Whiz-kid Mathematics

11 – 13 Years

Teach yourself

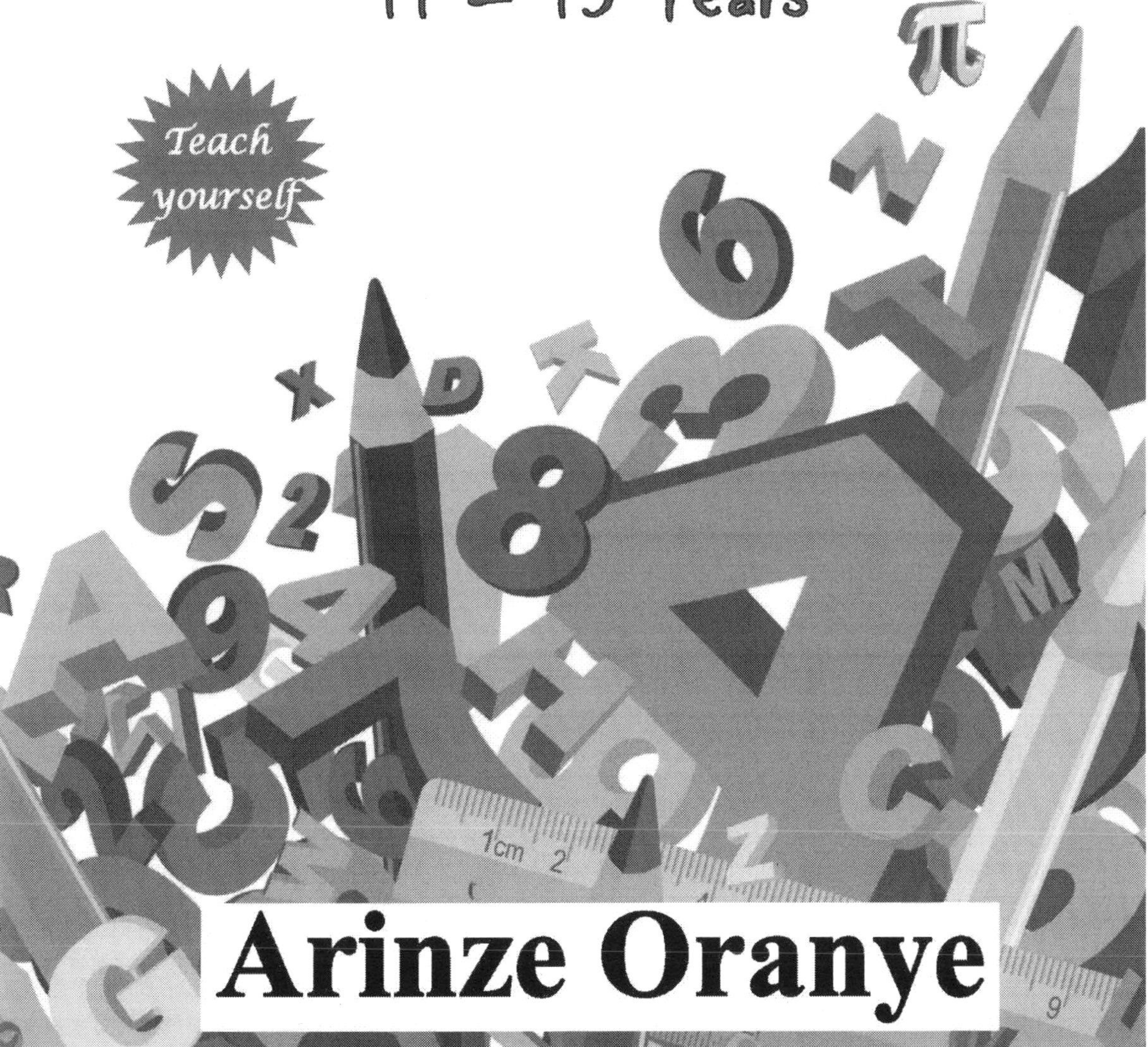

Arinze Oranye

ACKNOWLEDGEMENTS

The author would like to thank the following organisations for permission to reproduce photographs and table.

1) www.123rf.com for most of the pictures used in this book.
2) Pictures of students in chapter 9 algebra section, chapter 8 3d hemisphere, exercise 14a – cubes ....Designed by Creativeart/Freepik
3) www.surveyanalysis.org .... Statistical table on soft drinks - Chapter 18
4) Wikipedia – Picture of Fibonacci

Every effort was made to contact copyright owners/holders of materials reproduced in this book. If through oversight, any omissions will be rectified in future printings of this book if notice is given to the copyright holder.

CONTENTS

1 Number Work 1

This section covers the following topics:

- Types of Numbers
- Place Value
- International System of Units
- Days, Weeks and Months Calculations
- Time and Clock
- Timetables
- Chapter Review Section

LEARNING OBJECTIVES

By the end of this unit, you should be able to:

a) Know the difference between number types
b) Understand place values
c) Write numbers in words and figures
d) Work with the SI units
e) Perform calculations involving days, weeks, months and years
f) Know and read the time
g) Interpret timetables

KEYWORDS

- Integers
- Decimals
- Place value
- Significant figures

1.1 TYPES OF NUMBERS

Whole Numbers: These are numbers without decimal points or fractions.

Examples: 1, 2, 3, 4, 5, 6, 7…………..

Counting or Natural Numbers: They are numbers just like whole numbers.

Integers: These are positive or negative whole numbers including zero (0).

Examples: Negative (-1, -2, -3, -4, -5…), Positive (1, 2, 3, 4, 5…), or zero (0).

Positive Numbers: They are numbers greater than zero (0). They sometimes have a plus (+) sign or no sign attached to them.

Examples: 1, 2, 3, 4, 5… could be written as +1, +2, +3, +4, +5 ...

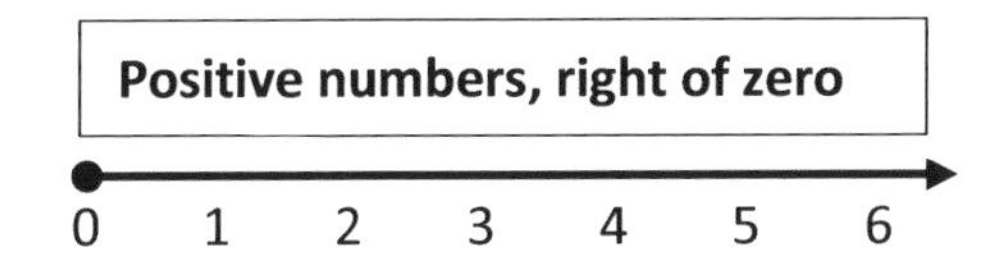

Negative Numbers: They are numbers less than zero and always written with a minus (-) sign.

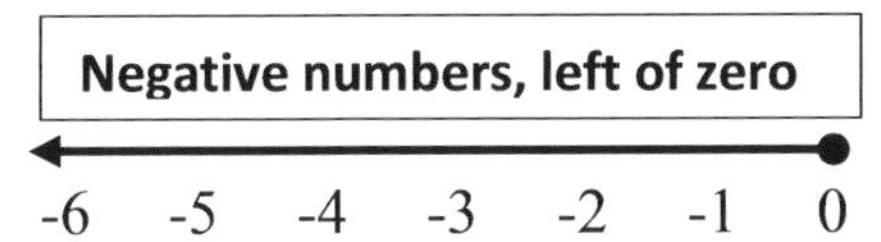

- Zero is neither negative nor positive. It is neutral.

Directed Numbers: They are numbers that have either a positive or a negative sign attached to them.

Examples: -3, -2, -1, +1, +2, +3 ……..

1.2 PLACE VALUE

The place occupied by a figure affects its value as we shall find out shortly.

In the number **567**, the digits have different values since they take up different places in the order of events. Using the place value table, the number 567 takes up different positions as shown:

Formerly called **units**

Hundreds	Tens	Ones
5	6	7

The value of the digit 5 is five hundred or **500**

The value of the digit 6 is 6 tens or **60**

The value of the digit 7 is 7 ones or simply, **7**

To explain the composition of the number 567, we say that it is made up of five hundred, six tens, and seven ones.

Example 1:

1 **4** 2 the 4, has value **40**

4 1 3 7 the 4, has value **4000**

4 5 **8** 6 the 8, has value 80

8 **0** 6 the 0, has value **0**

Example 2:

Anambra State has a population of 4,182,032 in 2016

Digit 3 represents 30 people

Digit 1 represents 100,000 people

Digit 4 represents 4,000,000 people

Example 3:

Using the digits **4 9 3**, write down all the three-digit numbers that can be made.

There could be six numbers which could be 493, 439, 943, 934, 349 or 394

A decimal place table can also include a decimal point to show varieties of numbers. The number 5973.862 can be written in the place value table below as

Thousands	Hundreds	Tens	Ones	•	tenths	hundredths	thousandths
5	9	7	3	•	8	6	2

The value of **5** is 5000

The value of **9** is 900

The value of **7** is 7 tens or 70

The value of **3** is 3 ones or 3

The value of **8** is 8 tenths or 8/10 or 0.8

The value of **6** is 6 hundredths or 6/100 or 0.06

The value of 2 is 2 thousandths or 2/1000 or 0.002

EXERCISE 1A

1) Write down the value of the numbers in bold.

a) **2** 5

b) 1 **4** 5

c) 7 0 **2** 0

d) 2 **1** 6 3

e) **4** 5 1 2

f) **8** 5 1 3

g) **8** 9 0

h) 1 **3** 9

i) 3 **5** 0 9

j) **3** 9 9

k) 3**7**

l) **3** 0 7

m) 1 **9** 1 3 5

n) 8 3 1 **2**

o) 3 6 8 **8**

p) 5 5 **5** 5

q) 8 **5**

r) 9 **3** 1 2

s) 1 **6** 1 6 1

t) 1 0 **3** 0

EXERCISE 1B

In questions 1 – 7, write down the value represented by **each** digit of the number.

1) 2537 people watched Wimbledon tennis in 2009.
2) 432 pupils passed their maths examination.
3) 1263 cats were caught in the bushes.
4) 3451 car accidents were reported in Sokoto State.
5) 15872 attended away games in 2007.
6) 1231 students passed their physics examination.

7) Only 753 lions left in the wild.

8) Using the digits 9 6 7 5, different four-digit numbers can be made.

   a) How many four-digit numbers can be made with 9 6 7 5?

   b) Write down the largest and smallest numbers from all the combinations.

   c) Write down the value of the second number in the largest four-digit number from question 8b.

9) Write the following numbers in a place value **table**.

   a) 45

   b) 135

   c) 5678

   d) 134.98

   e) 0.89

   f) 12345

10) Write down the value of **all** the digits in questions 9d, e and f above.

1.3 USING CODES

The numbers below are linked to different shapes.

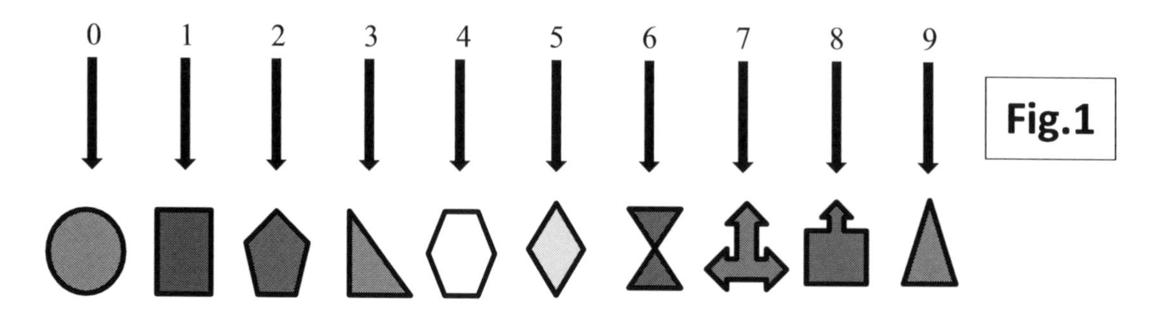

Example: Write down the value of the underlined symbol.

Solution: The symbols represent 4 <u>7</u> 1 0

Therefore, the value of 7 is **700.**

EXERCISE 1C

Using the codes and the symbols in fig.1, write down the value of each underlined symbol.

1)

2)

3)

4)

5)

6)

1.4 WRITING NUMBERS IN WORDS AND FIGURES

Accurate use of numbers is important in learning mathematics. As a positive step, we need to know these numbers in words.

1 ⟶ One

2 ⟶ Two

3 ⟶ Three

4 ⟶ Four

5 ⟶ Five

6 ⟶ Six

7 ⟶ Seven

8 ⟶ Eight

9 ⟶ Nine

10 ⟶ Ten

20 ⟶ Twenty

30 ⟶ Thirty

40 ⟶ Forty

50 ⟶ Fifty

60 ⟶ Sixty

70 ⟶ Seventy

80 ⟶ Eighty

90 ⟶ Ninety

100 ⟶ One hundred

950 ⟶ Nine hundred and fifty

Example 1: Write these numbers in figures.

a) Eighty-five

b) One hundred and twenty

c) Seven hundred and thirty-five

d) Eight hundred and ninety-nine

Answers: a) 85 **b)** 120 **c)** 735 d) 899

LARGE NUMBERS

There is no boundary of small and large numbers, but we have to recognise some numbers bigger than the usual ones we are used to. In higher mathematics lessons and real-life situations, large numbers come in handy. Some large numbers are displayed below.

Numbers	**Value**	**Power of ten**
One thousand	1 000	10^3
One hundred thousand	100 000	10^5
One million	1 000 000	10^6
One billion	1 000 000 000 ➡ a thousand multiply by a million	10^9
One trillion	1 000 000 000 000 ➡ a million multiply by a million	10^{12}

If you notice in the table above, the digits of the large numbers in the **value** column are grouped in threes from the decimal point (Remember: every whole number has an imaginary decimal point at the end) and separated by a space or gap between each group for easy identification.

For example, one thousand is written as 1 000 (1 ***then*** space ***then*** three zeros).

One million is written as 1 000 000 (1 ***then*** space ***then*** three zeros ***then*** another space ***then*** another three zeros).

From experience, some large numbers may be confusing to pronounce if the digits are not written properly. The number 761229 may be difficult to say in words. Is it still in the thousands or millions? However, if they are grouped in sets of three digits, it would be easy to read.

With the correct spacing, the number above should be written as 761 229. In words, it is seven hundred and sixty-one thousand two hundred and twenty-nine.

Note that commas between groups of digits are no longer in vogue. That is to avoid ambiguity as some educational establishments across the globe regard commas as decimal points.

EXERCISE 1D

Write these numbers in words.

1) 43

2) 6

3) 147

4) 222

5) 42 135

6) 8 020

7) 20 000

8) 4 619

9) 571 235

10) 2 315 200

11) 8 359

12) 751 350 210

13) 1 111

14) 88 888

15) 999 999

16) 8 997

17) 344 057

18) 1,888,300

19) 819, 270,000,000

20) 8 400 000 000,000

EXERCISE 1E

Write these numbers in figures.

1) Two

2) Eleven

3) Twenty-five

4) Two hundred and five

5) Eight hundred and three

6) One thousand and one

7) Ten thousand, nine hundred and fifty

8) Twenty-three thousand six hundred

9) Five hundred and one thousand, two hundred and sixty-five

10) One million, seven hundred and three thousand, nine hundred and 50

11) Six hundred and seven million, four hundred thousand

12) Four billion two hundred and ninety million Naira

EXERCISE 1F

CLASS ACTIVITY

In groups of threes, rewrite the numbers below using the correct standard of grouping in threes and not using commas. Remember to space the groups when applicable.

1) 4567
2) 45678
3) 1 thousand
4) 987654
5) 250000
6) 2 million
7) 12 million
8) 1 billion
9) 345000000000
10) 1 trillion
11) 3490034
12) 7623
13) 4 billion
14) 3.5 million
15) 200000000
16) 6789743
17) 0.3 billion
18) 5 trillion
19) 1564566
20) ₦7½ billion

21) Still, in the groups of threes (or any arrangement of your teacher's choice), say each number in activity 1 to your mate.

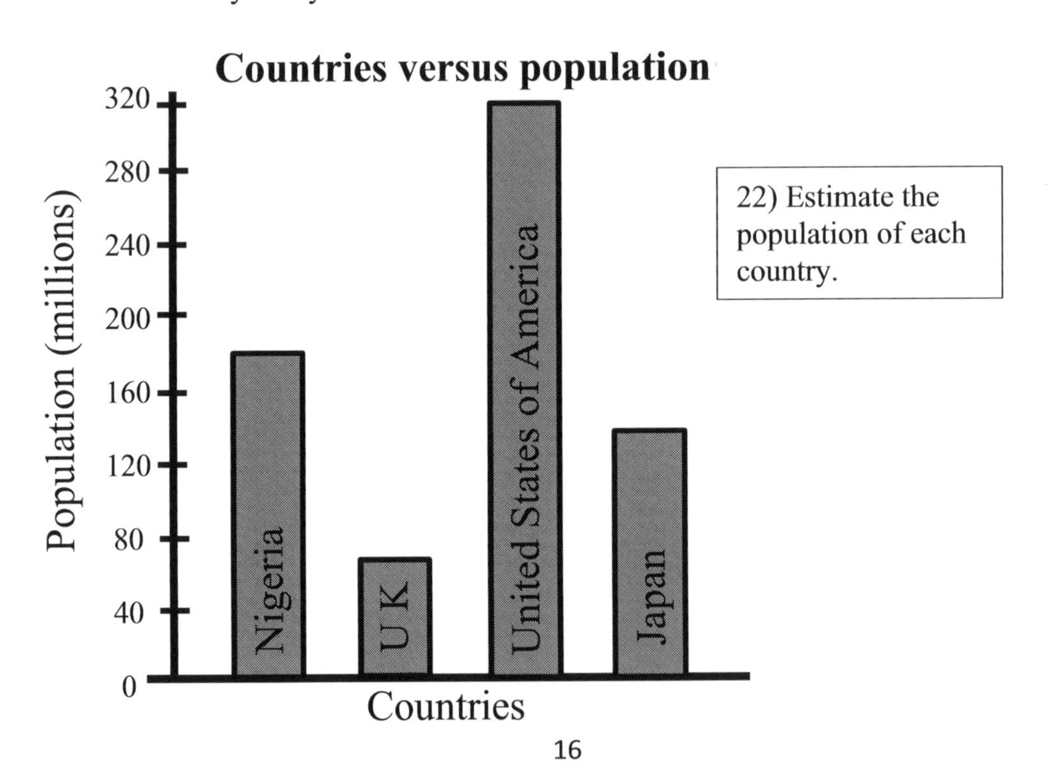

1.5 INTERNATIONAL SYSTEM OF UNITS (SI)

The International System of Units (SI) is the modern form of the metric system which is based on the decimal system. It is the most widely used system of units for length, mass, and capacity.

LENGTH

In the SI units, the metre is the basic unit of length. The conversions tables are listed below, and it is our duty to memorise these conversions. It will be handy in converting from one unit to another.

10 millimetres (mm)	= 1 centimetre (cm)
100 centimetres (cm)	= 1 metre (m)
1000 millimetres (mm)	= 1 metre (m)
1000 metres (m)	= 1 kilometre (km)

MASS

Mass is a measure of the amount of matter in an object. In the SI units, the gramme (gram) is the basic unit of mass.

1000 grams (g)	= 1 kilogramme (kg)
1000 kilogrammes (kg)	= 1 tonne (t)

CAPACITY

Capacity is the maximum amount that something can contain. In the SI units, a litre is the basic unit of capacity.

100 centilitres (cl)	= 1 litre (l)
1000 millilitres (ml)	= 1 litre (l)
1000 litres (l)	= 1 cubic metre (m^3)
1000 litres (l)	= 1 kilolitre (kl)

Conversions between units

Rule 1

To change to a smaller unit, multiply by 10, 100 or 1000 depending on the required unit.

Example 1: Convert 3 metres to centimetres.

Using the conversion 100 cm = 1 metre, multiply by 100.

Therefore, 3 metres = 3 × 100 = **300 cm**

Example 2: Convert 14 kilometres to metres.

Using the conversion 1000 m = 1 kilometres, multiply by 1000.

Therefore, 14 km = 14 × 1000 = **14000 m**

Example 3: Convert 4 tonnes to kilogrammes.

Using the conversion 1000 kg = 1 tonne, multiply by 1000.

Therefore, 4 tonnes = 4 × 1000 = **4000 kg**

RULE 2

To convert to a large metric unit, divide by 10, 100 or 1000 depending on the required unit.

Example 4: Convert 2000 metres to kilometres.

Using the conversion 1000 m = 1 km, divide by 1000.

Therefore, 2000 m = 2000 ÷ 1000 = **2 km**

Example 5: Convert 4800 millilitres to litres.

Using the conversion 1000 ml = 1 litre, divide by 1000.

Therefore, 48000 ml = 4800 ÷ 1000 = **4.8 litres**

Example 6: Change 14 millimetres to centimetres.

Using the conversion 10 millimetres = 1 centimetre, divide by 10

Therefore, 14 mm = 14 ÷ 10 = **1.4 cm**

EXERCISE 1G

Copy and Complete

1) 40 cm = ◯ mm

2) 17 cm = ◯ mm

3) 180 cm = ◯ mm

4) 4 m = ◯ mm

5) 13 m = ◯ mm

6) 876 m = ◯ mm

7) 13 m = ◯ cm

8) 330 m = ◯ cm

9) 29 m = ◯ cm

10) 330 m = ◯ cm

11) 2 km = ◯ m

12) 5.5 km = ◯ m

13) 45 km = ◯ m

14) 27 cm = ◯ mm

15) 19 m = ◯ mm

16) 412 m = ◯ mm

17) 16 cm = ◯ mm

18) 1.3 km = ◯ m

19) 82.5 km = ◯ m

20) 169.8 m = ◯ mm

EXERCISE 1H

Write as kilograms.

1) 20000 g **2)** 40 g **3)** 5 g **4)** 4000 g

5) 900 g **6)** 1.5 g **7)** 8.65 g **8)** 50 000 g

EXERCISE 1I

Copy and complete

1) 5 litres	=	☐	cl	
2) 2 tonnes	=	☐	kg	
3) 15 kg	=	☐	g	
4) 14 litres	=	☐	ml	
5) 4 m^3	=	☐	l	
6) 6 tonnes	=	☐	kg	
7) 20 kg	=	☐	g	
8) 100 kg	=	☐	g	
9) 7 litres	=	☐	ml	
10) 9.5 kg	=	☐	g	
11) 0.45 kg	=	☐	g	
12) 16 m^3	=	☐	l	
13) 6.9 tonnes	=	☐	kg	
14) 40 mm	=	☐	cm	
15) 17 mm	=	☐	cm	
16) 318 mm	=	☐	m	
17) 417 mm	=	☐	m	
18) 120 cm	=	☐	m	
19) 870 cm	=	☐	m	
20) 4000 m	=	☐	km	
21) 2000 kg	=	☐	t	
22) 8888 m	=	☐	km	
23) 8.88 litres	=	☐	ml	
24) 865 mm	=	☐	m	
25) 9 tonnes	=	☐	kg	
26) 6.3 litres	=	☐	cl	

EXERCISE 1J

1) Choose a more sensible weight for the weights below.

a) A tin of milk (400g or 5 kg)?

b) A pack of pasta (5kg or 500 g)?

c) An onion bulb (80 g or 2 kg)?

2) Afolarin buys 2kg of beans. 800 g were eaten. What is the weight of the beans that are left?

3) A cup of plain flour weighs 0.12 kg. What is the weight of 35 cups of the flour?

4) A carton of milk weighs 2 kg. What will 100 cartons weigh in grams?

5) Copy and complete the table below.

	Capacity(litres)	Capacity (ml)
Bottle of water	2litres	
Tin of Milk		300
Bottle of Coca Cola		1000
Petrol tank	30	

6)

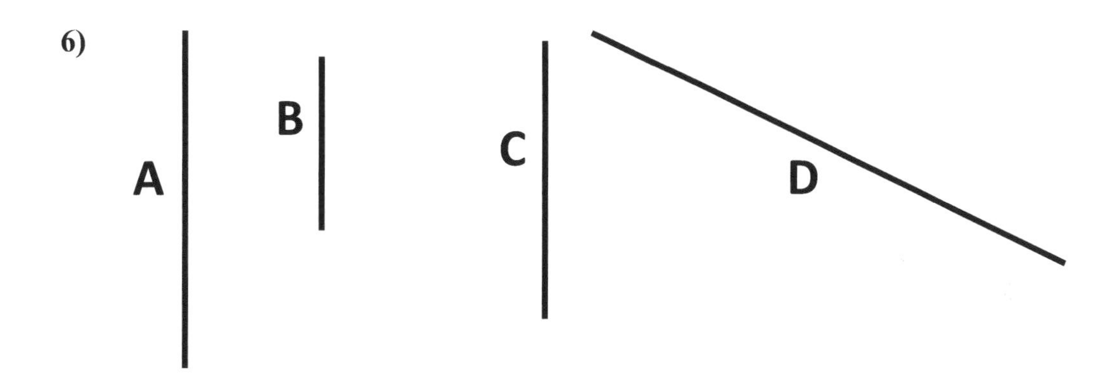

a) **Estimate** the lengths of lines A to D. Write down the units used.

b) **Measure** the lengths of the lines A to D in the same units used in part **a**.

c) Work out the differences between your answers in part **a** and part **b**

d) What is the actual length of line **C** + 113 cm?

1.6 TIME

The basic unit of time is the **second,** and it **does not** follow the decimal system like the previous SI units already mentioned. The tables below are vital and must be memorised.

1 minute	= 60 seconds
1 hour	= 60 minutes
	= 3600 seconds
1 day	= 24 hours
1 week	= 7 days

1 year	= 365 days
	= 366 days (leap year)
	= 52 weeks
	= 12 months

Example 1: What is two days in hours?

Since 1 day = 24 hours,

2 days must be 2 × 24 hours = **48 hours**

Example 2: How many weeks are there in 4 years?

Since one year = 52 weeks,

Four years must be 4 × 52 = **208 weeks**

Example 3: How many seconds are there in 20 minutes?

Since 60 seconds = 1 minute,

20 minutes must be 20 × 60 seconds = **1200 seconds**

Example 4: If a train takes 4 minutes and 34 seconds to reach another train station, Write this time in **seconds** only.

4 minutes is (4 × 60) seconds = 120 seconds

Therefore, the number of seconds in 4 min 34 s = 120 s + 34 s
= **154 seconds**

Example 5: Write three weeks two days, 11 weeks and eight days in weeks and days.

3 weeks + 11 weeks = **14 weeks** 2 days + 8 days = 10 days

But, 10 days = **1 week** + 3 days

Therefore, three weeks two days, 11 weeks and eight days is

15 weeks (14 weeks + 1 week) and 3 days

15 weeks and three days

EXERCISE 1K

Copy and complete

1) ---------- months = 2 years
2) ---------- months = 4 years
3) ---------- weeks = 1 year
4) ---------- days = 1 year
5) ---------- minutes = 3 hours
6) ---------- minutes = 120 seconds
7) ---------- weeks = 35 days
8) ---------- minutes = 180 seconds
9) ---------- minutes = 5 hours
10) ---------- weeks = 28 days

Use the Calendar below to answer questions 11 - 14

JANUARY

SUN	MON	TUE	WED	THUR	FRI	SAT
	1	2	3	4	5	6
7	8	9	10	11	12	13
14	15	16	17	18	19	20
21	22	23	24	25	26	27
28	29	30	31			

11) On which day of the week falls the following birthdays?

a) January 3rd b) January 8th c) January 28th d) January 13th

12) How many days have January?

13) What is the date of the second Friday in the month of January?

14) How many Saturdays are there in January?

15) Write the following as days.

a) 8 weeks b) 40 weeks c) 9 weeks d) 6 hours

16) Write the following as minutes.

a) 5 hours

b) 20 hours

c) 3.5 hours

d) 300 seconds

17) Write the following as years.

a) 156 weeks

b) 260 weeks

c) 365 days

d) 12 months

18) Write four weeks three days, 14 weeks and ten days in weeks and days.

19) How many seconds are there in 30 minutes?

20) How many weeks are there in 3 years?

21) Add the times below but remember to give your answer in hours and minutes.

300 minutes + 2 hours 42 minutes

22) Today is Monday. In 102 days, what day of the week will that be?

23) Find the total of 11 days and two weeks, seven days and nine weeks in weeks and days.

1.7 CLOCK

Ante Meridiem **(a.m.)**, is Latin for before midday (12 noon)

Post Meridiem **(p.m.)**, is Latin for after midday (12 noon)

There are two types of the clock:

a) Analogue clock

b) Digital clock

ANALOGUE CLOCKS

From 12 clockwise, we read the minutes as **PAST** (before 30 minutes)

After 30 minutes, we read as **TO**.

DIGITAL CLOCKS

Digital clocks have figures only.

8: 10 4: 42 7: 00

There are two ways of showing the time on the clock.

- 12 – hour clock
- 24 – hour clock

12-hour clock:

In a 12 – hour clock, there are two periods each of 12 hours' duration during each day. Between midnight and noon, the period is often referred to as a.m.

Between noon and midnight, the period is referred to as p.m.

Please note: 7.30 a.m. is a time in the morning.

7.30 p.m. is a time in the evening.

24-hour clock:

In a 24-hour clock, there is only one period of 24 hours. The 24-hour clock is mainly used in the aviation industry for airline timetables. They are also used in other fields and are written with four figures.

For example, 2.15 a.m. will be written as 02:15 hours using a 24-hour clock.
2.15 p.m. will be written as 14:15 hours. That is to avoid any confusion between a.m. and p.m.

Between midnight and noon, the times are stated as 00:00 hours and 12 hours respectively. Between noon and midnight, the times are stated as 12:00 hours and 24:00 hours respectively.

The 24 Hour Clock

EXERCISE 1L

1) Find the length of time between the following hours

a) 09:12 hours and 11:13 hours

b) 18:30 hours and 23:30 hours

c) 4.22 am and 9.47 pm

d) 3.12 a.m. and 3.12 p.m.

e) 1230 noon and 7.30 p.m.

f) 22:10 hours and 23:35 hours

g) 1.29 p.m. and 5.50 p.m.

h) 4.30 p.m. and 12 noon

2) Convert these a.m. and p.m. times into the 24-hour clock.

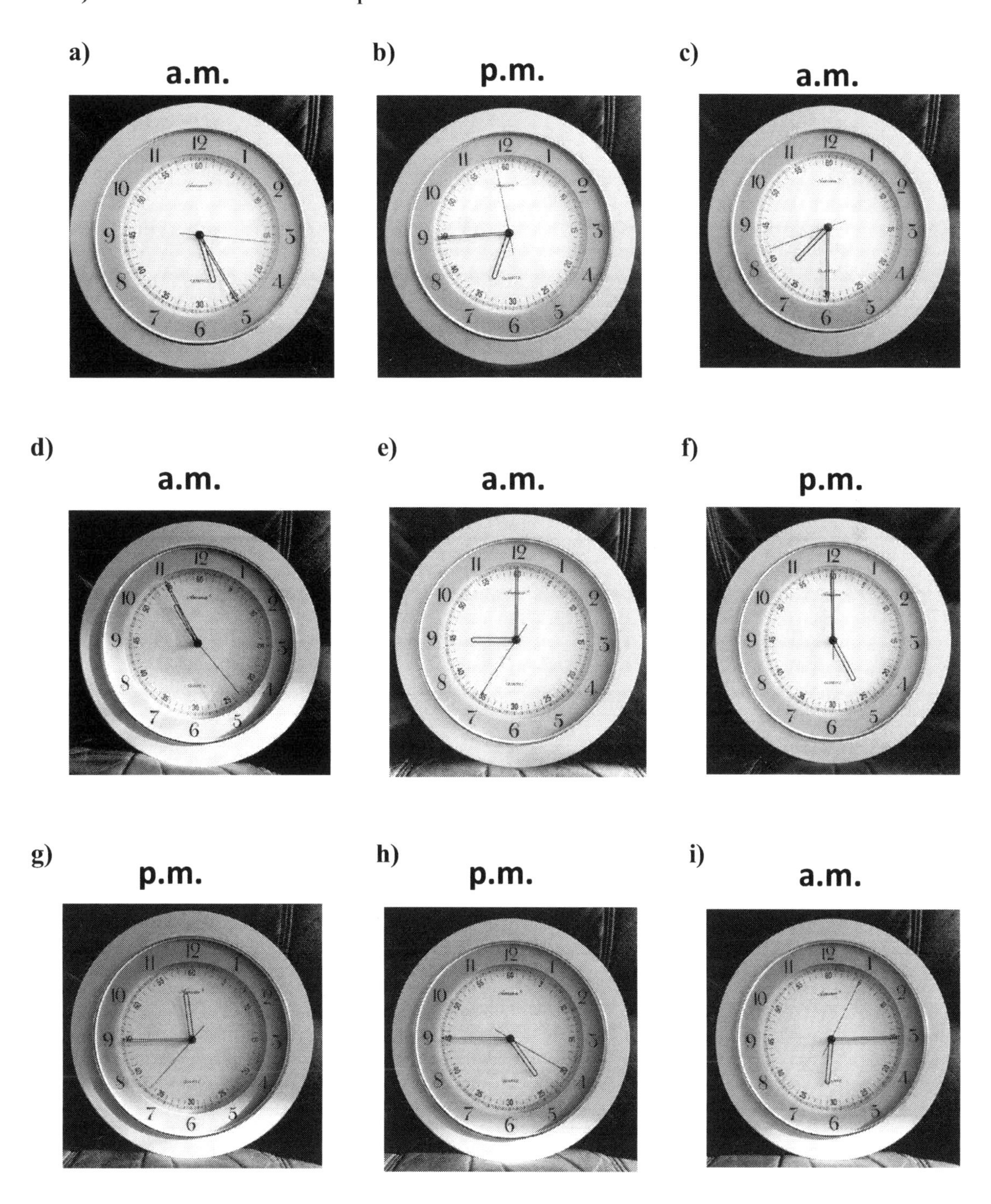

1.8 TIMETABLES

Most people travel daily by bus or train. It is, therefore, crucial that people can read and interpret timetables to avoid delays and other inconveniences. Timetables are usually written using the 24-hour clock.

The timetable below is for a journey from Lagos to Asaba by bus.

TABLE 1A

Lagos	Departure	0700	0730	0800	0830	0900	0930	1000	1030
Ore	Arrival	0930	1000	1100	1125	1155	1220	1250	1320
	Departure	0931	1002	1105	1125	1157	1220	1250	1321
Benin	Departure		1045	------	1146	1245	1259	1315	1350
Agbor	Arrival	1245	------	------	1230	------	------	------	1420
	Departure	------	------	------	1350	1420	1453	1528	1550
Asaba	Arrival	1320	1355	1420	1457	1525	1559	1630	1658

Example 1: A bus arrives at Agbor at 1420 from Lagos.

a) At what time did the bus leave Lagos?
b) How long was the journey?

Solution: a) The bus left Lagos at **10:30**

b)

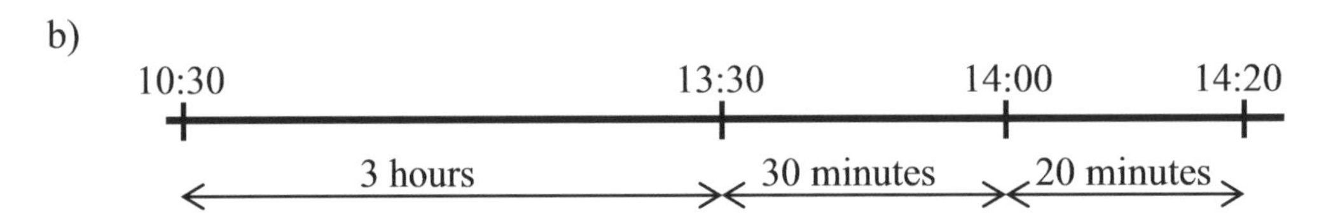

Altogether, 3 hrs + 30 mins + 20 mins = 3 hours 50 minutes

The journey took **3 hours 50 minutes**

EXERCISE 1M

Questions 1 to 3 relate to **table 1** above.

1a) Ikenna planned to travel from Lagos to Agbor. He must be in Agbor by 12:55 pm. What is the latest bus that he could catch from Lagos?

1b) How long was Ikenna's journey from Lagos to Agbor?

2) Chuba is travelling from Ore to Asaba and gets on the same bus as Ikenna.

- **a)** At what time does Chuba leave Ore?
- **b)** How long is Chuba's journey?
- **c)** Chuba says "My journey was longer than Ikenna's journey." Is he correct? Explain.

3) Awolowo gets on the 0800 bus from Lagos. He wants to travel to Asaba but will need to Change at Ore.

- a) How long is the wait at Ore?
- b) At what time will he arrive in Asaba?
- c) How long did the journey from Ore to Asaba take?
- d) How long is the entire journey from Lagos to Asaba **excluding** the wait at Ore?

4a) A journey from a bus stop to a school takes 15 minutes. Buses run every10 minutes. Copy and complete **table 1B** below,

TABLE 1B

	BUS A	BUS B	BUS C	BUS D	BUS E
BUS STOP	07:00	07:10			
SCHOOL		07: 25			

b) What is the latest bus a teacher will catch if he started work at

i) 7:20 ii) 7: 30 iii) 7: 40 iv) 7: 50?

Use the television time table in table 1C below to answer questions 5 to 7.

TABLE 1C

13:00	**News Summary**
13:20	**Sports**
13:55	**Weather**
14:00	**New Hour**
14:13	**Africa World**
15:00	**News Summary**
15:20	**Asia World**
15:45	**Are you there?**
16:15	**Football Matters**
17:00	**News Summary**

5) Which programme could you watch at these times?

a) 15:20 b) 16:00 c) 13:20 d) 14:30

6) How long are the following programmes?

a) Football matters

b) Are you there?

c) News Hour

d) Sports

7) Which programme has the shortest air time?

Chapter 1 Review Section

Assessment

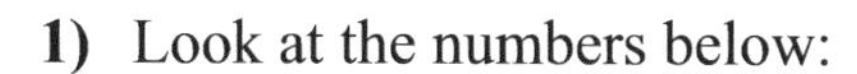

1) Look at the numbers below:

5 7 1 6

a) Write down the smallest number that can be made using all the four digits.

.... **1 mark**

b) Write down the largest number that can be made using all the four digits.

.... **1 mark**

c) Using the knowledge of place value, write down the value of **6** when the smallest number is formed in question 1a.

.... **1 mark**

d) Using the original numbers given, write down an extra number needed to make the number ten times bigger.

...............**2 marks**

2) Copy and complete the number machines below.

a) 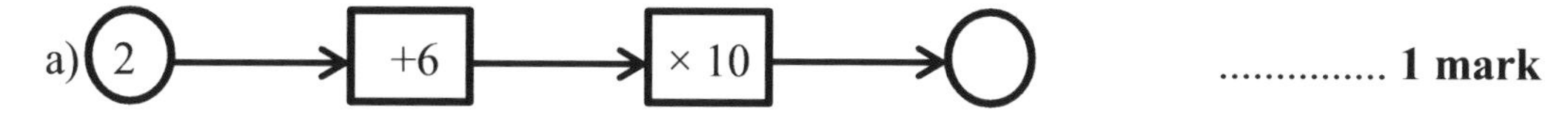

.... **1 mark**

b) 

.... **1 mark**

c) 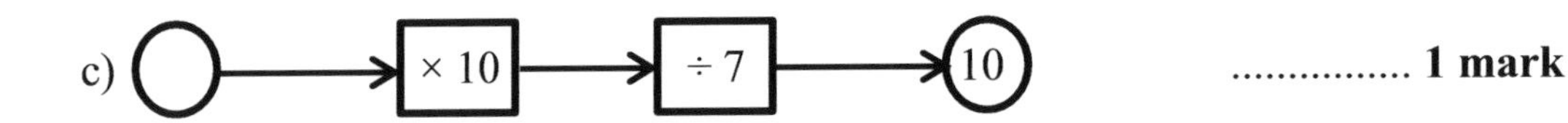

.... **1 mark**

3) Write down the missing number in each box.

a) 800 + ☐ = 950 .... **1 mark**

b) 60 × ☐ = 300 .... **1 mark**

c) 127 - ☐ = 83 .... **1 mark**

4) When a number is multiplied by itself, the answer is 81.
What is the number? .... **1 mark**

5) Chudi has 235 pens to share out amongst 50 pupils.

a) How many complete pens does each pupil receive? .... **2 marks**

b) How many pens does Chudi have left over? .... **1 mark**

6) An interview information is shown below:

Start time: 09:15

Written interview duration: 30 minutes

Break time: 10 minutes

Oral interview duration: 38 minutes

a) At what time will the written interview end? .... **1 mark**

b) At what time will the oral interview end? .... **1 mark**

c) What is total time spent on oral and written interview? .... **1 mark**

7) The clock shown could be showing two different times. Write down the times using

a) the 12-hour clock

b) the 24-hour clock.

.... **2 marks**

8) Using the 24-hour clock, write down two different times that each clock could be showing.

a) 

b) 

c) 

.... **3 marks**

9) Which months have 30 days exactly? .... **1 mark**

10) Convert each of the following into centimetres.

a) 70 mm　　c) 4 m　　e) 0.7 m

b) 87 mm　　d) 7.6 m　　f) 675 mm

.... **6 marks**

11) Work out the number of minutes in 540 seconds .... **1 mark**

12) Add together three weeks 3days, 11 days and six weeks two days.
Give your answer in days. .... **2 marks**

13) Add together 100 mm, 67 cm and 560 mm.

a) Give your answer in millimetres .... **2 marks**

b) Give your answer in centimetres .... **2 marks**

14) Hot chocolate drinks are packed in the 5kg bag shown. The weight of each chocolate drink is 500 g. How many chocolate drinks are needed to fill the bag?

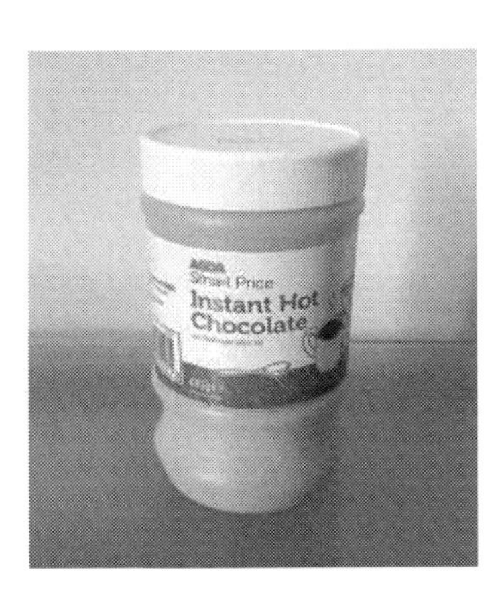

.... **3 marks**

15) The calendar below shows Okafor's itinerary for December 2017

MON	TUES	WED	THURS	FRI	SAT	SUN
						Evening Party
				Fishing 1.20 pm		
	Cinema 5.30 pm					
				Cycling 11.30 am		

a) How many days are there in December 2017? .... **1 mark**

b) On what day will Okafor go fishing? .... **1 mark**

c) After going to the cinema, how many days will Okafor wait before cycling? .... **1 mark**

d) Okafor will receive his monthly salary three weeks after going fishing. What date will he receive his salary? .... **1 mark**

16) Write these numbers in figures.

a) Three hundred and sixty-five

.... **1 mark**

b) Nine hundred and forty

.... **1 mark**

c) Four thousand, six hundred and eighty-seven

.... **1 mark**

d) Four hundred and forty thousand, nine hundred

.... **1 mark**

17) The population of three countries is shown in the table below.

Country	**Population**
Luxembourg	510 456
Latvia	1 967 500
Austria	8 456 398

a) Write the population of Latvia in words. .... **1 mark**

b) What is the difference between the population of Austria and Luxembourg?

.... **1 mark**

c) Write your answer to **part b** above in words.

.... **1 mark**

d) Bimbo says "The population of Latvia is double that of Luxembourg."
Is Bimbo correct? Explain

.... **2 marks**

2 Number Work 2

This section covers the following topics:

- Rounding
- Estimation & Approximation
- Adding and Subtracting Whole Numbers
- Multiplying and Dividing Whole Numbers
- Multiplying by 10, 100, 1000

LEARNING OBJECTIVES

By the end of this unit, you should be able to:

a) Round numbers to the nearest 10, 100, and 1000
b) Round numbers to a given number of decimal places
c) Estimate and approximate numbers
d) Add and subtract whole numbers
e) Multiply and divide whole numbers
f) Multiply by 10, 100, 1000...

KEYWORDS

- Round
- Estimate
- Approximate
- Add
- Subtract
- Whole numbers

2.1 ROUNDING

Numbers are rounded off to make them easier to handle and compare. It is used to give an approximate value of a number.

We round **down** if the number is less than halfway to the next rounding number.

We round **up** if the number is halfway or more to the next rounding number.

Rounding to the nearest 10

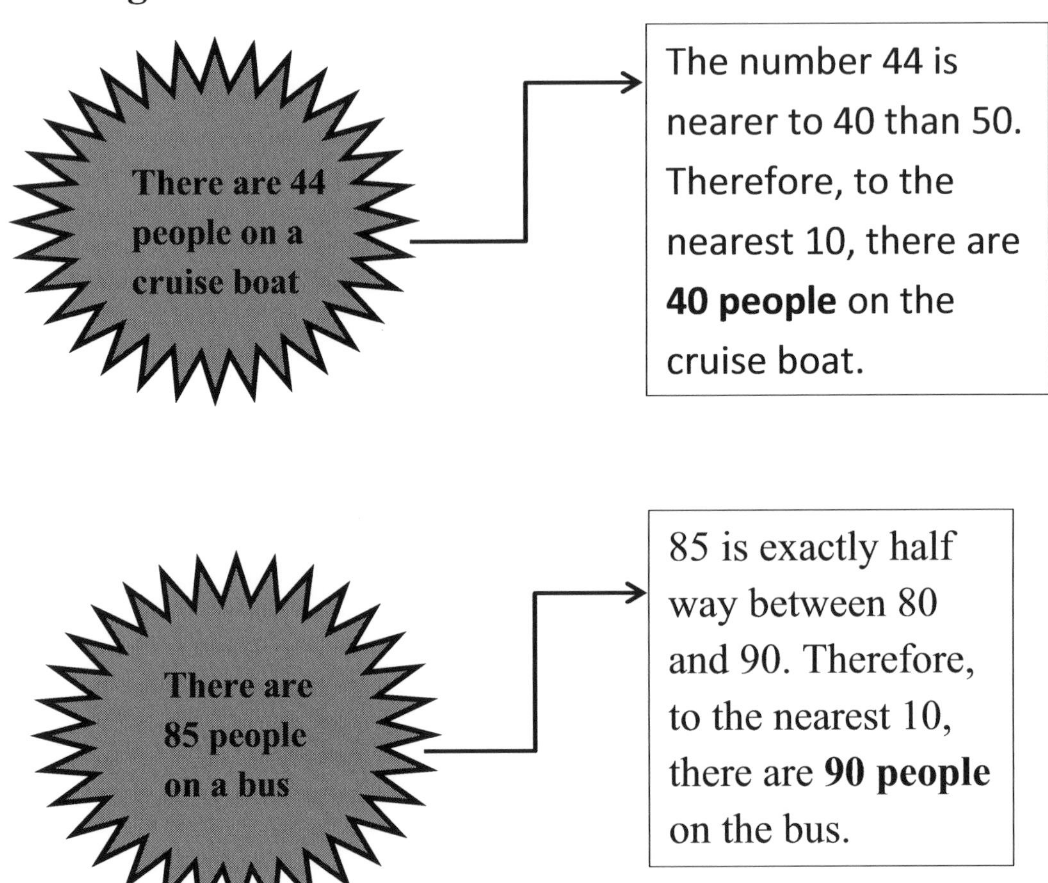

Another way to look at rounding is to consider the ones (units) column.

For example, to round 24 to the nearest 10, we must look at the digit in the ones column.

- If the digit is less than 5, round **down**. If that digit is exactly 5 or more, then round **up**.

Going back to the example; the digit in the ones column is 4, which is less than 5. We, therefore, round down. So, 24 to the nearest ten is **20**.

Example 2: 57 to the nearest ten rounds to **60** since seven is up to 5.

Rounding to the nearest 100

When rounding to the nearest 100, it is the same process except you look at the tens column instead of the ones column.

For example, 723 to the nearest 100 would be **700.** In the tens column, 2 is not up to 5, we, therefore, round down.

Also, 723 is closer to 700 than 800.

An accountant said, "To the nearest 100, we sold 700 black pens last month."
The exact number of black pens sold could be any number between 650 and 749.

Example 1: Round these numbers to the nearest 100.

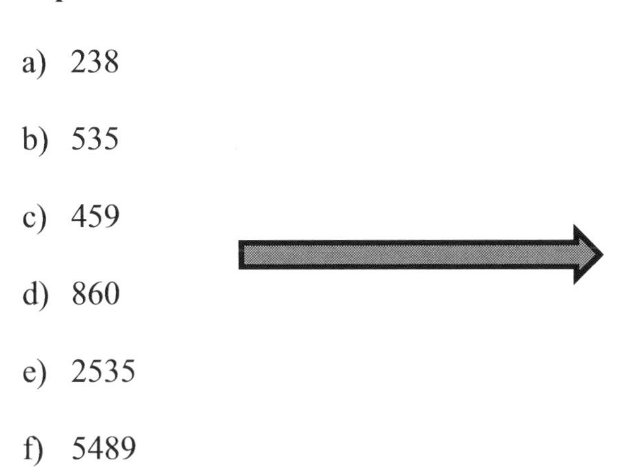

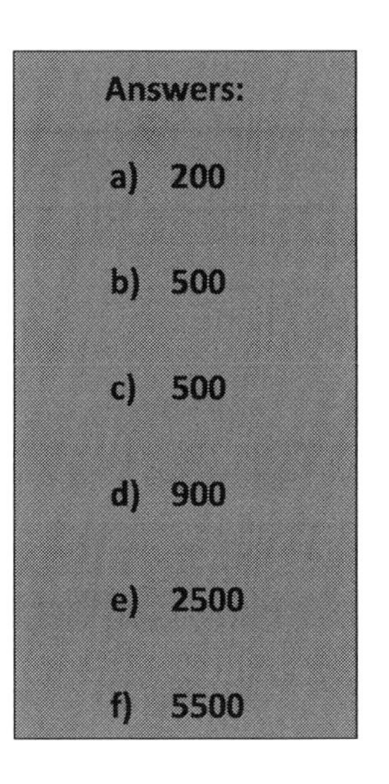

Rounding to the nearest 1000

Larger numbers can be rounded to the nearest 1000. It is the same principle as rounding to the nearest 10 or 100 except that we look at the digit in the hundreds column and apply the rules of rounding afterwards.

25732 to the nearest 1000 is 26000. The digit in the hundreds column is 7 which is up to 5. We, therefore, round up to get 26000.

68355 people attended the Arsenal Versus Manchester United game at the Emirates Stadium. A newspaper reports "68000 watched Arsenal Versus Man United game." 68355 is closer to 68000 than 69000. The number has been rounded to the nearest 1000.

Rounding to the nearest whole number

When rounding to the nearest whole number, look at the **first** number after the decimal point.

If the first number after the decimal point is less than 5, the answer will just be the left part - whole number(s) without the decimal number(s). You have rounded down.

If it is 5 or more, round up the whole number.

Example 1: Round to the nearest whole number.

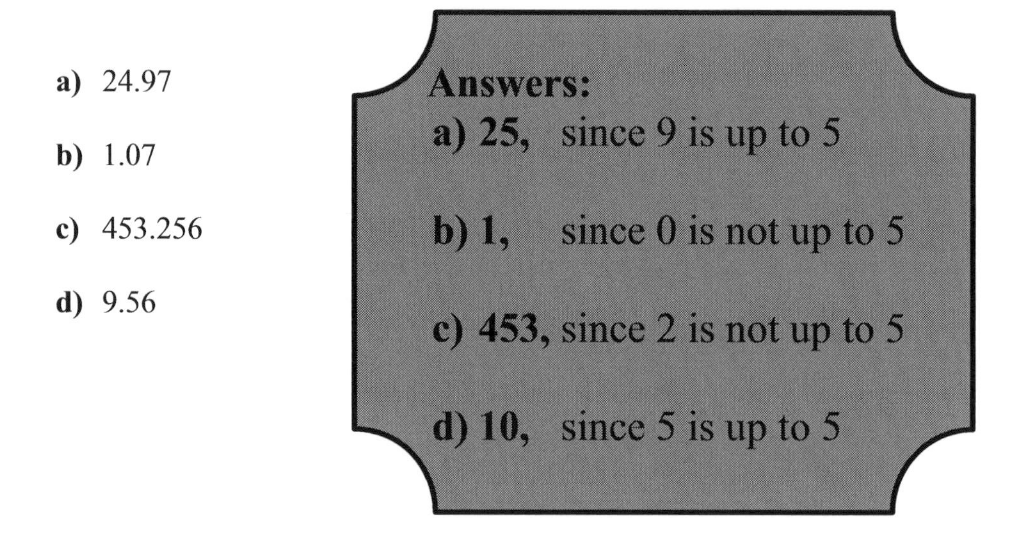

a) 24.97

b) 1.07

c) 453.256

d) 9.56

Answers:
a) 25, since 9 is up to 5

b) 1, since 0 is not up to 5

c) 453, since 2 is not up to 5

d) 10, since 5 is up to 5

EXERCISE 2A

1) Round each of these numbers to the nearest 10.

a) 7
b) 14
c) 25
d) 48
e) 92
f) 425

g) 525
h) 331
i) 586
j) 177
k) 216
l) 407

m) 586
n) 888
o) 409
p) 135
q) 198
r) 221

s) 923
t) 663
u) 88
v) 137
w) 5463
x) 1234

2) Round each of these numbers to the nearest **100**.

a) 74
b) 836
c) 208
d) 672

e) 469
f) 647
g) 345
h) 151

i) 850
j) 1425
k) 1387
l) 2208

m) 999
n) 11888
o) 6035
p) 1111

3) Round each of these numbers to the nearest 1000.

a) 4120
b) 3459
c) 3271
d) 8500

e) 1465
f) 2398
g) 876
h) 7365

i) 42 258
j) 76903
k) 13820
l) 619419
m) 506431

4) The attendance at Enugu Rangers FC Versus Enyimba Football Club was 23175. Write this figure to the nearest 100.

5) A conservatory is estimated to coat N450 340. Write this quotation to the nearest ₦100.

6) Round the following numbers to the nearest whole number.

a) 1.69 b) 24.396 c) 103.8kg d) 349.50

7) To the nearest 10, I own 70 pens.
   a) Write down how many pens I could have.
   b) Write down the range of numbers (integers) the statement could represent.

ROUNDING TO DECIMAL PLACES

Numbers are often rounded to decimal places. The number of figures after the decimal point gives the number of decimal places in a decimal number.

Note: Zeros will not be counted if they occur at the end of the decimal number.

14.253 has 3 decimal places
14.203 has 3 decimal places
14.230 has **2** decimal places (the zero at the end is not counted)

EXERCISE 2B

Write down how many decimal places each of the decimal numbers below have.

1) 0.2
2) 0.45
3) 0.05
4) 0.372
5) 0.750
6) 0.305
7) 0.097
8) 3.52
9) 8.307
10) 9.2070
11) 5.897
12) 15.808
13) 42.3
14) 85.2074
15) 45.9821
16) 1.789543
17) 329.2
18) 498.073
19) 186.23
20) 0.0435
21) 9.234
22) 1471.2678954
23) 59.9070
24) 764.128

Sometimes, numbers need to be written to a given number of decimal places. We can also round up or down.

Example 1:
Round 1.367 to one decimal place

Solution:
The answer (number) **must** have one number after the decimal point. That number we are interested in is **3**. However, we must look at the next number immediately after 3, which in this case, is 6. We then round up since 6 is up to 5 (rounding rules).

So, 1.367 to one decimal place is **1.4**

Example 2:
Round 48.53176 to two decimal places

Solution:
The answer (solution) **must** have two numbers after the decimal point. The numbers are 5 and 3. However, we look at the next number immediately after 3. That number is 1 which is not up to 5. We then round down (rounding rules).

So, 48.53176 to 2 decimal places is **48.53**

EXERCISE 2C

1) Round off these numbers correct to one decimal place.

a) 2.323

b) 0.3791

c) 4.533

d) 0.0757

e) 0.1598

f) 23.762

g) 153.997

h) 32.0572

i) 100.5397

j) 1234.77216

k) 1.04212

l) 72.12331

2) Round off the numbers in question one above, correct to two decimal places.

3) Round to three decimal places

a) 3.7522

b) 2.635386

c) 10.23781

d) 11.98602

e) 0.1258764

f) 17.98999

g) 879.12344

h) 0.76285

i) 10.087654

j) 76.1298

k) 100.67512

l) 9.651287

4) The volume of a sphere is 456.783256 cm^3. Round the volume to

a) One decimal place

b) 2 decimal places

c) 3 decimal places

d) 4 decimal places

5) The area of a football pitch is 153 786.45362 m^2. Round the area to

a) One decimal place

b) Two decimal places

c) Three decimal places

d) Four decimal places

ROUNDING TO SIGNIFICANT FIGURES

Writing a number to a given number of significant figures is almost the same as writing a number to a given number of decimal places. The **only exception** is that we must count all the figures in the number, not just the decimal parts.

The first significant figure is the first **non-zero** digit of the number. Significant figures are often abbreviated to **s.f.**

Example 1:

Consider the number 4 2 5 9

1st s.f **2nd s.f.** **3rd s.f.** **4th s.f.**

Example 2:

Consider the number

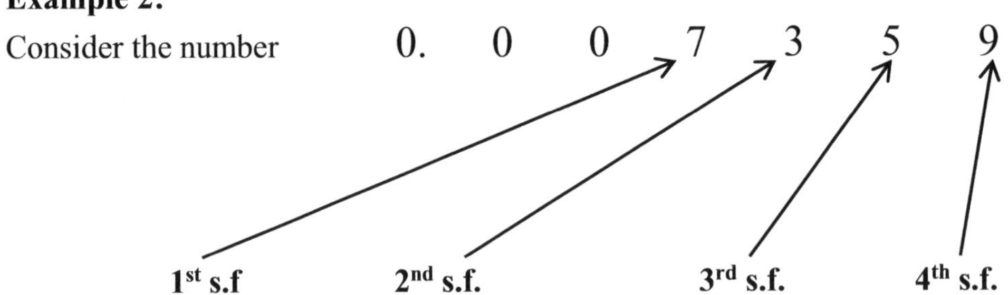

Example 3:

Consider the number

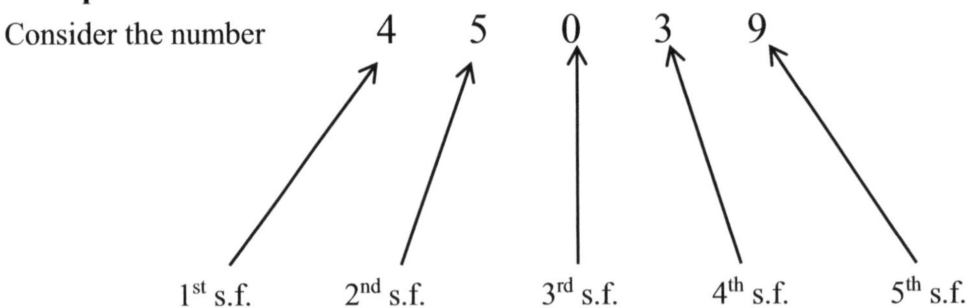

Example 4:

Write 423 to one significant figure

Solution: The first significant figure from 423 is 4. The number 2, which is after the number 4 is not up to 5, so we round down the remaining numbers to zero.

Therefore, 423 to 1 s.f. = **400**

Example 5:

Write 423 to 2 significant figures

Solution: The second significant figure from 423 is 2. The number 3, which is after the number 2 is not up to 5, so we round down to zero.

Therefore, 423 to 2 s.f. = **420**

Example 6:

Write 0.003756 to 2 significant figures

Solution: Remember that the first significant figure is the first non-zero digit. The number 7 is the second significant figure in the number 0.003756. Since the next number after 7 is 5, which is up to 5 (rounding rules), we round up.

Therefore, 0.003756 to 2 s.f. = **0.0038**

Example 7:

Write 84.375 to 3 significant figures

Solution: The 3rd significant figure from 84.375 is 3. However, 7 is up to 5 (rounding rules), so we round up.

Therefore, 84.375 to 3 significant figures = **84.4**

EXERCISE 2D

1) Write the first significant figure for the numbers given below.

a) 237

b) 34

c) 0.05

d) 0.954

e) 45876

f) 306

g) 5.67

h) 54.908

2) Write the second significant figure for the numbers given below.

a) 347

b) 0.187

c) 56.897

d) 0.107

e) 90.789

f) 78949

g) 106.9

h) 2.93

3) Round each of the numbers below to one significant figure.

a) 68

b) 786

c) 2.56

d) 1.987

e) 0.378

f) 72.99

g) 36778

h) 0.083

4) Write each of the numbers below to the given number of significant figures.

a) 4035 1 s.f.

b) 8.63 1 s.f.

c) 5.654 2 s.f.

d) 3.167 2 s.f.

e) 8.907 3 s.f.

f) 12.599 3 s.f.

5) Some lengths of leisure parks in Nigeria are given below with their lengths (not to scale).

Amusement Parks	Lengths (m)
Dreamland Africana (Lekki)	1234
Hi-Impact Planet(Lagos)	867
The Wild Bunch (Jos)	1234
Klubdelag (Lagos)	790
Fun Factory (Ibadan)	1041
Omu Resort (Lekki)	498

a) Round each length to one significant figure.

b) Round the length of *The Wild Bunch* park to 3 significant figures.

c) If the lengths of the Dreamland Africana and Fun Factory are rounded to 2 significant figures, what is the difference in their lengths?

6) A container of chocolates contains 70 chocolates to one significant figure. What is the smallest number of chocolates that could be in the container?

7) Write each number to the given number of significant figures.

a) 39 1 s.f.

b) 54.9 2 s.f.

c) 0.87 1 s.f

d) 89.54 2 s.f.

e) 6788 3 s.f

f) 56.23 1 s.f.

g) 812 1 s.f.

h) 0.8764 2 s.f.

i) 543.2 3 s.f.

j) 1.098 2 s.f.

k) 367 1 s.f.

l) 4.903 2 s.f

2.2 ESTIMATION AND APPROXIMATION

Estimation only means to make a guess very close to the correct answer. However, it might be impossible to estimate accurately. Estimating an answer to a given calculation gives an approximate answer which is close to the real answer.

We use estimation to check that the answer is about right. It is useful when shopping on a tight budget. Mental estimation of selected goods is vital so that we do not get embarrassed when trying to pay at the counter.

Rounding all the numbers to **one significant figure** is the most convenient way to estimate calculations.

Example 1: Estimate the answers to the problems below.

a) 4.2×5.3

b) 0.9×0.9

c) 48×9.6

d) $\sqrt{99} \times \sqrt{50}$

e) 30.2×8.7

f) $27.8 \div 0.98$

g) 9% of ₦89.95

Solutions: We use approximately equal to sign ($\approx$) when estimating calculations. Therefore, we convert all to one significant figure.

a) $4.2 \times 5.3 \approx 4 \times 5 = \mathbf{20}$

b) $0.9 \times 0.9 \approx 1 \times 1 = \mathbf{1}$

c) $48 \times 9.6 \approx 50 \times 10 = \mathbf{500}$

d) $\sqrt{99} \times \sqrt{50} \approx \sqrt{100} \times \sqrt{49} = 10 \times 7 = \mathbf{70}$

e) $30.2 \times 8.7 \approx 30 \times 9 = \mathbf{270}$

f) $27.8 \div 0.98 \approx 30 \div 1 = \mathbf{30}$

g) 9% of N89.95 $\approx$ 10% of ₦90 = **₦9**

EXERCISE 2E

1) Give an estimate for each of the following calculations.

a) 6.83×9.53

b) 18.99×10.09

c) 42.97×9.87

d) 48% of ₦88.90

e) $110 + 95 + 131$

2) The rent for a house is ₦995 per week. Estimate the total amount spent on rent in one year.

3) A size seven men's shoes cost ₦924. Estimate the cost of 47 shoes.

In questions 4 and 5, there are four calculations and four answers. Write down the correct answer from the list given for each calculation. The answers are estimates.

4) a) 3.2×9.3

b) $20.25 \div 4.89$

c) 42×1.97

d) 3.5×4.8

Answers: 4 20 30 80

5) a) $202.5 \div 99.73$

b) 19×1.83

c) 51% of 499.80

d) 3×0.98

Answers: 3 250 2 40

6) Estimate the answer/value the arrow is pointing at in the diagrams below.

a)

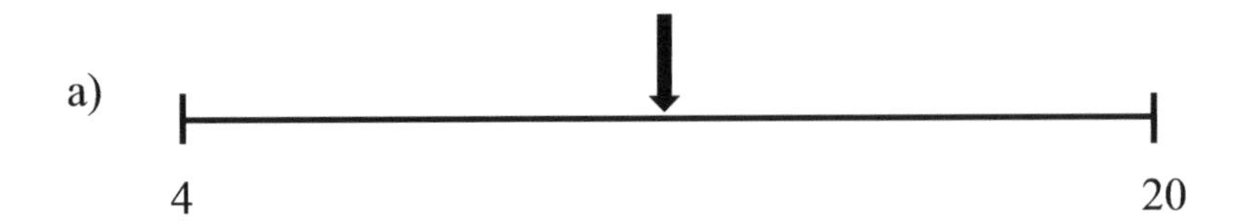

b)

0.0 1.0

c)

7) Estimate the answers to the following calculations.

a) $\dfrac{19 \text{ x } 1.8}{0.89 + 3.9}$

b) 8.93^2

c) $\sqrt{48.88} \text{ x } 1.7$

d) A tenth of 693

e) 13% of 517

f) 0.89% of 1083.45

g) In a grocery shop in London, Chukwudi bought eight tins of sardines at 95 pence each. He paid with a ten-pound note. Describe a quick way for Chukwudi to know how much he should expect as change.

2.3 ADDITION OF WHOLE NUMBERS

There are different ways of adding whole numbers. You can either work out calculations mentally (in your head), work it out on paper or use a calculator which is the easier method.

However, you will not be able to use a calculator all the time as some methods of assessment(s) do not require the utilisation of a calculator.

METHOD 1	METHOD 2
72 + 63 = 70 + 2 + 60 + 3 = 70 + 60 + 2 + 3 = **135**	72 + 63 = 72 7 + 6 = 13 + 63 2 + 3 = 5 **135**
465 + 183 = 400 + 60 + 5 + 100 + 80 + 3 = 400 + 100 + 60 + 80 + 5 + 3 = **648**	465 + 183 465 4+1 = 5 + 183 5 + 3 = 8 5+1 = **6** → **648** **1** 6 + 8 = **14**
3865 + 129 = 3000 + 800 + 60 + 5 + 100 + 20 + 9 = 3000 + 800 + 100 + 60 + 20 + 5 + 9 = **3994**	3865 + 129 3865 + 0129 **3994**

Add 343 + 19

$$\begin{array}{r} 343 \\ +\ \ 19 \\ \hline \mathbf{362} \\ \hline \mathbf{1} \end{array} \checkmark$$

EXERCISE 2F

1) Add the numbers below.

a) 8 + 6

b) 11 + 3

c) 2 + 98

d) 16 + 9

e) 19 + 34

f) 13 + 78

g) 54 + 30

h) 22 + 123

i) 56 + 346

j) 123 + 452

k) 984 + 128

l) 100 + 548

m) 751 + 831

n) 703 + 358

2) Find the total of the numbers below.

a) 10 and 45

b) 13 and 89

c) 76 and 98

d) 123 and 897

e) 786 and 178

f) 987 and 109

g) 564 and 834

3) Add the following numbers below.

a) 123 + 345

b) 17 + 19 + 57

c) 234 + 87 + 987

d) 675 + 432

e) 876 + 368 + 1098

f) 5674 + 390 + 2378

g) 1234 + 5678 + 1890

4) A restaurant displays the sign.
What is the cost of

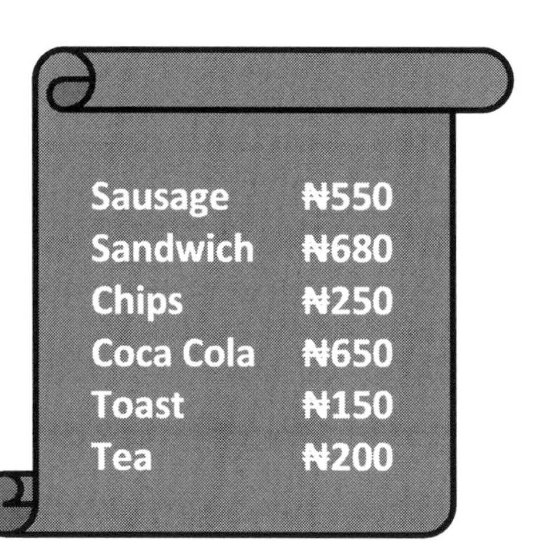

a) two toasts

b) a slice of toast, a sandwich and a cup of tea

c) 5 cups of tea

d) a Coca Cola and three toasts

e) three chips

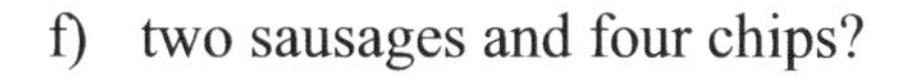

f) two sausages and four chips?

5) Try these

a) $\begin{array}{r} 4357 \\ +\ \ 179 \\ \hline \\ \hline \end{array}$

b) $\begin{array}{r} 2659 \\ +\ \ 137 \\ \hline \\ \hline \end{array}$

c) $\begin{array}{r} 66395 \\ +\ \ 1786 \\ \hline \\ \hline \end{array}$

6) The total of 382 and 7139 is $\boxed{}$

7) Kolade bought a pair of socks for ₦650, a table for ₦3,675 and a car for ₦750,000. How much did he spend altogether?

8) Copy and complete the blanks.

a) $19 + \square = 87$ b) $\square + 765 = 4532$ c) $\square + 54 + 62 = 510$

9) Copy and complete the following.

a) $\begin{array}{r} 6\ \square \\ +\ \ \ \ \ \ \\ 1\ 9 \\ \hline \square\ 1 \\ \hline \end{array}$

b) $\begin{array}{r} 2\ 6\ \square \\ +\ \ \ \ \ \ \ \ \ \\ 1\ 4\ 9 \\ \hline \square\ 1\ 4 \\ \hline \end{array}$

c) $\begin{array}{r} 1\ 2\ 0\ \square \\ +\ \ \ \ \ \ \ \ \ \ \ \ \\ \square\ 4\ 3\ 7 \\ \hline 3\ 6\ 3\ 7 \\ \hline \end{array}$

2.4 SUBTRACTING WHOLE NUMBERS

Some calculations can be done mentally.
Examples

$6 - 2 = 4$ $\qquad$ $100 - 8 = 92$

$10 - 7 = 3$

When subtractions are difficult to do mentally, we can do them on paper.

Example 1: 486 – 43

$$\begin{array}{rccc} & 4 & 8 & 6 \\ - & 0 & 4 & 3 \\ \hline & \mathbf{4} & \mathbf{4} & \mathbf{3} \\ \hline \end{array}$$

Always remember to start subtracting from the right side (ones end)

$6 - 3 = 3$
$8 - 4 = 4$
$4 - 0 = 4$

Example 2: 573 – 19

$$\begin{array}{rccc} & & \mathbf{6} & \mathbf{1} \\ & 5 & \not{7} & 3 \\ - & 0 & 1 & 9 \\ \hline & \mathbf{5} & \mathbf{5} & \mathbf{4} \\ \hline \end{array}$$

3 is smaller than 9, so we take 1 from the next column (tens) and 7 reduces to 6. Place that 1 in front of 3 to make 13.
$13 - 9 = 4$
$6 - 1 = 5$
$5 - 0 = 5$

Example 3: 800 - 253

$$\begin{array}{rccc} & \mathbf{7} & \mathbf{9} & \mathbf{1} \\ & \not{8} & \not{0} & 0 \\ - & 2 & 5 & 3 \\ \hline & \mathbf{5} & \mathbf{4} & \mathbf{7} \\ \hline \end{array}$$

Example 4: 1134 - 129

		2	1
1	1	~~3~~	4
- 0	1	2	9
1	**0**	**0**	**5**

Take 1 from 3 to make 14.
14 – 9 = **5**

3 in the tens column reduces to 2.
2 – 2 = **0**

1 – 1 = **0**

1 - 0 = **1**

EXERCISE 2G

1)

a) 7 – 3
b) 19 – 12
c) 70 – 13
d) 56 – 12
e) 29 – 17
f) 87 – 35
g) 32 – 19
h) 45 – 15
i) 53 – 17
j) 59 – 28
k) 92 – 67
l) 47 – 28
m) 48 - 23
n) 97 - 29
o) 94 - 70
p) 78 - 29
q) 43 - 17
r) 90 - 15

2)

a) 130 – 80
b) 269 – 128
c) 170 – 129
d) 249 – 177
e) 602 – 182
f) 876 – 120
g) 750 – 230
h) 889 – 677
i) 571 – 123
j) 417 – 209
k) 300 – 132
l) 780 – 333
m) 905 - 178
n) 873 - 125
o) 736 - 94
p) 18235 - 1375
q) 11111 - 1075
r) 2980 - 2345

3) College of the Immaculate Conception Volley ball team scored 89 points in a game. How many did they score in the first half if they scored 43 in the second?

4) In an election, People's Democratic Party (PDP) polled 46,235 votes in Anambra State while All Progressives Grand Alliance (APGA) polled 67,789 votes.

a) Who won the election?

b) By how many votes?

5) Find the difference between 19 and 543

6) What is 780 take away 526?

7) Work out

a) $\begin{array}{r} 558 \\ -\ 135 \\ \hline \\ \hline \end{array}$

b) $\begin{array}{r} 6007 \\ -\ 4895 \\ \hline \\ \hline \end{array}$

c) $\begin{array}{r} 8902 \\ -\ 3616 \\ \hline \\ \hline \end{array}$

8) Ifeoma has 165 pens. She gave 73 to her best friend Chikaodi and bought another 33 pens. How many pens does Ifeoma now have?

9) Copy and complete the calculations below.

a) $\begin{array}{r} 3\ 7 \\ -\ 1\ \square \\ \hline \square\ 5 \\ \hline \end{array}$

b) $\begin{array}{r} 6\ \square \\ -\ \square\ 9 \\ \hline 4\ 2 \\ \hline \end{array}$

c) $\begin{array}{r} 8\ 6\ \square \\ -\ \square\ \square\ 4 \\ \hline 6\ 8\ 1 \\ \hline \end{array}$

10) Work out

a) 78 – 34 + 78

b) 987 + 7 – 379

c) 564 – 342 + 876

d) 9452 – 2190 – 456

e) 187 – 98 - 43

f) 1000 – 456 + 7

2.5 MULTIPLYING WHOLE NUMBERS

Some multiplications can be done straight away in your head. However, some cannot.

$5 \times 3 = 15$
$8 \times 2 = 16$
$4 \times 9 = 36$
$7 \times 7 = 49$
$10 \times 10 = 100$

} can be done mentally

257×389 needs to be worked out on paper unless you are a genius or a walking computer. Multiplication table can help when multiplying whole numbers.

TABLE 2A

×	**1**	**2**	**3**	**4**	**5**	**6**	**7**	**8**	**9**	**10**
1	1	2	3	4	5	6	7	8	9	10
2	2	4	6	8	10	12	14	16	18	20
3	3	6	9	12	15	18	21	24	27	30
4	4	8	12	16	20	24	28	32	36	40
5	5	10	15	20	25	30	35	40	45	50
6	6	12	18	24	30	36	42	48	54	60
7	7	14	21	28	35	42	49	56	63	70
8	8	16	24	32	40	48	56	64	72	80
9	9	18	27	36	45	54	63	72	81	90
10	10	20	30	40	50	60	70	80	90	100

$2 \times 8 = 16$

$7 \times 9 = 63$

Three numbers can also be multiplied together.

$1 \times 3 \times 5 = 15$

$2 \times 4 \times 8 = 64$

$3 \times 7 \times 10 = 210$

Example 1: 13 × 4

$$\begin{array}{r} 13 \\ \times \underline{\quad 4} \\ \underline{52} \\ \textcircled{1} \end{array}$$

Always start your multiplication from the ones column.
3 × 4 = 12
Write down the digit **2** in the ones column and move the **1** over to the tens column. 4 × 1 = 1 plus the 1 (moved over) = 5.

Example 2: 24 × 12

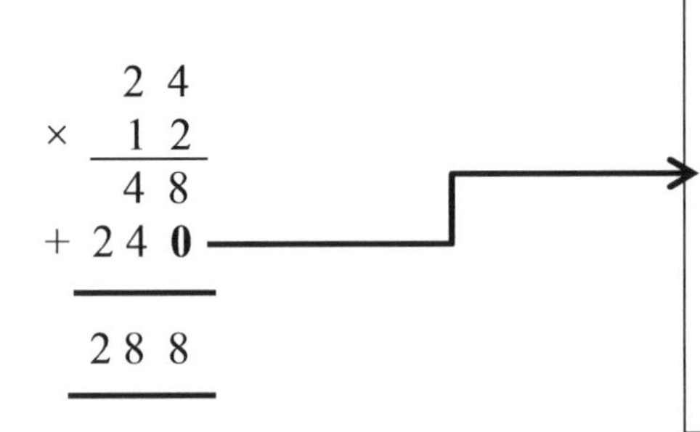

2 × 4 = 8
2 × 2 = 4

Add zero before multiplying by 1

1 × 4 = 4
1 × 2 = 2
Therefore, 24 × 12 = **288**

Example 3: 135 × 45

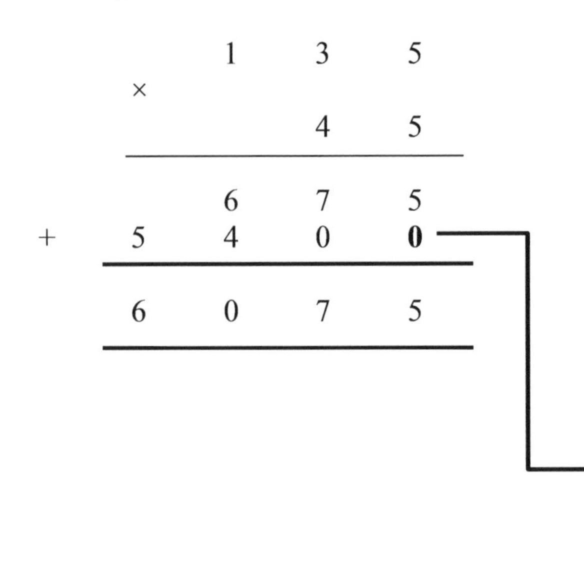

We first multiply 135 by 5 which gives 675.

Before multiplying by 4, add zero to the ones column.

Then, 135 x 4 = 540

Adding 675 and 5400 gives **6075**

THE GRID METHOD

EXAMPLE 4: Multiply 152 by 17

Split 152 into hundreds, tens, and ones. That gives 100 + 50 + 2

Split 17 into tens and ones. That gives 10 + 7

Put all the numbers in a table form.

×	100	50	2
10	**1000**	**500**	**20**
7	**700**	**350**	**14**

100 × 10 = **1000**, 100 × 7 = **700**, 50 × 10 = **500**, 50 × 7 = **350**

2 × 10 = **20** and 2 × 7 = **14**

Add up all the multiplications (blue numbers) to give

1000 + 500 + 20 + 700 + 350 + 14 = 2584

Therefore, 152 × 17 = **2584**

EXERCISE 2H

1) Without using a calculator, work out

a) 3×8
b) 14×9
c) 22×8
d) 61×7
e) 95×4
f) 12×14
g) 5×32
h) 76×3
i) 12×7
j) 98×5
k) 11×7
l) 67×9
m) 34×12
n) 57×13
o) 93×76

2)

a) $6 \times 5 \times 10$
b) $8 \times 3 \times 10$
c) $3 \times 4 \times 5$
d) $3 \times 8 \times 5$
e) $9 \times 2 \times 10$
f) $10 \times 3 \times 5$
g) $5 \times 5 \times 5$
h) $3 \times 7 \times 10$
i) $6 \times 7 \times 5$
j) $2 \times 4 \times 10$
k) $9 \times 2 \times 10$
l) $7 \times 8 \times 10$
m) $10 \times 2 \times 3$
n) $9 \times 2 \times 4$
o) $11 \times 10 \times 2$

3) What is 5 times 78?

4) Multiply 14 by 237

5) Tochukwu has five books. Ikechi has fifteen times as many. How many books does Ikechi have?

6) There are 19 cubes in a block. There are 120 blocks. How many cubes are there?

7) Answer all the questions without using a calculator.

a) 15×16
b) 67×57
c) 65×765
d) 786×12
e) 890×12
f) 34×198

8) Work out the product of 130 and 14

9) A fridge costs ₦25,500. How much would 12 fridges cost?

10) Copy and complete the grid below.

×	3	6	7	9	10
2					
			56		
7		42			
					110

11) The area of a flower garden is rectangular as shown below.

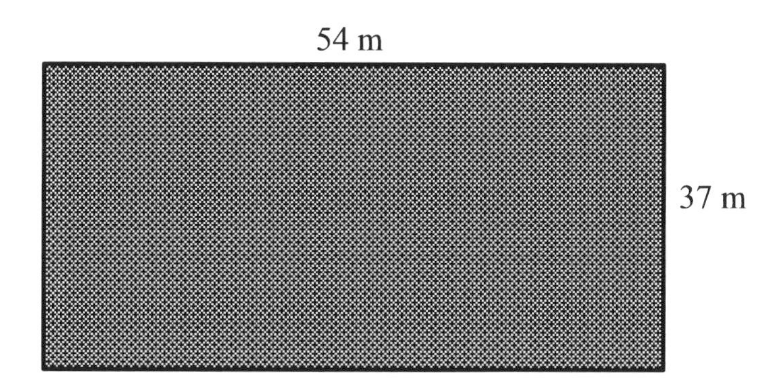

What is the area of the flower garden?

12) When you multiply 368 by 17, what is the value of the digit **2** in your answer?

13) Chinyere paid ₦23,000 into her current account which already had ₦13,400. A week later, her employer paid in three times the amount she had in her bank account. How much is in Chinyere's bank account now?

14) Copy and complete the multiplications below.

a) $13 \times \square = 91$ b) $\square \times 30 = 300$ c) $45 \times 67 = \square$

15) Kolade saves ₦1,350 every week from his salary. How much will he save after

a) 2 weeks b) 5 weeks c) 47 weeks d) 50 weeks?

16) There are 12 eggs in a box. There are 25 boxes. How many eggs are there altogether?

17) Look at the numbers in the box below.

a) What is the largest number multiply by 12?

b) Multiply the third smallest number by the smallest number

c) The product of two numbers in the box above is equal to another number in the box. Find the three numbers.

d) Identify a square number from the box above

18) To watch Super Eagles of Nigeria play football at the National Stadium, Aminu took his two brothers, a cousin, and three friends. The average ticket costs ₦2500. How much would Aminu pay to watch the game with his entourage?

19) An Arts Theatre in Ebonyi State has 47 rows and 35 seats. How many seats are there in the Arts theatre altogether?

20) Copy and complete the multiplication table below.

×	**4**			**9**
2				
		35	40	45
7		49		
	32			

2.6 DIVIDING WHOLE NUMBERS

Division is the opposite of multiplication.

Example 1: 6 ÷ 3

6 ÷ 3 simply means how many times 3 go into 6. In this case, the answer is 2

Example 2: 48 ÷ 12

Knowing your times table helps in division. 48 ÷ 12 means how many times 12 go into 48. In this case, the answer is 4.

Example 3: 6290 ÷ 5

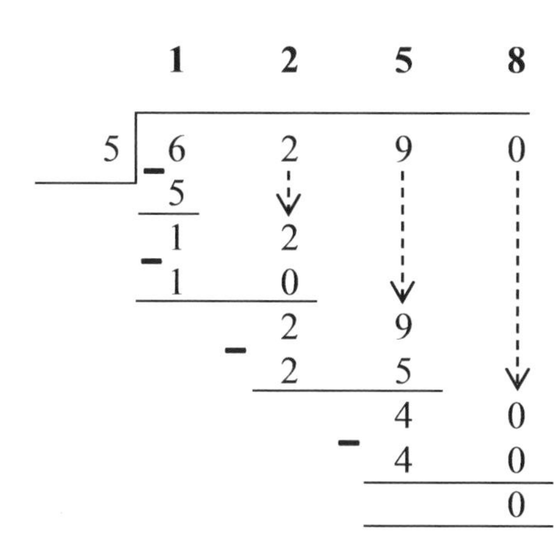

Using long multiplication, 5 goes into 6 once, so we write the **1** on top of 6 as shown.

1 × 5 = 5, so we write the 5 underneath the 6 and take it away. 6 – 5 = 1.

5 cannot go into 1, so we carry down the 2 from the next column to make 12. Now, 5 goes into 12, twice, so we write the 2 on top.

2 × 5 = 10, we take away 10 from 12 to be left with 2.

Again, 5 cannot go into 2, so we carry down the 9 from the next column to make 29. 5 can go into 29 five times, so we write 5 on to and multiply by 5 to give 25.
29 – 25 = 4. Also, 5 cannot go into 4 so we carry down the 0 from the next column to make 40. At this point, 5 can go into 40 exactly to give 8.
We therefore write the 8 on top.
6290 ÷ 5 = **1258**

Alternatively, there is a shorter method.

```
     1    2    5    8
   ____________________
  |       1    2    4
5 | 6     2    9    0
```

> 6 divide by 5 = 1, remainder 1.
> We write the **1** on top and move the remainder (1) to the top of the next number to make 12.
>
> 12 ÷ 5 = 2 remainder 2. We then write the **2** on top and move the remainder (2) to the top of the next number to make 29.
>
> 29 ÷ 5 = 5 remainder 4. We then write the **5** on top and move the remainder (4) to the top of the next number to make 40.
>
> 40 ÷ 5 = **8** exactly, we then write 8 on top.
>
> Therefore, 6290 ÷ 5 = **1258**

Example 4: 25 ÷ 4

Some division are not whole numbers.

```
     0  6.  2  5
    ______________
4 | 2  5
   -0  ↓
   ----
    2  5
   -2  4
   -----
       1  0
      -   8
      -----
          2  0
         -2  0
         -----
             0
```

> 4 cannot go into 2 (using the system of long multiplication), so we write zero (0) on top.
>
> 0 × 4 = 0 and then take zero away from 2 to give 2.
>
> 4 cannot go into 2, so we bring down the 5 from the next column to make 25. Four (4) then goes into 25, six (6) times. We write 6 on top and multiply by 4 to give 24. 25 – 24 = 1.
> We must now put a decimal point as there are no numbers left to bring down. Also, 4 cannot go into 1, we therefore bring down an imaginary zero (0) to make it 10. 4 then goes into 10 twice. We then write 2 on top. 2 × 4 = 8. 10 – 8 = 2. Again, we bring down an imaginary zero to make 20. At that point, 4 will go into 20, 5 times. We then write the 5 on top. 5 × 4 = 20, and 20 – 20 = 0
> Therefore, 25 ÷ 4 = **6.25**

EXERCISE 2I

1) Work out

a) $40 \div 2$	f) $182 \div 14$	k) $225 \div 3$
b) $9 \div 3$	g) $200 \div 20$	l) $2000 \div 5$
c) $44 \div 11$	h) $68 \div 2$	m) $3500 \div 5$
d) $100 \div 5$	i) $645 \div 5$	n) $9028 \div 122$
e) $72 \div 6$	j) $144 \div 9$	o) $15554 \div 7$

2) Share ₦4800 equally between two people. How much will each receive?

3) Divide 144 kg by 6

4) How many 5's make 600?

5) Three people win a prize of ₦1950 between them equally. How much does each person receive?

6) A calculator factory has nine workers to produce 6741 calculators every week. How many calculators does each worker produce per week if they all work at the same rate?

7) Twenty people won a competition. They shared the ₦19,740 prize equally between them. How much did each person receive?

8) A group of 140 tourists travels by coach. Each coach holds 24 people.

a) How many coaches are needed for the tourists?

b) If 60 more tourists joined the group of travellers, how many coaches are needed?

9) Three friends, Tokunbo, Chibogu, and Halima, received ₦9,250 from their Principal for hard work in school.

How much could each person receive?

2.7 MULTIPLYING WHOLE NUMBERS BY 10, 100, 1000

To multiply a number by 10, move the entire digit(s) one place to the left and add a zero (0). Simply put, for whole numbers; just add a zero (because there is only one zero from 10) at the end of the number.

Example 1: 3×10

Add a zero at the end of 3 to make 30.
Therefore, $\mathbf{3 \times 10 = 30}$ ✓

OR

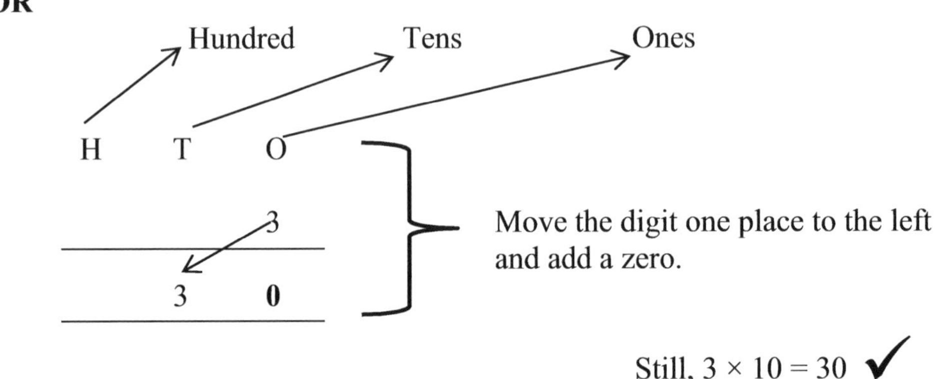

Example 2: 12×10

Just add a zero at the end of 12 to make 120.
Therefore, **12 x 10 = 120**

OR

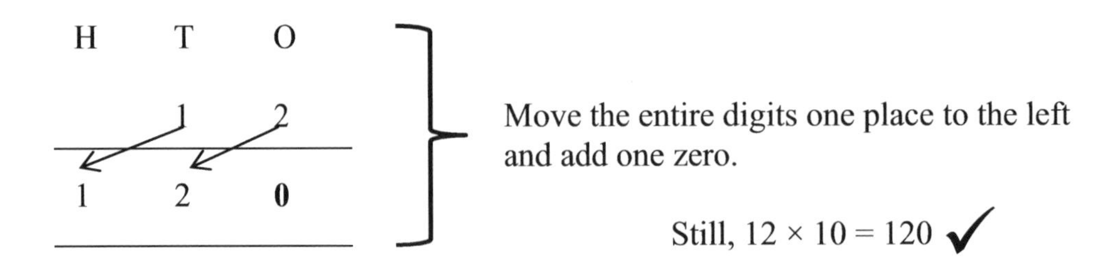

Example 3: 14×100

Add two zeros (because a hundred has two zeros) at the end of 14 to make 1400. Therefore, **14 x 100 = 1400**

OR

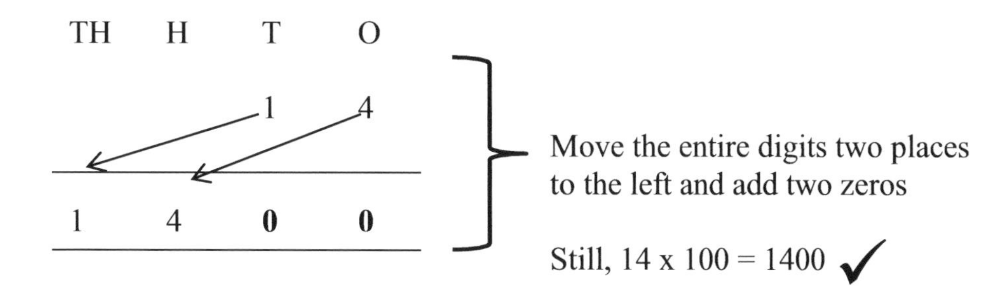

2.8 DIVIDING WHOLE NUMBERS BY 10, 100, 1000....

When dividing numbers by 10, move the entire digits one place to the right
When dividing numbers by 100, move the entire digits two places to the right
When dividing numbers by 1000, move the entire digits three places to the right

Likewise, to divide by 10,000, 100,000... Move all the digits to the right, according to the number of zeros.

Example 1: 70 ÷ 10

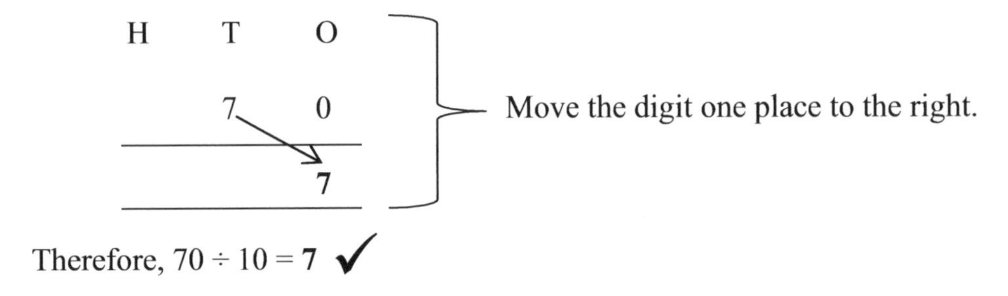

Therefore, 70 ÷ 10 = **7** ✓

Alternatively, you may cancel out each zero to leave you with 7 since you are dividing both sides by 10.

Example 2: 679 ÷ 10

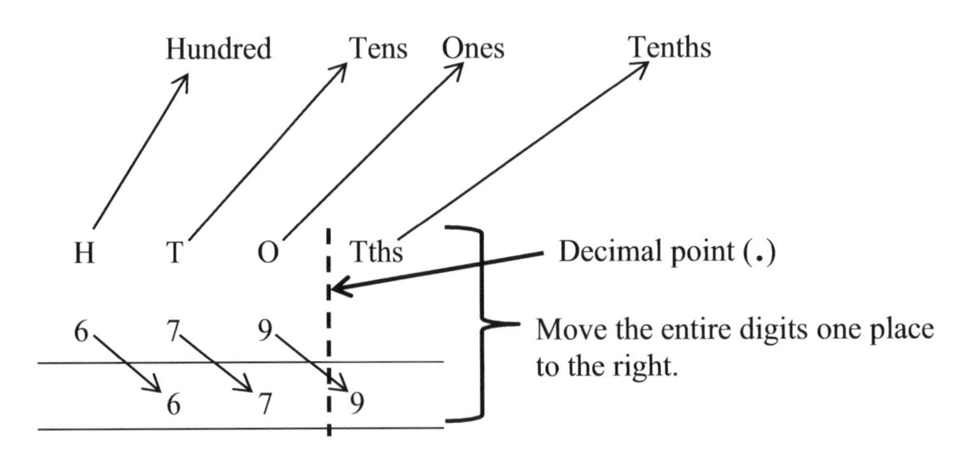

Therefore, 679 ÷ 10 = **67.9** ✓

Example 3: 546 ÷ 100

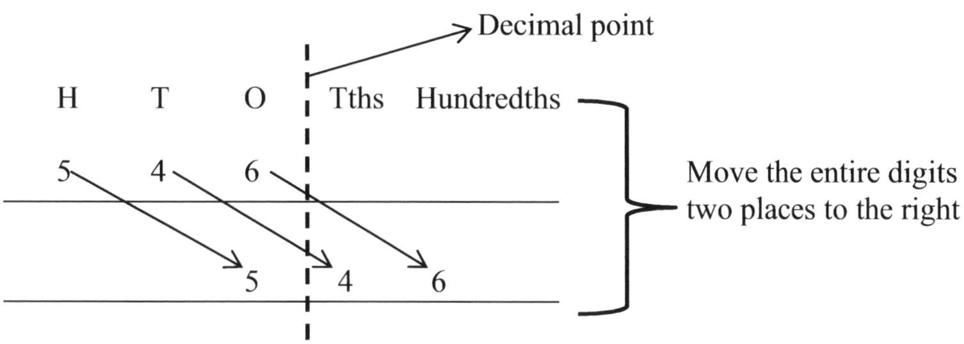

Therefore, 546 ÷ 100 = **5.46** ✓

Example 4: 762 ÷ 1000

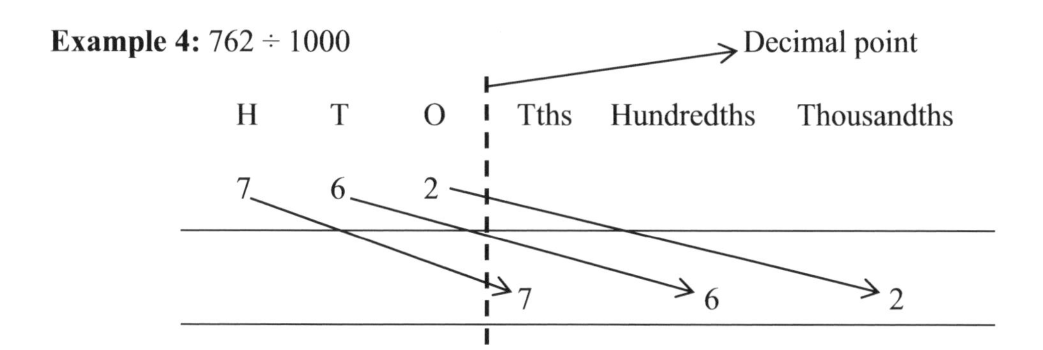

Move the entire digits three places (because there are three zeros in a thousand) to the right.

Therefore, 762 ÷ 1000 = **0.762** ✓

EXERCISE 2J

1) Work out without a calculator.

a) 3×10
b) 43×10
c) 77×10
d) 319×10
e) 400×10
f) 519×10
g) 76×10
h) 431×10
i) 780×10
j) 234×10
k) 980×10
l) 134×10
m) 555×10
n) 1111×10
o) 78965×10

2) Work out without using a calculator.

a) 30×100
b) 42×100
c) 6×100
d) 75×100
e) 319×100
f) 50×100
g) 419×100
h) 917×100
i) 7×100
j) 47×100
k) 9288×100
l) 1112×100
m) 976×100
n) 59×100
o) 86697×100

3) Work out without using a calculator.

a) 3×1000
b) 5×1000
c) 77×1000
d) 318×1000
e) 96×1000
f) 7×1000
g) 431×1000
h) 96×1000
i) 419×1000
j) 60×1000
k) 8776×1000
l) 92×1000
m) 1112×1000
n) 59×1000
o) 486×1000

4) Work out without a calculator.

a) $30 \div 10$
b) $70 \div 10$
c) $460 \div 100$
d) $96 \div 10$
e) $100 \div 100$
f) $465 \div 10$
g) $54321 \div 100$
h) $678 \div 10$
i) $12800 \div 100$
j) $4000 \div 100$
k) $83100 \div 100$
l) $57000 \div 10$
m) $1100 \div 100$
n) $500 \div 100$
o) $5000 \div 100$

5) Work out without a calculator.

a) $9000 \div 1000$
b) $86000 \div 1000$
c) $32450 \div 1000$
d) $10000 \div 100$
e) $6540 \div 10$
f) $4590000 \div 1000$
g) $700000 \div 100$
h) $2340 \div 10$
i) $87000 \div 100$
j) $318 \div 10$
k) $83 \div 100$
l) $89 \div 10$
m) $452 \div 1000$
n) $9874 \div 100$
o) $408 \div 10$

FUNCTIONAL MATHEMATICS

Some promotional items are on display as shown below.

1) What is the cost of three bottles of Nescafe?

2) How much will Blake pay for 3 cartons of Diet Coke?

3) Blake picked a packet of Tetley tea, a carton of Diet Coke and 5 packs of Pepsi Max.
   a) How much will he pay for the items?
   b) He paid with a £20 note. How much change would he receive?

4) Original Tetley tea contains 160 tea bags. In the promotion, 50% extra tea bags were added. How many tea bags are in the promotional pack as shown above?

5) Helen picked 4 bottles of Nescafe. a) How many grams altogether would that be?
   b) She gave the attendant £14. How much change would she receive?

6) Estimate the cost of 15 bottles of Nescafe.

Chapter 2 Review Section

Assessment

1) To the nearest 10, underline the numbers that would **round up**.

12 35 43 57 76 81 89 134 789

.... **2 marks**

2) To the nearest 100, underline the numbers that would **round down**.

123 245 353 397 578 621 705 892 947

.... **2 marks**

3) Round these number to the nearest 10

a) 43

b) 51

c) 79

d) 135

e) 278

f) 312

g) 456

h) 791

i) 837

j) 1456

k) 4567

l) 5555

.... **12 marks**

4) Below are some mountain heights in metres.

Mountain	Height (m)
Mount Elbrus	5642
Lhotse I	8501
Manaslu	8156
Aconcagua	6959

a) Round the height of **Lhotse I** mountain to the nearest 1000 m .... **1 mark**

b) Round the height of Aconcagua mountain to the nearest 100 m .... **1 mark**

c) Round the height of Mount Elbrus to one significant figure .... **1 mark**

d) Round the height of Manaslu mountain to two significant figures .... **1 mark**

5) A number rounded to one significant figure is 4000.
Mbakwe thinks the number could be 3490.
Is Mbakwe correct? Explain fully. .... **2 marks**

6) Copy and complete.

a) 30 + ☐ = 80 .... **1 mark**

b) 47 + ☐ = 134 .... **1 mark**

c) 54 + 78 + ☐ = 564 .... **1 mark**

d) ☐ - 78 = 655 .... **1 mark**

7) Eastern shop sold 15 paintings on Friday, 25 on Sunday and 10 the following Monday.

a) How many paintings will the shop need to order to replace the sold paintings?

b) If the cost of one painting is ₦ 5 450, what is the cost of the paintings sold on Friday?

c) How much was extra money realised on Sunday than on Monday?

…………. **6 marks**

8)

One Tuesday, Asda supermarket had 370 tins of baked beans in stock. However, they sold 49 tins on the same day and took delivery of another 670 tins.

a) How many tins of baked beans were still in the shop at the end of the day?

…..... **2 marks**

b) The cost of a tin of beans is £1.05. How much baked beans was sold on Tuesday?

……. **2 marks**

9) Work out

a) 23×100	……… 1 mark	d) 12×45	……... **1 mark**
b) 768×1000	……… 1 mark	e) 763×9	……… **2 marks**
c) 90089×10	……… 1 mark	f) 421×56	……… **2 marks**

10) The cost of an annual ticket to watch Enugu Rangers football club is ₦55 120.

How much would a ten-year ticket cost? ……….. **1 mark**

11) Complete the multiplication grid.

×	2	5		7	9
3			18		
					18
7					

……….. **3 marks**

12) Below is a magic square. The sum of the entries of any row, any column or any diagonals is the same.

Complete the magic square

11		
12	10	
		9

…………. **3 marks**

13) College of the Immaculate Conception Alumni Association has ₦12 500 000 in the bank. ₦345 786 was spent on painting the walls of the school. How much money is left in the account? …………. **2 marks**

14) Udoka earns ₦135 456 per month. How much does she earn in 9 months? …………. **2 marks**

15) This ceiling light casing holds three bulbs. The cost without the bulbs is ₦3 450.

Amaka wants to buy 9 of the ceiling light casings for her new house project.

a) How much will Amaka pay for the ceiling light casings? .... **2 marks**

The cost of a bulb is ₦430.

b) How much will she pay for bulbs **only** to fit into all the nine casings? .... **2 marks**

c) How much will she pay altogether to fit all the bulbs and casings in her new house if the cost of labour is ₦20 000? .... **2 marks**

16) A container is worth **£25** and can take up to 30 indoor footballs.

a) Find the cost of 5 indoor footballs. ...... **1 mark**

b) If the container is full of footballs, how much is the container worth? .... **2 marks**

c) To the nearest £10, what is the price of a container? ....**1 mark**

d) Round your answer to **part a** to the nearest whole number. .... **1 mark**

e) Round the price of three footballs to one decimal place. .... **1 mark**

17) 2023 students are to be divided into seven equal groups. How many students will be in each group? .... **2 marks**

18) Work out 245 ÷ 5 .... **1 mark**

19) The diagram shown journeys between some towns in Nigeria.

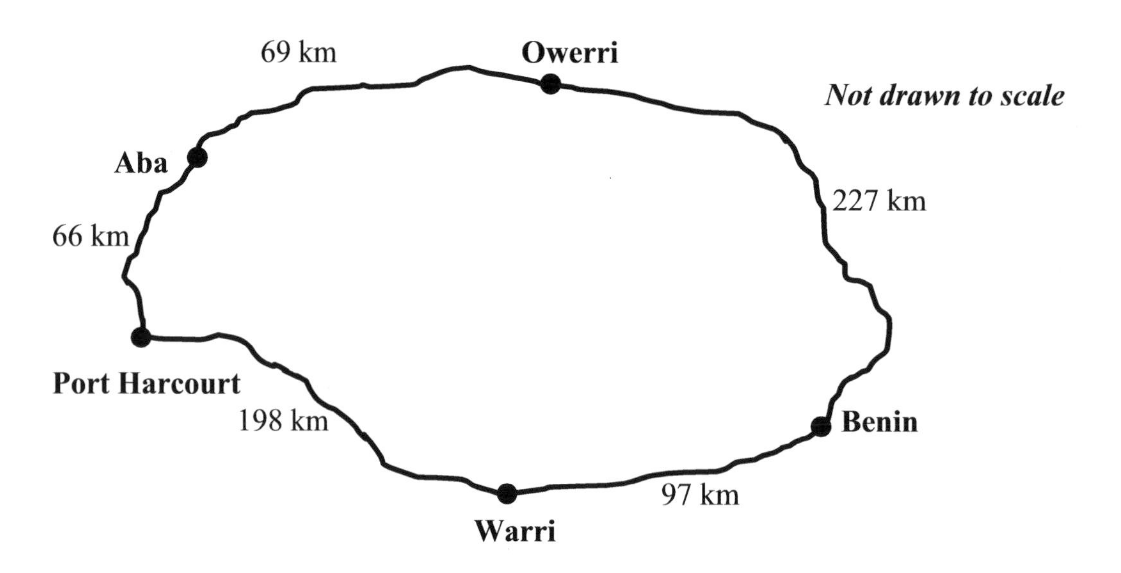

a) Which is the quicker way from Port Harcourt to Benin? Show all working out. .... **2 marks**

b) What is the distance from Aba to Benin via Port Harcourt? .... **1 mark**

c) Wole left Warri in the morning and returned in the evening. What was his total distance for the entire journey? .... **2 marks**

d) George's residence is in Aba, and he wants to travel to Benin. His friend, Achike advised that his shortest route would be to pass through Owerri.
Is Achike correct? Explain fully. .... **3 marks**

20) Round the answers to these calculations to **one** significant figure.

a) $34 + 87$

b) $123 + 456$

c) $987 – 453$

d) $452 - 98$

e) $1000 - 675$

f) $3590 – 1999$

……… **12 marks**

21) Round the answers to the calculations in question 20 to **two** significant figures

……… **12 marks**

22) Round to one decimal place.

a) 1.26

b) 23.53

c) 0.95

d) 897.345

e) 3.111

f) 98.567

……… **6 marks**

23) Estimate the answers to these calculations.

a) $23 + 45.7$ ……… **2 marks**

b) $603 + 789.9$ ……… **2 marks**

c) $\frac{6.789 \times 4.8}{5.06 + 1.6}$ ……… **2 marks**

24) Work out the following **without** using a calculator. Show all working out.

a) $270 \div 6$ ……… **2 marks**

b) $120 \div 5$ ……… **2 marks**

c) $3222 \div 9$ ……… **2 marks**

d) $7896 \div 12$ ……… **2 marks**

3 Number Work 3

This section covers the following topics:

- Adding and Subtracting Decimal Numbers
- Multiplying and Dividing Decimal Numbers
- Multiplying Decimals by 10, 100, 1000
- Writing as decimal numbers
- Ordering Decimal

LEARNING OBJECTIVES

By the end of this unit, you should be able to

a) add and subtract decimal numbers
b) multiply decimals
c) order decimals

KEYWORDS

Decimal numbers
Add
Subtract
Decimal point
Order
Multiply
Divide

In its simplicity, decimal numbers are numbers with decimal points. As we move right from the decimal point, each place value is divided by 10 to have tenths, hundredths, thousandths, etc.

Examples of decimal numbers are: 6.2, 5.75, 0.9 789.45

3.1 ADDING AND SUBTRACTING DECIMALS

Care must be taken when adding and subtracting decimals. We must line up the decimal points on top each other and then do the calculations.

Example 1: 2.5 + 3.3

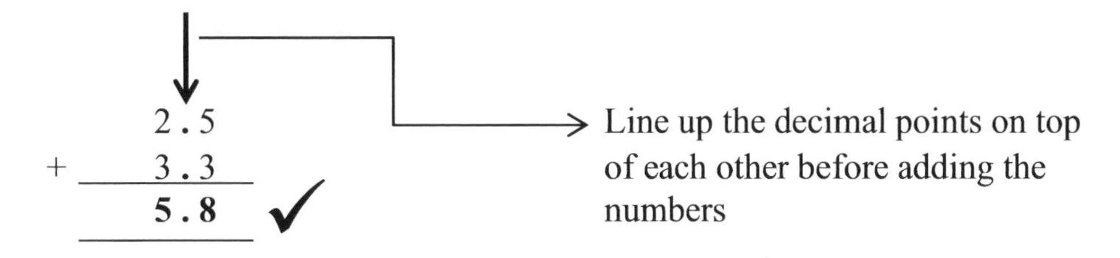

Example 2: 1.79 + 0.65

```
     1 . 7 9
+    0 . 6 5
     -------
     2 . 4 4  ✓
     -------
     1   1
```

Example 3: ₦20.50 + 20 kobo

Convert 20 kobo to Naira
100 kobo = 1₦

So, 20 kobo = 20 ÷ 100 = ₦0.20

```
     2 0 . 5 0
+      0 . 2 0
     ---------
     2 0 . 7 0  ———>  ₦20.70 ✓
     ---------
```

Example 4: 568. 457 – 45.89

```
              1
          7   3 1
     5 6 8 . 4 5 7
-      4 5 . 8 9 0
     -------------
     5 2 2 . 5 6 7  ✓
     -------------
```

Example 5: 600 – 253.4

```
    5 9 9   1
    6 0 0 . 0
-   2 5 3 . 4
   -----------
    3 4 6 . 6  ✓
   -----------
```

Example 6: 0.076 – 0.009

```
          6 1
    0 . 0 7 6
    0 . 0 0 9
   -----------
    0 . 0 6 7  ✓
   -----------
```

EXERCISE 3A

Addition questions

1) 4.43 + 1.25 = ______

2) 8.6 + 1.9 = ______

3) 0.35 + 7.19 = ______

4) ₦18.37 + ₦17.09 = ______

5) 1.9 + 3.5 = ______

6) 15.35 + 0.39 = ______

7) 139.05 + 17.99 = ______

8) 82.50 + 2.75 = ______

9) 56.35 + 1.2 = ______

10) 00.84 + 18.10 = ______

11) 6.2 + 3.5

12) 8.03 + 3.9

13) 0.3 + 0.5 + 1.2

14) 100.35 + 0.79 + 1.75

15) 18.34 + 1.7 + 0.87

16) 8.09 + 0.57 + 12.874

17) 222.39 + 100.1 + 523.23

18) 654.37 + 12.88 + 0.076

19) Gbenga bought a box for £15.87. He had £20.75 left after buying the box. How much money did he have before buying the box?

20) Three iron rods are joined. Each length is 18.56 cm, 45.9 cm and 81.13 cm. What is the total length of the three iron rods?

21) Work out these decimals and leave your answer in kilometres.

a) 4.35 km + 98.07 km + 56 m

b) 7000m + 54.34 km + 65.3km

22) Work out the perimeter of the rectangular garden below in metres.

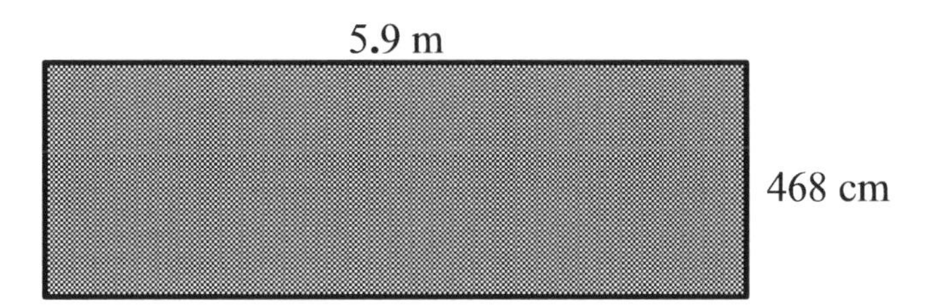

23) In a chemistry experiment, a container full of water has a mass of 2.298kg. The mass of the container is 0.21 kg. What is the mass of water in the container?

EXERCISE 3B

Subtraction questions

1) 5.63
 - 3.52

7) 1.00
 - 0.86

2) 3.68
 - 2.97

8) 12.35
 - 7.89

3) 31.93
 - 4.52

9) 200.5
 - 129.4

4) 18.05
 - 13.98

10) 11.7
 - 2.3

5) 7.29
 - 3.18

11) 8.29
 - 2.43

6) 0.423
 - 0.38

12) 745.078
 - 45.99

13) 86.3 – 11.5

14) ₦657.89 - ₦34.45

15) 169kg – 34.89kg

16) 25.89 – 2.09

17) James has ₦1759.35 in his bank account. He withdraws ₦1590.50. How much money is left in his bank account?

18) Ifeoma bought a watch costing ₦2030.56 and paid with a cheque of ₦3000.00 How much change did she get?

19) Copy and complete.

a) 54.43 - ☐ = 34.72

b) 900.8 - ☐ = 769.73

c) ☐ - 67.1 = 86.34

d) ☐ - 0.089 = 0.987

e) 87.654 - ☐ = 45.5

f) ☐ - 100.345 = 131.76

Write down two possible values for questions 20 - 23

20) ☐ - ☐ = 7.25

21) ☐ - ☐ = 102.57

22) ☐ - ☐ = 66.66

23) ☐ - ☐ = 993.8

24) Taiwo added two amounts, ₦345.98 and ₦23.93. Find the difference between his total and ₦11 701.

3.2 MULTIPLICATION OF DECIMAL NUMBERS

Multiplying with decimal numbers is the same as multiplying with whole numbers. The only difference is where the decimal point would be.

When decimal numbers are multiplied, the answer must have the same number of decimal places as in the original question. There are numerous methods of multiplying decimals, but only three methods will be explained.

Method 1 for multiplying decimal numbers

- Make sure you line up the numbers on the right and do not try to align the decimal points.
- Then multiply in the same way you would **whole numbers.**
- Finally, starting from the right, place the decimal point in the answer by moving the number of places that are equal to the sum of the decimal places in the question.

Example 1: 4.23 × 1.5

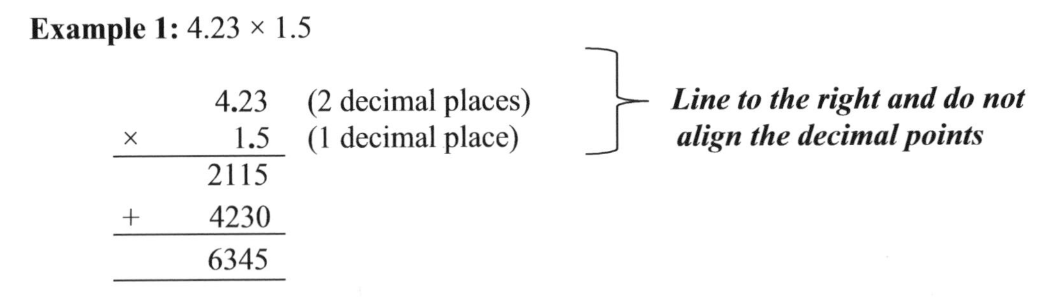

We must at this point, put back the decimal point. From the original question, **2** decimal places + **1** decimal place = 3 decimal places. Therefore, the answer **must** have three decimal places. 4.23 × 1.5 = **6.345** ✓

Check your answer: You may check your answer for the correct decimal place value by looking at the two numbers you multiplied. 4.23 is not up to 5, so take it as **4.** Also, 1.5 is not up to 2, so take it as **2** (use sensible rounding).

Therefore, 4 × 2 = 8 this means that your answer must be close to 8.
6.345 is closer to 8 than any other position you may put the decimal point, and hence the right answer. 0.6345 is too small and far away from 8. 63.45 is too far away from 8. 634.5 is far too away from 8.

Method 2 of multiplying decimal numbers

- Take away the decimal points
- Multiply the whole numbers together
- Put the decimal point back by counting the number of decimal points from the question given.

Example 2: 0.34×465.1

Take away the decimal points. 34×4651

USING THE COLUMN METHOD

			4	6	5	1
		×			3	4
		1	8	6	0	4
+	1	3	9	5	3	0
	1	5	8	1	3	4
		1	1			

Altogether, 0.34 and 465.1 have 3 decimal places.

Therefore
$0.34 \times 465.1 =$ **158.134** ✓

Example 3: 0.2×0.5
Firstly, $2 \times 5 = 10$

Secondly, 0.2 and 0.5 have two decimal places altogether. Therefore, the answer must have two decimal places. 0.2×0.5 then equals **0.10** ✓

Example 4: 7×0.3
$7 \times 3 = 21$

Since there is only one decimal place in the question (7×0.3), the answer must have one decimal place as well. $7 \times 0.3 =$ **2.1** ✓

Method 3: Grid method

Example 5: 1.8×24.5

Split 1.8 into 1 + 0.8

Split 24.5 into 20 + 4 + 0.5

Put all the numbers in a table form.

×	1	0.8
20	**20**	**16**
4	**4**	**3.2**
0.5	**0.5**	**0.4**

$1 \times 20 =$ **20**, $1 \times 4 =$ **4**, $1 \times 0.5 =$ **0.5**,

$0.8 \times 20 =$ **16** $0.8 \times 4 =$ **3.2** $0.8 \times 0.5 =$ **0.4**

Add up all the multiplications (blue numbers) to give

$20 + 16 + 4 + 3.2 + 0.5 + 0.4 = 44.1$

Therefore, $1.8 \times 24.5 =$ **44.1**

MULTIPLYING DECIMAL NUMBERS BY 10, 100, 1000 …

Example 5: 4.5 × 10

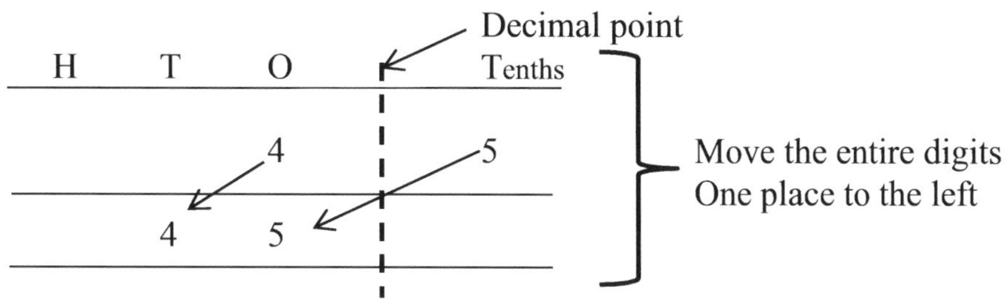

Remember that the tenths column is on the right side of the one's column.
Therefore, 4.5 × 10 = **45** ✓

* There is an **unconventional method** which you might find quicker and useful, though not recommended. That is where the decimal point moves to the right the same number of zeros when multiplying. When multiplying by 10, the decimal point moves to the right once.

So, 4.5 × 10

4 . 5

45. = **45**

Likewise, to multiply a number by 100, move the entire digits two places to the left and add two zeros. Do same for 1000, 10,000 and so on.

Example 6: 15.3 × 100

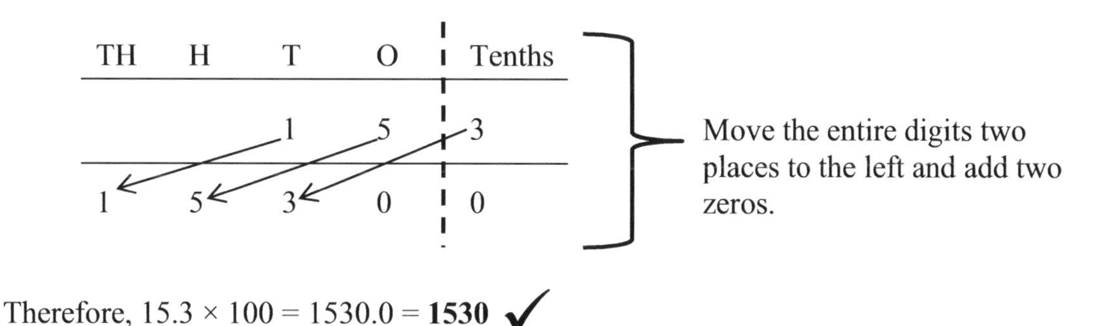

Therefore, 15.3 × 100 = 1530.0 = **1530** ✓

EXERCISE 3C

Work out and show all working out.

1) 2.8×4.5
2) 1.9×0.2
3) 0.08×3.4
4) 0.5×19
5) 8.76×2.1
6) 79.76×0.5
7) 5.55×5
8) 760.07×1.2
9) 0.2×0.2
10) 0.4×0.4
11) 5.43×3.2
12) 0.9×56
13) 21.3×4.3
14) 4.20×0.5
15) 0.9×0.9
16) 4.21×7.6
17) 213.76×0.2
18) 67.23×2.1
19) 100.3×7
20) 5.90×3.45

EXERCISE 3D

Work out

1) 3.2×10
2) 4.56×10
3) 100×176.34
4) 5.123×1000
5) 0.82×10
6) 13.9×1000
7) 800.34×100
8) 0.005×1000
9) 1.897×10
10) 100×18.39

EXERCISE 3E PROBLEM SOLVING (1 – 9)

1) Paving slabs cost ₦654.50 each. Work out the total cost of 64 slabs.

2) Anthony has a part-time job as a bouncer in a South London nightclub. He is paid £7.25 for each hour he works. Last week, Anthony worked for 35 hours. Work out Anthony's total pay for the week he worked.

3) A British charity shop buys 123 literature books, each costing 15 pence. What is the total cost of the books in British Pound Sterling?

4) Cameron got three prescriptions for stomach upset from his doctor. He is not entitled to free medicals. The cost of each prescription is £8.60. Work out how much Cameron would pay for his prescriptions.

5) A girl buys 100 pens each costing £0.31. What is the total cost of the pens?

6) A bottle of Olive oil costs £3.79. Adam bought 19 bottles. Work out the total cost.

7) In question 6 above, Adam gave the cashier at the shop, £80. What change would Adam receive?

8) Work out the total cost of 10 calculators at £4.50 each, 17 shirts at £7.25 each and two laptops at £399.50 each.

9) A private lawn tennis court has rectangular dimensions as shown below.

Work out the area of the lawn tennis court.

10) Write down the product of 23.45×1.2

11) Express 6.53 kg in tonnes.

12) Work out:

a) $(4.5)^2$

b) $(0.7)^2$

c) $(0.03)^2 + (0.4)^2$

d) $43.56 - (1.3)^2$

WRITING AS DECIMAL NUMBERS

Remember, $10^2 = 10 \times 10 = \mathbf{100}$

$10^3 = 10 \times 10 \times 10 = \mathbf{1000}$

$10^6 = 10 \times 10 \times 10 \times 10 \times 10 \times 10 = \mathbf{1000000}$

Example 1: Write 4.56×1000 as a decimal number.

$4.56 \times 1000 = \mathbf{4560}$

Example 2: Write 0.73×10^2 as a decimal number.

$0.73 \times 100 = \mathbf{73}$

Example 3: Write 923.67×10^6 as a decimal number.

$923.67 \times 1000000 = \mathbf{923670000}$

Example 4: Write $\frac{651}{1000}$ as a decimal number.

$651 \div 1000 = \mathbf{0.651}$

EXERCISE 3F

Write your answer as decimal numbers.

1) 0.78×10^4
2) 34.78×10^2
3) $9.07 \times 10^2 \times 10^3$
4) 4312.89×10000
5) 100×10^6
6) 0.54×100
7) 12.145×10^5
8) 0.007×10^3
9) 8.88×100
10) $10^3 \times 456.54$

3.3 DIVIDING DECIMALS

There are different ways of dividing two decimal numbers, depending on which one you prefer. However, the method outlined below is straightforward and easy to remember when dividing decimals.

STEPS FOR DIVIDING DECIMAL NUMBERS

- Write the question as a fraction
- The numerator and denominator **must** be multiplied by the same number using 10, 100, 1 000 or any multiples of 10 appropriate to the given question. The effect of this is that the two numbers give a whole number which is easy to divide.
- Using any preferred method of your choice, divide the numerator by the denominator.

Example 1: Work out $46 \div 0.2$

Using the steps above, write as a fraction: $\frac{46}{0.2}$

It is appropriate to multiply by ten since it will make both the denominator and numerator, whole numbers. The fraction becomes

$$= \frac{46 \times \mathbf{10}}{0.2 \times \mathbf{10}}$$

$$= \frac{460}{2}$$

$$= \mathbf{230} \checkmark$$

> Always remember that if we divide by a number greater than 1, the answer is less than the original number. However, if we divide by a number between 0 and 1, the answer is greater than the original number.

Example 2: 9.75 ÷ 0.01

Write as a fraction $\frac{9.75}{0.01}$

Multiply numerator and denominator by 100

$$= \frac{9.75 \times \mathbf{100}}{0.01 \times \mathbf{100}} = \frac{975}{1} = \mathbf{975}$$ ✓

Example 3: Divide 4.86 by 0.0162

Write as a fraction $\frac{4.86}{0.0162}$

Now multiply numerator and denominator by 10 000

$$= \frac{4.86 \times \mathbf{10\ 000}}{0.0162 \times \mathbf{10\ 000}}$$

$$= \frac{48600}{162}$$

You may consider cancelling down at this point.
Divide top and bottom by 2.

$$= \frac{24300}{81}$$

Divide top and bottom by 27

$$= \frac{900}{3}$$

Divide top and bottom by 3

$$= \mathbf{300}$$ ✓

Other things to know about dividing decimal

- Dividing by 0.1 is the same as multiplying by 10
- Dividing by 0.01 is the same as multiplying by 100
- Dividing by 0.001 is the same as multiplying by 1000 and so on.

$4.3 \div 0.1 = 4.3 \times 10 = 43$

$8.635 \div 0.01 = 8.635 \times 100 = 863.5$

EXERCISE 3G

1) Work out:

a) $17.4 \div 0.2$

b) $1.5 \div 0.03$

c) $32 \div 0.4$

d) $31 \div 6.2$

e) $0.96 \div 3$

f) $1.3 \div 0.1$

g) $4.9 \div 0.01$

h) $5.706 \div 0.001$

i) $42.5 \div 100$

j) $0.5 \div 0.01$

2) Work out:

a) $0.22 \div 11$

b) $0.055 \div 0.11$

c) $0.0257 \div 0.005$

d) $23.4 \div 9$

e) $1.844 \div 0.2$

f) $9.9 \div 100$

g) $3.74 \div 5$

h) $8.7 \div 0.4$

i) $16.8 \div 7$

j) $34.5 \div 10$

In questions 3 to 7, copy and fill in the missing numbers.

3) $1.1 \times \square = 3.3$

4) $1.4 \times \square = 2.94$

5) $0.2 \times \square = 0.06$

6) $0.5 \times \square = 0.001$

7) $8.6 \times \square = 27.52$

8) Nwachukwu bought 3.5 metres of cloth for ₦2 303. How much did he pay for a metre of cloth?

9) Chuma multiplied 56 and 72 and got 4032.

   Work out the value of $40.32 \div 7.2$

Work out the value of:

10) $558 \div 3$

11) $5.58 \div 0.3$

12) $55.8 \div 0.03$

13) $558 \div 0.3$

14) $5.58 \div 0.03$

15) $5.58 \div 0.003$

16) $5.58 \div 3$

17) $55.8 \div 3$

3.4 ORDERING DECIMALS

Ordering decimals simply mean arranging the decimal numbers from ascending or descending order taking into account the place values.

Some people can order decimals by mere inspection while others require a stringent rule to guide them.

Below are some rules to follow when ordering decimals.

- Line the numbers on top of each other in a place value table.
- Identify the first significant figure of each number and note its value
- Starting from the column farthest left and moving to the right, the number with the highest first significant **figure and value** is the biggest number.
- Repeat the process to obtain the next largest number and so on.

Example 1: Order the following numbers from highest to lowest
4.89   48.9   489   4.98

Hundreds	Tens	Ones	•	Tenths	hundredths
		4	•	8	9
	4	8	•	9	
4	8	9	•		
		4	•	9	8

From the above table, it is glaring that the number with the highest first significant figure value is 489. The first significant figure is 4 with a value of 400. It is then followed by 48.9 as 4 is the first significant figure with a value of 40. We are left with 4.89 and 4.98 both with the first significant figure of 4 with a value of 4. At this point, we look at the next column to the right, which is the tenths column. 9 tenths is bigger than 8 tenths, so 4.98 is greater than 4.89.

Therefore, in descending order (highest to lowest): **489, 48.9, 4.98, and 4.89** ✓

Example 2: Write these numbers in ascending order:

245.7   2.457   24.57   0.2457   0.25

Again, draw the place value table and insert the numbers in their correct positions.

Hundreds	Tens	Ones	•	tenths	hundredths	thousandths	ten thousandths
2	4	5	•	7			
		2	•	4	5	7	
	2	4	•	5	7		
		0	•	2	4	5	7
		0	•	2	5		

From the above table, it is glaring that the number with the highest first significant figure value is 245.7. The first significant figure is 2 with a value of 200. It is then followed by 24.57 as 2 is the first significant figure with a value of 20. From the ***Ones*** column of the remaining numbers, 2.457 is the next higher number since the first significant figure is 2 with a value of 2. We are then left with 0.2457 and 0.25. They have the same values in the ones and tenths column, so we can **only** compare them from the hundredths column. We can see that 5 is higher than 4, so 0.25 is bigger than 0.2457.

Therefore, in ascending order (lowest to highest):

0.2457, 0.25, 2.457, 24.57 and 245.7 ✓

EXERCISE 3H

Write these decimal numbers in ascending order.

1) 2.5	0.41	4.01	2.14	
2) 40.54	4.054	4.05	4.504	
3) 4.2	2.4	0.24	42.4	
4) 9.6	6.90	96	9.69	
5) 5	0.53	5.35	1.53	0.953
6) 0.5	0.7	0.43	0.6	0.57
7) 2.6	6.2	26.2	20.6	21.62

Sort these numbers in descending order.

8) 1.3	3.1	1.43	4.31
9) 6.85	86.5	80.65	8.65
10) 1.3	1.5	0.9	2.7
11) 8.9	8	8.8	9.88
12) 10.3	30.1	1.3	13.3
13) 0.032	0.32	0.023	0.23
14) 1.87	8.17	0.187	0.78

In questions 15 – 20, copy and complete by writing =, < or > in the box.

15) 1.7 ☐ 7.1

17) 17.0 ☐ 17

19) 89.6 ☐ 98.6

16) 11.8 ☐ 1.8

18) 0.3 ☐ 0.03

20) 0.42 ☐ 0.421

Chapter 3 Review Section

Assessment

1) Work out.

a) 4.3 + 2.2 = ______

b) 9.6 + 7.8 = ______

c) 34. 8 - 2.6 = ______

.... **3 marks**

2) Work out the following.

a) 0.67 + 4.7 .... **1 mark**

b) 1.2 + 6.4 .... **1 mark**

c) 0.78 + 0.12 .... **1 mark**

d) 678.43 – 45.98 .... **1 mark**

3) Put each number in order of size, from lowest to highest.

a) 0.7, 7.2, 0.07, 0.712 .... **1 mark**

b) 4.67, 4.6, 4.76, 4.7 .... **1 mark**

c) -0.3, -0.55, -0.4, -6 .... **1 mark**

d) 3.08, 2.83, 3.83, 2.38 .... **1 mark**

4) Copy and complete

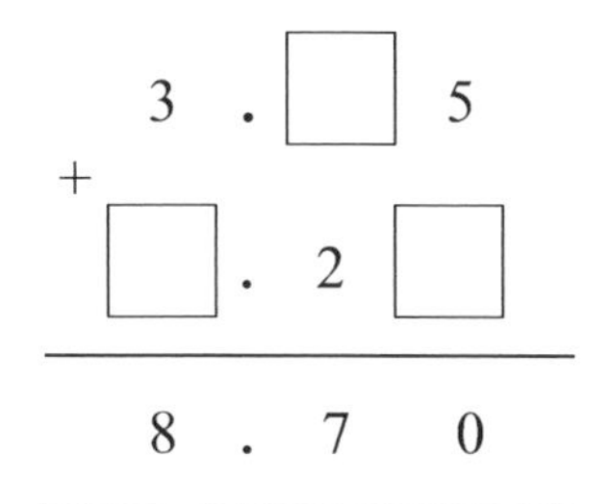

.... **2 marks**

5)

Biodun and his family went for a vacation in London and bought two maths books as shown above for his two sons, Kola and Ade.

a) How much did he spend altogether? .... **1 mark**

b) What is the difference in price for a copy of **each** maths book?
.... **1 mark**

c) The Bunny maths was for Kola while Algebra Secrets was for Ade.
Ade's friend liked his book and informed his mother. The mum went to a shop and bought **three** copies. How much did she spend on the three books?
.... **2 marks**

d) Ade's mum paid for the maths books with a fifty-pound note. How much did she get back as change? …………… **1 mark**

6) In a 100 m sprint, Chukwudi's time for his first run was 14.25 seconds. He tried again and was 0.36 seconds quicker. What was Chukwudi's time for the second trial/run?

7) Work out

   a) 43.9×100 ………. **1 mark**

   b) $76.7 \div 10$ ………. **1 mark**

   c) 0.078×1000 ………. **1 mark**

   d) $239.34 \div 100$ ………. **1 mark**

8) Work out

   a) 4.3×7 ………. **1 mark**

   b) 2.1×5 ………. **1 mark**

   c) 34×1.2 ………. **1 mark**

   d) 78.8×0.5 ………. **2 marks**

9) Work out

   a) $400 \div 0.8$ ………. **2 marks**

   b) $27.2 \div 6.8$ ………. **2 marks**

   c) 56.7×0.04 ………. **2 marks**

   d) 0.023×3.3 ………. **2 marks**

10) A crate of free range eggs costs £2.50. A box of sweets costs £3.75. Ndubuisi buys three crates of eggs and four boxes of sweets.

   a) How much does he spend altogether?
   ……….. **3 marks**

   b) What is the difference in cost between a crate of eggs and a box of sweets?

………… **1 mark**

11) A field is represented by the diagram below.

What is the perimeter of the field?

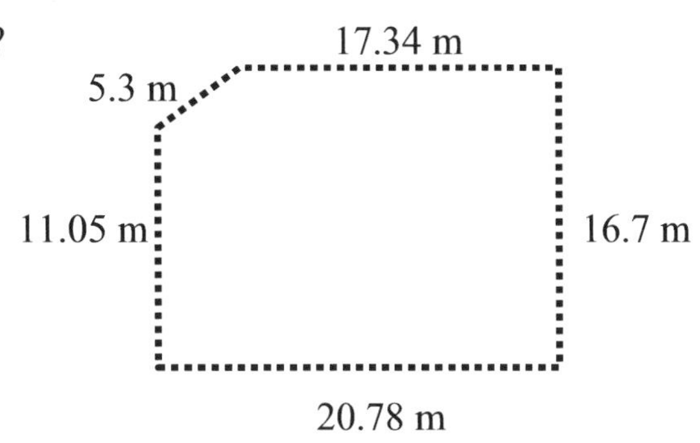

……………..… **2 marks**

12) In May, Hassan weighed 83.4 kg. He was advised to go on a diet and he did. In October that same year, he weighed 77.8 kg.

a) How much weight did Hassan lose? …………………. **1 mark**

b) In April the next year, his weight was 94.5 kg.
How much weight did he gain since May last year?

…………………. **2 marks**

13) Work out

a) 278 + 8.87 + 123.09 …………………. **1 mark**

b) 18.95 – 6.12 + 57.3 …………………. **1 mark**

c) 9 – 0.06 …………………. **1 mark**

d) 32 – 0.5 …………………. **1 mark**

e) 0.0534 ÷ 0.006 …………………. **2 marks**

14) A male tailor buys 8.9 metres of cloth for making trousers at £18.30 per metre.
How much did he pay for the cloth?

…………………. **3 marks**

15) Offiong went to a restaurant in Akwa Ibom with Ekaete.

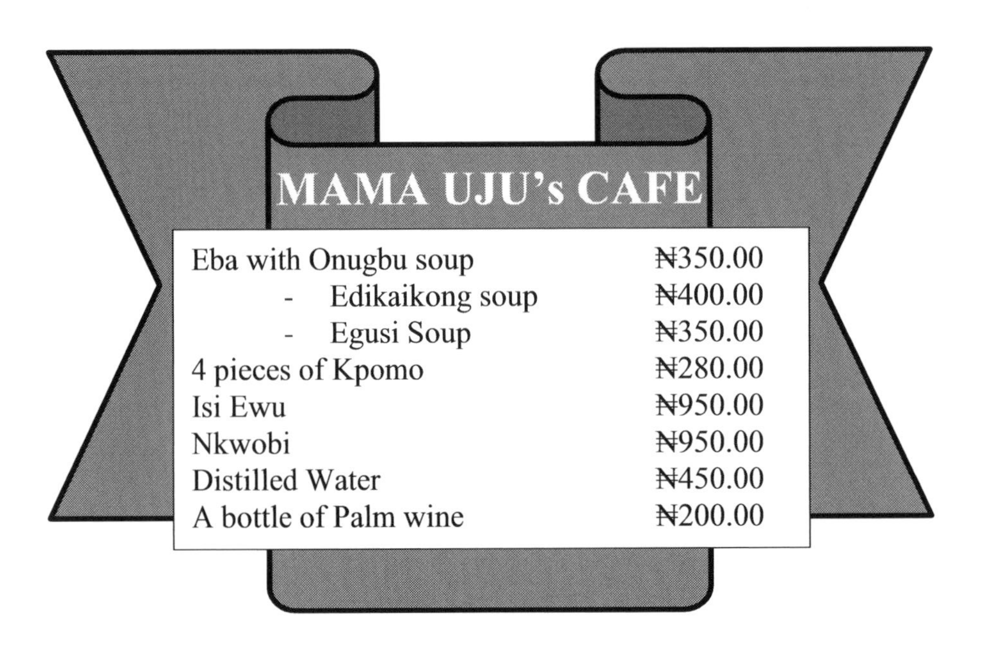

a) Offiong ordered Eba with Edikaikong soup and palm wine. How much will he pay for his orders?

.... **1 mark**

b) Ekaete requested for Nkwobi and distilled water. How much will she pay for her orders?

.... **1 mark**

c) Offiong gave the sales lady a ₦1 000 note.
How much change did he receive?

.... **1 mark**

d) How much is a piece of kpomo? .... **1 mark**

e) As a takeaway, Offiong ordered ten pieces of kpomo, three bottles of palm wine, Eba with egusi soup, and a plate of Isi ewu.

How much will he pay for his orders?

.... **3 marks**

4 Number Work 4

This section covers the following topics:

- Multiples
- Factors
- Primes
- Test for divisibility

LEARNING OBJECTIVES

By the end of this unit, you should be able to:

a) Understand the meaning of multiples
b) Understand and be able to write factors of numbers
c) Understand and identify prime numbers
d) Find the Highest Common Factor (HCF) of two or more numbers
e) Find the Lowest Common Multiple (LCM) of two or more numbers
f) Write or express numbers as a product of its factors
g) I can test if a number is divisible by 2, 3, 4, 5, 6, 7, 8, 9 and 10

KEYWORDS

- Factors
- Multiples
- Prime numbers
- Prime factors
- Product of prime factors
- Highest common factor
- Lowest common multiple
- Divisibility

4.1 MULTIPLES

By now, you must be familiar with the necessary time's tables as it will be vital in setting out multiples of numbers.

Multiples of a number are all the numbers in the multiplication table of that number. The most important fact to remember is that the first multiple of a number is the **number itself.**

Also, multiples of a number goes on forever (infinite).

The multiples of 3 are the answers to
$(1 \times 3), (2 \times 3), (3 \times 3), (4 \times 3), (5 \times 3), (6 \times 3), (7 \times 3)$.........

Therefore, the multiples of 3 are **3, 6, 9, 12, 15, 18, 21.........**

EXERCISE 4A

1) Write down the first four multiples of:

a) 1 d) 7 g) 13

b) 2 e) 9 h) 17

c) 5 f) 10 i) 20

2) Look at the list of numbers below.

2	5	7	12	16	22	30	36	49	54	70

a) Which numbers are multiples of 2?

b) Which numbers are multiples of 3?

c) Which numbers are multiples of 4?

d) Which numbers are multiples of 5?

e) Which numbers are multiples of 7?

3) Write down a number that is odd and a multiple of 3

4) Write down a number that is odd and a multiple of 2

5) Write down a number that is even and a multiple of 5?

6) Write down a number that is a square number and a multiple of 7

7) Write down two multiples of 4 with a sum of 28

8) Look at the list of numbers below.

a) Which numbers in the list are multiples of 11?

b) Which numbers in the list are multiples of 9?

c) Which number in the list is a square number?

9) Is 18 a multiple of 6? Explain fully.

10) Sanusi says "the first five multiples of 4 are 8, 12, 16, 20, 24."
Is Sanusi correct? Explain fully.

11) Write down:

a) the first five multiples of 7

b) the first ten multiples of 3

c) a common multiple of 3 and 7

12) Find two multiples of 6 that have a difference 18 but with a product of 360.

13) Three multiples of 9 add up to 117. Find the three numbers.

4.1.1 LOWEST COMMON MULTIPLE (LCM)

The lowest common multiple of two or more numbers is the smallest (first) of their common multiples.

Example 1: Find the LCM of 3 and 5

List some multiples of 3 and 5 as follows:

3:	3	6	9	12	15	18	21…….
5:	5	10	15	20	25	30	35…….

The lowest number which is in both lists is 15.

Therefore, **15** is the LCM of 3 and 5.

Example 2: Find the LCM of 4, 5 and 6

List some multiples of 4, 5 and 6 as follows:

4: 4 8 12 16 20 24 28 32 36 40 44 48 52 56 60 64……

5: 5 10 15 20 25 30 35 40 45 50 55 60 65…….

6: 6 12 18 24 30 36 42 48 54 60 66……..

The lowest number which is on all the lists is 60.

Therefore, **60** is the LCM of 4, 5 and 6.

USING PRODUCT OF PRIME FACTORS TO WORK OUT THE LOWEST COMMON MULTIPLE (LCM)

When listing multiples of numbers to find LCM becomes cumbersome, it is advisable to use the prime factor method.

Example 1: Find the LCM of 16 and 20

Find all the prime factors of 16 and 20 (refer to section 4.5)

16: $2 \times 2 \times 2 \times 2 \longrightarrow 2^4$
20: $2 \times 2 \times 5 \longrightarrow 2^2 \times 5$

In this system, the LCM contains the highest powers of **each** factor/number.

The highest power of 2 is $\mathbf{2^4}$
The highest power of 5 is $5^1 = \mathbf{5}$

The LCM is the product of all the highest powers

$2^4 \times 5 = 2 \times 2 \times 2 \times 2 \times 5 = \mathbf{80}$ ✓

Example 2: Find the LCM of 20, 24 and 30 using the product of prime method.

Find all the prime factors of 20, 24 and 30 using the factor tree method (Refer to section 4.6)

20: $2 \times 2 \times 5 \longrightarrow 2^2 \times 5$

24: $2 \times 2 \times 2 \times 3 \longrightarrow 2^3 \times 3$

30: $2 \times 3 \times 5 \longrightarrow 2 \times 3 \times 5$

The LCM contains the highest powers of **each** number/factor.

The highest power of 2 is $\mathbf{2^3}$
The highest power of 3 is $\mathbf{3^1} = \mathbf{3}$
The highest power of 5 is **5**

Therefore, the LCM is the product of all the highest factors

$2^3 \times 3 \times 5 = 2 \times 2 \times 2 \times 3 \times 5 = \mathbf{120}$ ✓

USING THE VENN DIAGRAM TO WORK OUT THE LOWEST COMMON MULTIPLE (LCM)

Example 1: Find the LCM of 16 and 20 using the Venn diagram.

Find the prime factors of 16 and 20 (refer to section 4.5)

16: $2 \times 2 \times 2 \times 2$

20: $2 \times 2 \times 5$

Draw a Venn diagram as shown below. The common prime factors are listed in the intersection part.

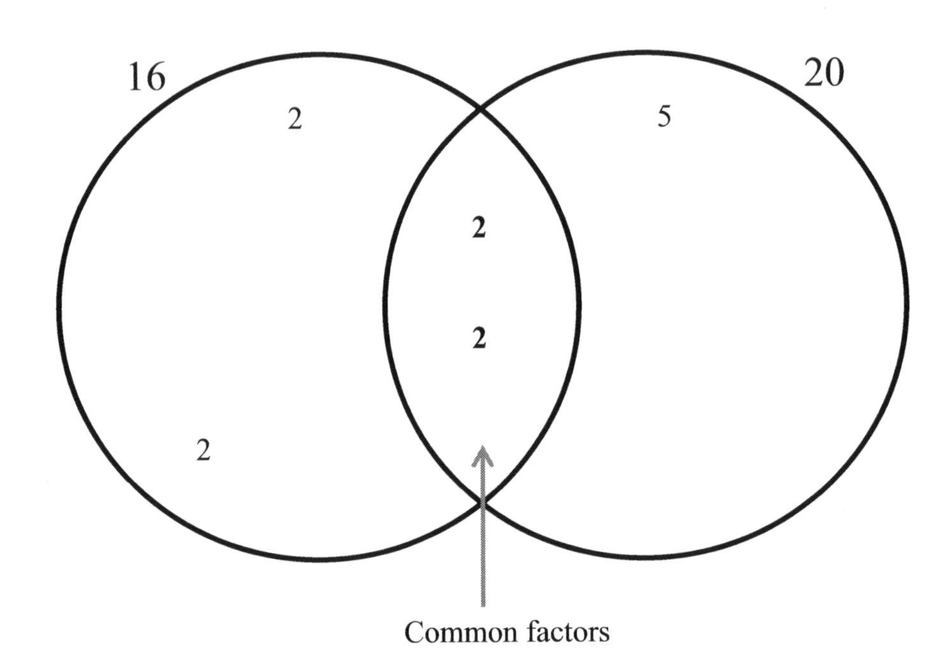

The lowest common multiple is the product of **all** the prime factors in the diagram.

LCM = $2 \times 2 \times 2 \times 2 \times 5 =$ **80** ✓

In chapter 4.6, we shall learn how to use the Venn diagram to find the highest common factor (HCF) of two or more numbers. It is simply performing the steps above and multiplying only the common factors in the intersection part of which is $2 \times 2 = 4$. Therefore, the HCF of 16 and 20 is **4**.

EXERCISE 4B

1) Find the LCM of:

a) 2 and 4

b) 3 and 6

c) 5 and 7

d) 7 and 10

e) 6 and 8

f) 10 and 15

g) 15 and 20

h) 4 and 5

2) Find the LCM of these sets of numbers.

a) 2, 3 and 4

b) 5, 6 and 7

c) 10, 12 and 16

d) 2, 7 and 8

e) 5, 7 and 9

f) 4, 6 and 7

g) 30 and 45

h) 50 and 60

3) Tony and Emma took part in a jogging experiment. Tony took 40 seconds to jog round the athletics track once and Emma 45 seconds. If they both start from the same point at the same time, how long will it take in minutes before they cross the start line together?

4) By leaving your answers in prime factors, find the LCM of the following:

a) $3 \times 3 \times 3 \times 2 \times 2$ and $2 \times 3 \times 3$

b) $2 \times 2 \times 3 \times 3 \times 5$ and $2 \times 2 \times 5 \times 5$

c) $2 \times 2 \times 3 \times 3 \times 3$ and $5 \times 5 \times 2 \times 2 \times 2 \times 3$

d) $2 \times 3 \times 7 \times 7$ and $2 \times 2 \times 2 \times 7 \times 7 \times 7$

e) $5 \times 5 \times 7 \times 11 \times 11$ and $7 \times 7 \times 7 \times 11$

f) $2 \times 3 \times 3 \times 3 \times 3 \times 4 \times 4 \times 5 \times 5 \times 5$ and $2 \times 2 \times 3 \times 3 \times 5 \times 5$

5) Jude says "The LCM of two numbers is always the product of the numbers."
Is Jude correct?
Explain fully with an example.

6) Find the LCM of 70 and 80

7) Find the LCM of 10, 20 and 40

8) Okoye and Chidi have the same number of drinking cups. Okoye placed his cups into 4 equal parts while Chidi arranged his own in 2 equal parts. Find the least number of cups they could each have.

4.2 FACTORS

Like in multiples, the knowledge of multiplications is important.

A factor of a number will divide **exactly** into that number. *Exactly* means there will be no remainder(s).

A fact about factors is that every number has at least two factors: **1 and itself**.

Any number that divides exactly into 6 is called a factor of 6.

The factors of 6 are: **1, 2, 3** and **6**

We can also find factor as pairs

What are the factors of 20?

Using factor pairs: (1 and 20) because $1 \times 20 = 20$
(2 and 10) because $2 \times 10 = 20$
(4 and 5) because $4 \times 5 = 20$

Putting them in order 1 2 4 5 10 20

Therefore, the factors of 20 are 1 2 4 5 10 and 20

Example 1: Find all the factors of 12.

To start with, 1 and the number itself are factors. So, 1 and 12 are factors of 12.

2 and 6 are also factors because $2 \times 6 = 12$

3 and 4 are also factors because $3 \times 4 = 12$

Putting all in order, the factors of 12 are **1 2 3 4 6 and 12**

Note: Factors of a number start with **1** and end with the number itself.

EXERCISE 4C

1) Find all the factors of:

a) 3
b) 4
c) 14
d) 24
e) 30
f) 35
g) 36
h) 48
i) 56
j) 84
k) 100
l) 120

2) Look at the following list of numbers:

3 5 6 7 9

From the list above, choose a number or numbers that are factors of the following:

a) 48
b) 21
c) 18
d) 27
e) 30
f) 108

3) Write down a number that is odd and a factor of 40

4) Write down even numbers that are factors of 18

5) Write down three factors of 50 that have a sum of 40

6) Write down two factors of 35 that have a sum of 6

7) Write down two factors of 72 that have a difference of 14

8) Anthony Joshua says that 7 is a factor of 56.
Is Anthony correct?
Explain fully

9) Add together all the factors of 30

4.3 PRIME NUMBERS

A prime number is a number that has only **two** factors, **1** and **itself**. This statement means that once a whole number has more than two factors, the number is **not** a prime number.

Example 1: a) list all the factors of 1, 2, 3, 4, 5, 6, 7, 8 and 9

b) Which of the numbers are prime numbers?

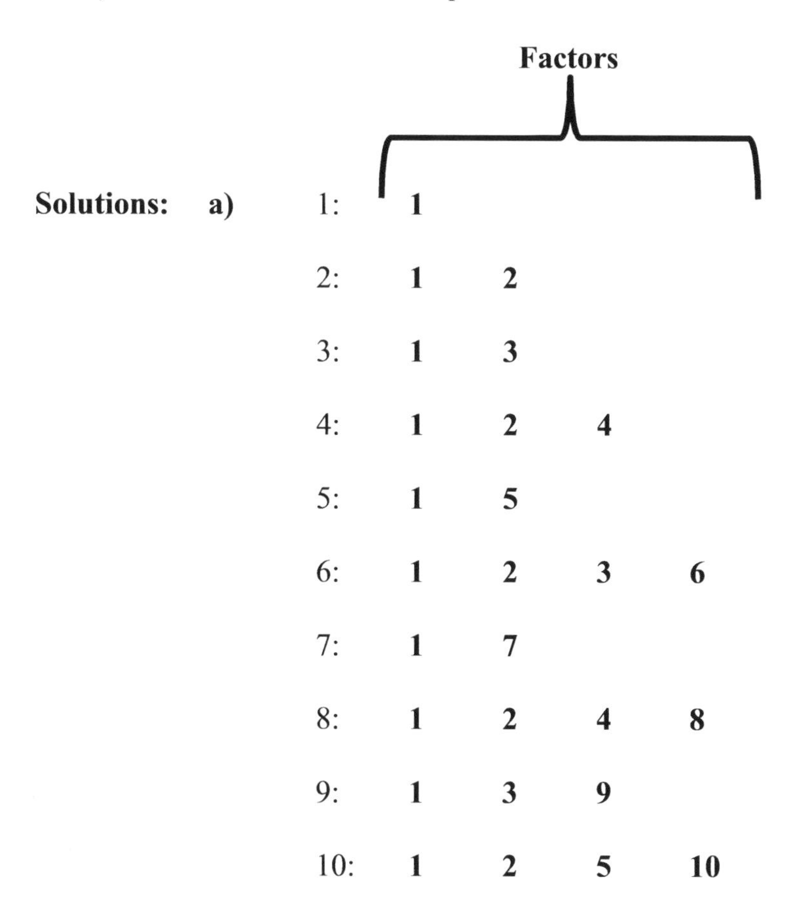

b) Since prime numbers have only two factors, the prime numbers from the list above are **2, 3, 5,** and **7**

Notice that the number **1** is **not** a prime number as it has only **one** factor.

Also, the number **2** is the **only even prime number.**

Example 2: Ifeanyi says: "**9** is a prime number because it is an odd number." Is he correct? Explain fully

Solution: List all the factors of 9. They are 1, 3 and 9.
However, a prime number is a number that has only two factors. 9 has three factors, and cannot be a prime number.

Therefore, **9 is not** a prime number because it has **more** than two factors, 1, 3 and 9 and a prime number must have only two factors.

The first 20 prime numbers are:

2	3	5	7	11	13	17	19	23	29
31	37	41	43	47	53	59	61	67	71

4.4 PRIME FACTORS

Prime factors of a number are the factors of that number that are prime numbers.

Example: The factors of 20 are: 1 (2) 4 (5) 10 20

From the factors, only 2 and 5 are prime numbers.

Therefore, **2** and **5** are the **prime factors** of 20.

POINTS TO NOTE

- A number is prime if the only factors are 1 and itself
- 1 **is not** a prime number since it has only one factor, 1
- 2 is the only even prime number
- Prime factors of a number are the prime numbers from factors of that number

EXERCISE 4D

1) Identify all the prime numbers from the numbers below:

a) 1	3	7	15	19	27
b) 2	5	8	13	18	33
c) 6	9	14	28	31	41
d) 10	20	25	37	59	97
e) 4	11	17	60	78	121

2) Find the prime factors of

a) 4 b) 12 c) 34 d) 50

3) Identify a prime number that is also a factor of 6

4) Identify a prime number that is also a factor of 24

5) a) Copy and complete the 7 by 7 square grid following the existing pattern.

1	2	3	4	5	6	7
8	9	10	11	12	13	14

b) Colour in all the prime numbers in the table above.

c) Add up all the prime numbers in the 5th column.

d) Is the number obtained in **part c** above a prime number? Explain fully.

4.5 PRODUCT OF PRIME FACTORS

In section 4.4, prime factors were discussed. Please revisit if you need extra help.

Any whole number which is not a prime number can be broken down into prime factors. It is called prime factor decomposition. The product of the prime factors obtained will always give the original number.

There are different ways to write numbers as the product of prime factors:

Method 1: Factor tree method

Example 1: Write 20 as the product of prime factors.

Step1: Find any two numbers that multiply to give 20 →
If any of the factors are prime, circle it.

Step 2: For the factors **not circled, in this case, 4**, repeat step 1.

Step 3: Since all the numbers are circled, it means **no more decomposition** (breaking down). At this point, you have decomposed the number, 20 into its prime factors.

Multiply all the circled numbers $2 \times 2 \times 5$

Therefore, as a product of prime factors, $20 = 2 \times 2 \times 5$ or $\mathbf{2^2 \times 5}$ in index forms.

Example 2: Write 320 as the product of prime factors.

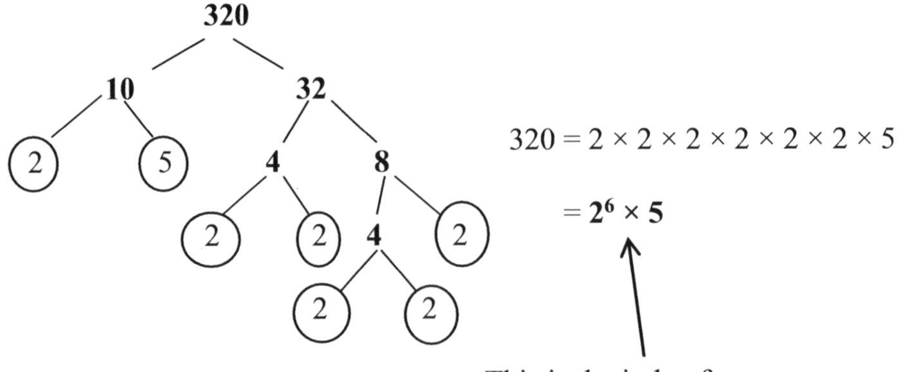

This is the index form

Method 2 of finding product of prime factors

This is a method of dividing the number by the smallest (first) prime number that divides into the number. The process is repeated until the answer becomes 1.

Example 3: Express 60 as the product of prime factors

The first prime number that divides into 60 is 2. Use 2 to divide. Repeat the process until the answer becomes 1.

2	60
2	30
3	15
5	5
	1

Therefore, 60 written as the product of prime factors is $2 \times 2 \times 3 \times 5 = \mathbf{2^2 \times 3 \times 5}$

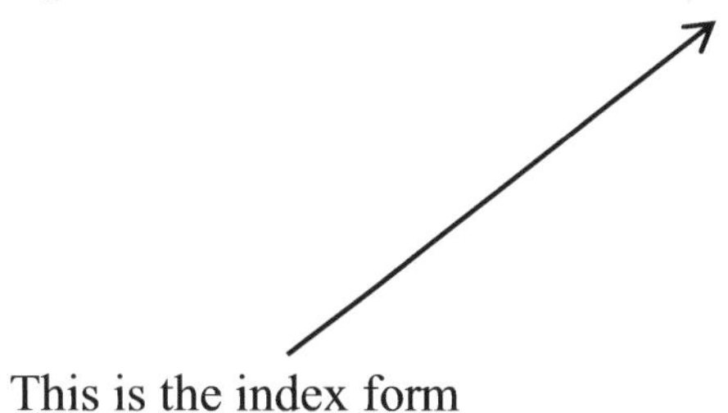

EXERCISE 4E

1) Express each of the numbers below as a product of prime factors

a) 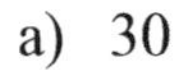 30	c)  70	e) 720
b) 55	d) 144	f) 940

2) a) Complete the factor tree.

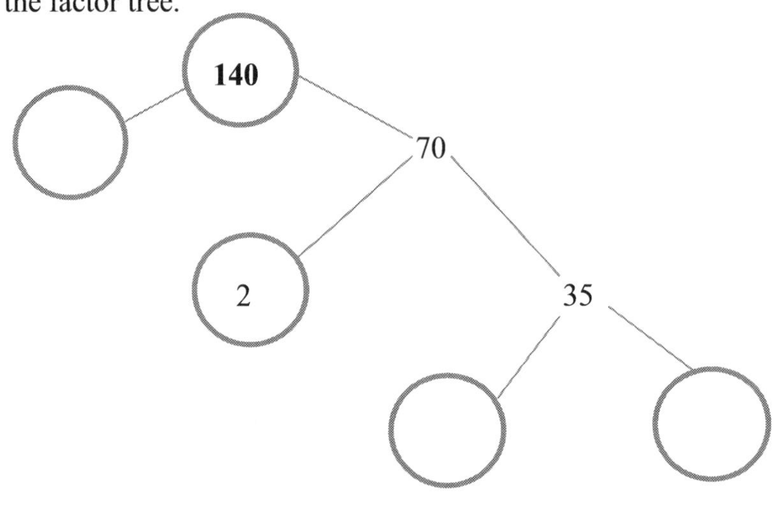

b) Write down the prime factors of 140 in index form.

3) Write each number as a product of prime factors in index form.

a) 150 b) 500 c) 510

4) Write the following as the product of two prime factors.

a) 6 b) 21 143

5) Copy and complete the factor trees.

a)

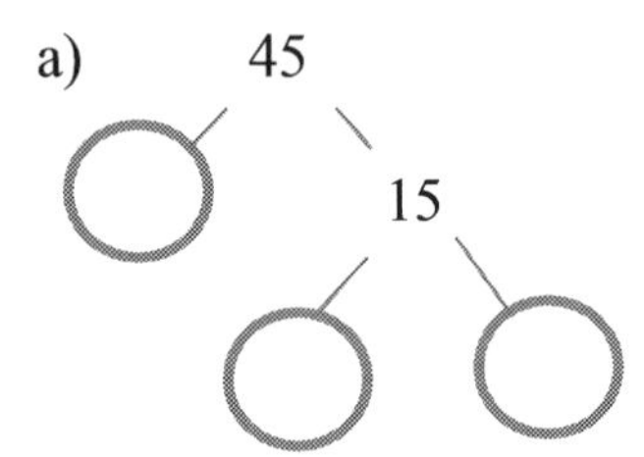

b)

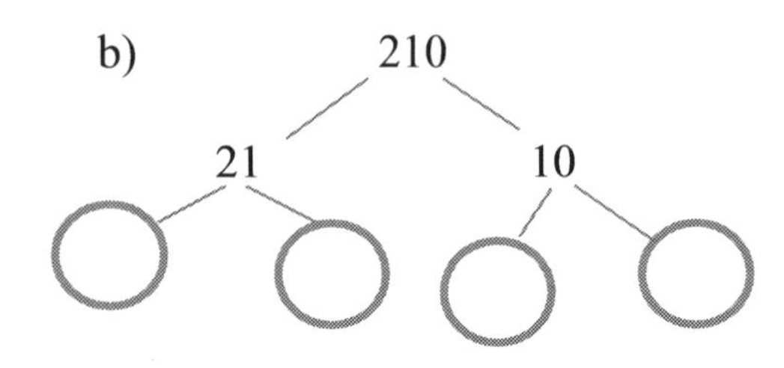

6) Write in index form.

a) $2 \times 2 \times 2 \times 3 \times 3 \times 3$

b) $2 \times 2 \times 5 \times 7 \times 7 \times 7 \times 7$

c) $3 \times 3 \times 3 \times 3 \times 5 \times 5$

7) Maduaburochukwu says "Prime factors will be different if you chose a different factor pair."
Is he correct?

Explain fully

8) Write each number as a product of prime factors in index form.

a) 3300

b) 1540

4.6 HIGHEST COMMON FACTOR (HCF)

The highest common factor of two or more numbers can be obtained. The first method is to find all the factors of the numbers and identify the **common factors**. The highest number of the common factors will be the HCF of the numbers.

Refer to **section 4.2** for finding factors of numbers.

Example 1: Find the HCF of 4 and 8

Identify all the factors of 4 and 8 as follows:

4:	**1**	**2**	**4**	
8:	**1**	**2**	**4**	8

The common factors are 1, 2 and 4.

However, 4 is the largest number of the common factors. Therefore, **4** is the highest common factor (HCF) of 4 and 8.

It is also the highest number that can divide into both numbers.

Example 2: Find the HCF of 24 and 36

Identify all the factors of 24 and 36.

24:	**1**	**2**	**3**	**4**	**6**	8	**12**	24	
36:	**1**	**2**	**3**	**4**	**6**	9	**12**	18	36

The common factors are 1, 2, 3, 4, 6 and 12.

However, the highest number common to both numbers is 12.

Therefore, **12** is the highest common factor (HCF) of 24 and 36.

It is the highest number that can divide into both numbers.

USING PRIME FACTORS TO FIND HCF

When listing factors of numbers in other to find HCF becomes cumbersome, it is advisable to use the prime factor method.

Example 1: Find the HCF of 24 and 36

Find all the prime factors of 24 and 36 (refer to section 4.5)

24: $2 \times 2 \times 2 \times 3 \longrightarrow 2^3 \times 3$

36: $2 \times 2 \times 3 \times 3 \longrightarrow 2^2 \times 3^2$

In this system, find the pairs of common prime factors and multiply them.

24:	**2**	**2**	2	**3**
36:	**2**	**2**	3	**3**
	↓	↓		↓
	2	2		3

++The HCF would be $2 \times 2 \times 3 =$ **12** ✓

ALTERNATIVELY

After finding the product of prime factors, we multiply the lowest power of **each common** prime factor to give the HCF.

24: $2 \times 2 \times 2 \times 3 = 2^3 \times 3$

36: $2 \times 2 \times 3 \times 3 = 2^2 \times 3^2$

LOWEST POWER OF 2 is $\mathbf{2^2}$ while the LOWEST POWER OF 3 is **3.**

Therefore, HCF of 24 and 36 $= 2^2 \times 3 = 2 \times 2 \times 3 =$ **12** ✓

Example 2: Find the HCF of $3 \times 3 \times 3 \times 3 \times 5 \times 5 \times 7 \times 7 \times 7$ **and** $3 \times 3 \times 3 \times 7 \times 7$
Leave your answer in prime factors in index form.

$3 \times 3 \times 3 \times 3 \times 5 \times 5 \times 7 \times 7 \times 7 = 3^4 \times 5^2 \times 7^3$ and

$3 \times 3 \times 3 \times 7 \times 7 = \mathbf{3^3} \times \mathbf{7^2}$

The lowest power of 3 contained in the two numbers is 3^3

The lowest power of 7 contained in the two numbers is 7^2

** Notice that the number **5** is **not** common to both numbers and as such is **not included**.*

Therefore, in prime factor index form, the HCF is $\mathbf{3^3 \times 7^2}$ ✓

Please note: The above index answer is acceptable if the question says so.
If not, work out the real answers to $3^3 \times 7^2$ as detailed below.

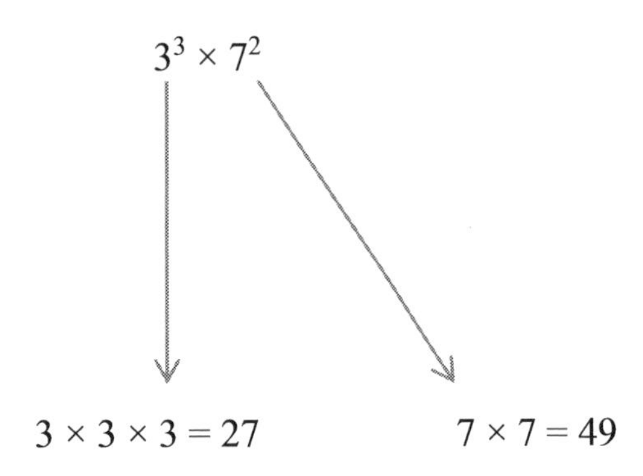

Highest common factor (HCF) $= 27 \times 49$

$= \mathbf{1323}$

USING THE VENN DIAGRAM TO WORK OUT THE HIGHEST COMMON FACTOR (HCF)

Example 1: Find the HCF of 16 and 20 using the Venn diagram.

Find the prime factors of 16 and 20 (refer to section 4.5)

16: $2 \times 2 \times 2 \times 2$

20: $2 \times 2 \times 5$

Draw a Venn diagram as shown below. The common prime factors are listed in the intersection part.

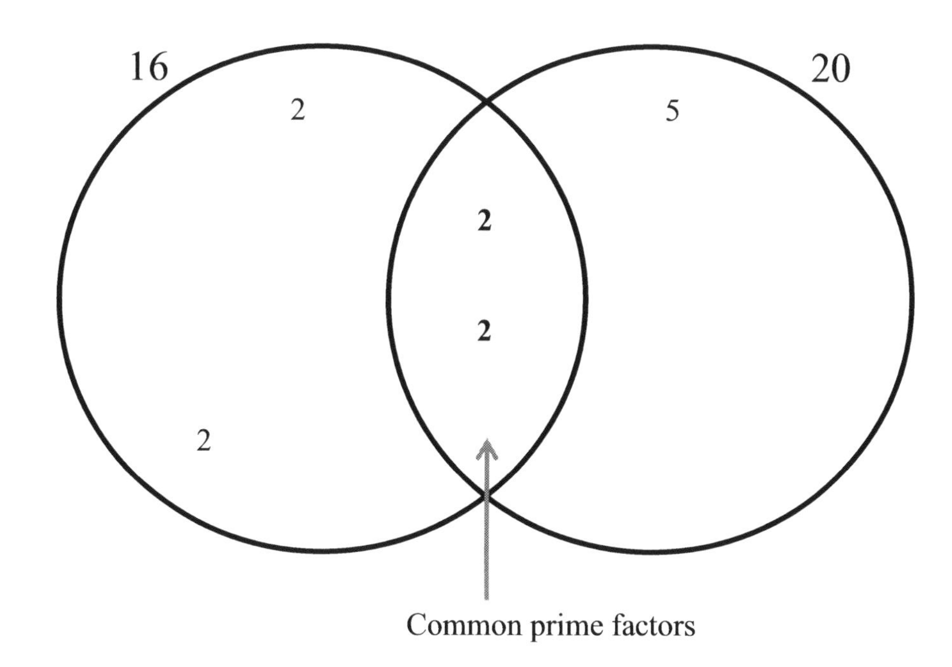

The highest common factor is obtained by multiplying the numbers in the intersection part.

$$2 \times 2 = 4$$

Therefore, the HCF of 16 and 20 is **4**.

In all, you may use any method of your choice to find HCF unless told otherwise.

EXERCISE 4F

1) i) Write down all the factors of the numbers below.

   ii) Circle the common factors of each pair.

   iii) Pick out the highest common factor.

   a) 6 and 10
   b) 12 and 18
   c) 5 and 25
   d) 27 and 49
   e) 30 and 76
   f) 36 and 48

2) Find the highest common factor using the prime factor method for the following pairs of numbers.

   a) 40 and 60
   b) 120 and 150
   c) 420 and 700

3) Find the HCF of the following numbers, leaving your answers in prime factor forms using index notation.

   a) $3 \times 3 \times 5 \times 5 \times 5 \times 5 \times 7 \times 7 \times 7$ and $3 \times 5 \times 5 \times 5 \times 7 \times 7$
   b) $2 \times 2 \times 2 \times 2 \times 3 \times 3 \times 3 \times 3 \times 3 \times 3 \times 11$ 11×11 and $2 \times 2 \times 3 \times 3 \times 3$
   c) $5 \times 5 \times 7 \times 7 \times 7 \times 13$ and $5 \times 7 \times 13$

4) Find the HCF of the following numbers using the Venn diagram method.

   a) 18 and 60
   b) 20 and 75

5) Using the index notation, find the HCF of the numbers below.

a) 8 and 40

b) 63 and 270

c) 30, 60 and 90

d) 25, 70 and 130

6) Write out the common factors of

a) 4 and 16

b) 12 and 20

c) 45 and 63

d) 8, 20 and 52

7) Nwatu needed to cut out congruent squares for making envelopes from a piece of rectangular brown cardboard paper as shown.

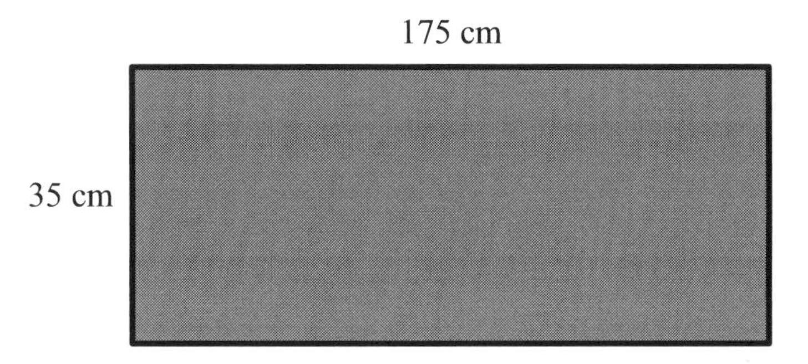

Without wasting any cardboard paper, what is the maximum square size Nwatu can use for the envelopes?

8)

Kenechukwu says "The HCF of 25 and 50 is 5."

Is she correct? Explain fully.

4.7 DIVISIBILITY TEST

To test if a whole number is divisible by 2, 3, 4, 5, 6, 7, 8, 9 or 10, use the divisibility rules which are set out below.

DIVISIBILITY BY 2

A number is divisible by 2 if the last digit is 0 or even.

Look at the numbers below.

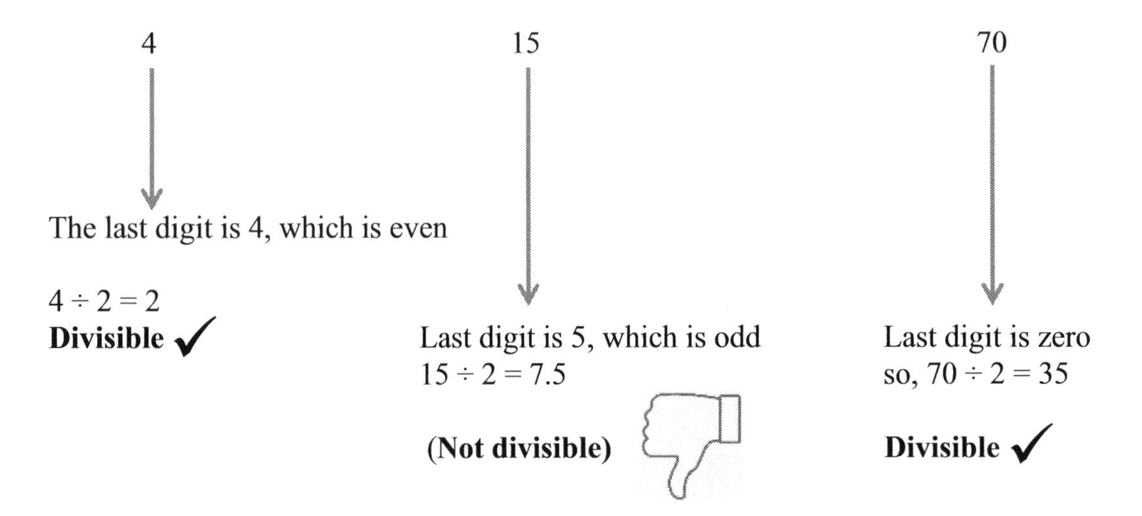

DIVISIBILITY BY 3

To find out if a number is divisible by 3, add up all the digits. If the sum is divisible by 3, then the number is divisible by 3.

If you cannot figure out if the added numbers are divisible by 3, add together the digits of the added numbers and divide by 3 to check.

Example 1: Check if 453 is divisible by 3.

$4 + 5 + 3 = 12$
Clearly, 12 is divisible by 3 because $12 \div 3 = 4$
Therefore, 453 **is divisible** by 3

$453 \div 3 = 151$ ✓

Example 2: Is 79796223 divisible by 3?

$7 + 9 + 7 + 9 + 6 + 2 + 2 + 3 = 45$

If you are familiar with your 3 times table, then 45 is divisible by 3 since $15 \times 3 = 45$

Yes, 79796223 **is divisible** by 3. ✓

You may use a calculator to check. $79796223 \div 3 = 26\ 598\ 741$

Example 3: Is 265 divisible by 3?

$2 + 6 + 5 = 13$

$13 \div 3 = 4.3333333.....$ (**Not** a whole number)

13 is **not** divisible by 3. Therefore, 265 **is not divisible** by 3.

DIVISIBILITY BY 4

If the **last two digits** of a number are divisible by 4, then the number is divisible by 4

Example 1: Is 459 divisible by 4?

The last two digits are 59

$59 \div 4 = 14.75$ (**Not** a whole number)

No, 459 **is not divisible** by 4

Example 2: Is 2792 divisible by 4?

The last two digits are 92

$92 \div 4 = 23$ (Whole number)

Yes, 2792 **is divisible** by 4 ✓

DIVISIBILITY BY 5

A number is divisible by 5 if the last digit is 0 or 5.

Example 1: Is 345 divisible by 5?

The last digit is 5; therefore it **is divisible** by 5. ✓

$345 \div 5 = 69$ (whole number)

Example 2: Is 677 divisible by 5?

The last digit is 7; therefore it **is not divisible** by 5

$677 \div 5 = 135.4$ (**Not** a whole number)

Example 3: Is 120 divisible by 5?

The last digit is 0; therefore it **is divisible** by 5 ✓

$120 \div 5 = 24$ (whole number)

DIVISIBILITY BY 6

A number is divisible by 6 if it is divisible by both 2 and 3. Check the rules above!

Example 1: Is 156 divisible by 6?

Step 1: Check with divisibility by 2 rules:
It is even; therefore it can be divided by 2 (Yes)

Step 2: Check with divisibility by 3 rules:
$1 + 5 + 6 = 12$
…and 12 is divisible by 3;
therefore it can be divided by 3 (Yes)

Since 156 is divisible by 2 and 3, it is divisible by 6 ✓

Example 2: Is 86 divisible by 6?

Step 1: 86 is even, so it is divisible by 2 (Yes)

Step 2: Check for divisibility by 3.

$8 + 6 = 14$
$(14 \div 3 = 4.666666\ldots)$ (X)

14 is not divisible by 3. Therefore, 86 is not divisible by 3 as well.

Since 86 failed divisibility by 3 rules, it cannot be divided by 6.

DIVISIBILITY BY 7

For a number to be divisible by 7, double the last digit and subtract it from a number made by the other digits. If the outcome/result is divisible by 7, then the number itself is also divisible by 7.

Example 1: Is 238 divisible by 7?

Double the last digit $8 \times 2 =$ **16**

Subtract 16 from the remaining numbers $23 - 16 =$ **7**

7 is divisible by 7 $(7 \div 7 = 1)$ …..(Whole number)

Therefore, 238 **is divisible** by 7 ✓

Example 2: Is 957 divisible by 7?

Double the last digit $7 \times 2 = 14$

Subtract 14 from the remaining numbers $95 - 14 = 81$

$81 \div 7 = 11.57\ldots.$ (Not a whole number)

Therefore, 957 is **not divisible** by 7

DIVISIBILITY BY 8

If the **last three** digits are divisible by 8, then the number is divisible by 8

Example 1: Is 1 547 divisible by 8?

The last three numbers are 547.
547 ÷ 8 = 68.375 (Not a whole number)

Therefore, 1547 is **not divisible** by 8

Example 2: Is 32 984 divisible by 8?

The last three numbers are 984
984 ÷ 8 = 123 (Whole number)

Therefore, 32 984 **is divisible** by 8 

Points to note: *At times the last three numbers might be a problem to divide. Halve it three times, and if the result is still a whole number, then that number is divisible by 8.*

DIVISIBILITY BY 9

A number is divisible by 9 if the sum of the digits is divisible by 9.

Example 1: Is 154 divisible by 9?

1 + 5 + 4 = 10

10 ÷ 9 = 1.1111111 (**Not** a whole number)

Therefore, 154 is **not divisible** by 9

Example 2: Is 322 866 divisible by 9?

$3 + 2 + 2 + 8 + 6 + 6 = 27$

$27 \div 9 = 3$ (Whole number)

Therefore, 322 866 **is divisible** by 9 

DIVISIBILITY BY 10

A number is divisible by 10 is the last digit is zero (if the number ends in zero (0).

Example 1: Is 4 567 divisible by 10?

The last number is 7

4 567 **is not divisible** by 10 because the last number is not zero

Example 2: Is 56 430 divisible by 10?

The last number (digit) is 0

Therefore, 56 430 **is divisible** by 10

EXERCISE 4G

1) Are these numbers divisible by **2**? Give a reason.

a) 12

b) 26

c) 27

d) 69

e) 780

f) 7865

2) Are these numbers divisible by 3? Give a reason using divisibility tests.

a) 12

b) 21

c) 34

d) 46

e) 180

f) 654

3) Obinna says "287945 is divisible by 7." Is he correct? Explain using the divisibility rule.

4) Are these numbers divisible by 9? Perform divisibility tests to check.

a) 81

b) 108

c) 1 553

d) 2 345

e) 5 166

f) 142 911

5) Is 65432 divisible by 5? Explain fully.

6) Is 7893451 divisible by 8? Explain fully

7) Is 6754320 divisible by 2 and 10? Explain fully.

8) Using the divisibility test, check whether the following numbers are divisible by 4.

a) 876

b) 1258

Chapter 4 Review Section
Assessment

1) Write down the first five multiples of

a) 3 .... **1 mark**

b) 4 …………**1 mark**

c) 14 ………... **1 mark**

2) Write down all the factors of

a) 13 …………**1 mark**

b) 36 ………... **1 mark**

c) 104 ………... **1 mark**

3) From the cloud, write down the number(s) that are

a) factors of 35

b) factors of 63

c) multiples of 6

d) prime numbers

e) multiples of 92

23 2 72 7 6 9

………. **5 marks**

4) Write 250 as a product of prime factors in index form. .... **3 marks**

5) Work out the HCF of

a) 27 and 63 .... **2 marks**

b) 400 and 500 .... **3 marks**

6) Work out the LCM of

a) 2 and 7 .... **2 marks**

b) 35 and 50 .... **2 marks**

c) 182 and 420 .... **3 marks**

7) Using the Venn diagram method, find the LCM and HCF of

a) 66 and 132 .... **5 marks**

b) 420 and 560 .... **5 marks**

8) Find the HCF of the following and leave your answers in prime factors and index notation.

a) $2 \times 2 \times 3 \times 3 \times 3 \times 5 \times 5 \times 5$ and $2 \times 3 \times 3 \times 3 \times 3 \times 5$ .... **1 mark**

b) $7 \times 7 \times 7 \times 7 \times 11 \times 11$, $7 \times 7 \times 13 \times 13 \times 13 \times 13$ and

$7 \times 11 \times 13 \times 13$ .... **1 mark**

9) Azubuike says "The Lowest common multiple of two numbers cannot be one of the numbers."

Is Azubuike correct? Explain fully.

.... **2 marks**

10) The Lowest common multiple of 8 and another number is 72. If the other number is more than 8 but less than 15, what is the other number?

.... **2 marks**

11) Tolu, Nnamdi, and Okechukwu started at the same time to ring a school bell.

Tolu rings a bell every 3 seconds.

Nnamdi rings a bell every 5 seconds.

Okechukwu rings a bell every 11 seconds.

How long will it take before they ring the bell at the same time?

.... **3 marks**

12) a) Write 54 as a product of its prime factors. .... **2 marks**

b) Express your answer in index form .... **1 mark**

c) Find the HCF of 54 and 126 .... **2 marks**

13)

Dictionaries are sold in packs of 12.

Scrabbles are sold in packs of 20.

A shop wants to buy the same number of dictionaries and scrabbles.

What is the lowest number of packs of dictionaries and scrabbles they could buy?

.... **3 marks**

14) Henry says "1 is a prime number and 2 is not."
Is Henry correct?

Explain fully. .... **3 marks**

15) Write down the next **five** prime numbers larger than 36.

.... **2 marks**

16) Draw a factor tree for the following numbers.

a) 16 .... **2 marks**

b) 144 .... **2 marks**

17) Is the number 189357 divisible by 3?

Explain your answer.

.... **3 marks**

18) Write down the prime factor of 750. .... **2 marks**

19) From these set of numbers

12 **16** **32** **33** **47** **54**

Write down

a) a prime number .... **1 mark**

b) a multiple of 9 .... **1 mark**

c) a multiple of 8 .... **1 mark**

d) a factor of 144 .... **1 mark**

e) a square number .... **1 mark**

5 Number Work 5

In this section, we shall consider the following:

- Types of fractions
- Equivalent fractions
- Simplifying fractions
- Adding and subtraction fractions
- Multiplying and dividing fractions
- Fraction of an amount

LEARNING OBJECTIVES

By the end of this unit, you should be able to

a) Understand what fractions are

b) Work out equivalent fractions

c) Simplify fractions to their lowest form

d) Add and subtract fractions including mixed numbers

e) Multiply and divide fractions including mixed numbers

f) Work out fraction of an amount

KEYWORDS

- Fractions
- Numerator
- Denominator
- Add and Subtract
- Mixed number
- Simplify
- Divide and multiply

5.1 UNDERSTANDING FRACTIONS

Fractions usually have two parts, the **numerator,** and **denominator**.

The numerator is the top number while the denominator is the bottom number in a fraction.

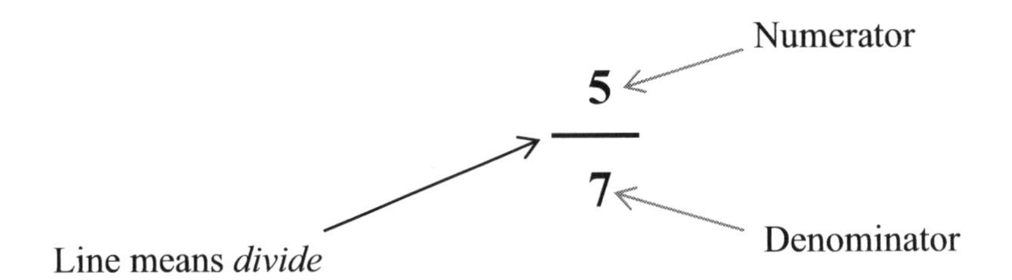

When a whole number is divided into equal parts, each of the parts is a fraction of the whole. Fractions could also be written with a slash (/) instead of a horizontal line (**—**) like 2/3, 4/7, 5/7.

5.2 TYPES OF FRACTIONS

There are three types of fractions at this level.

a) Proper Fractions

b) Improper fractions

c) Mixed Numbers

PROPER FRACTIONS

In a proper fraction, the numerator (top number) is less than the denominator (bottom number).

Examples of proper fractions: $\frac{2}{3}$ $\frac{4}{7}$ $\frac{12}{15}$

IMPROPER FRACTIONS

In an improper fraction, the numerator is bigger than the denominator.

Examples of improper fractions: $\frac{4}{3}$ $\frac{15}{4}$ $\frac{24}{7}$

MIXED NUMBERS

A whole number combined with a fraction is called a mixed number.

Examples of mixed numbers: $1\frac{2}{3}$ $7\frac{3}{4}$ $11\frac{1}{7}$

5.3 CONVERSION FROM MIXED NUMBERS TO IMPROPER FRACTIONS

Example 1: Convert $\mathbf{2\frac{3}{4}}$ to improper fraction.

- Multiply the denominator (bottom number) by the whole number
- Add the outcome to the numerator
- Finally, divide the result by the denominator **(which stays the same)**

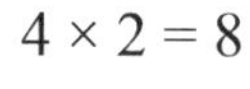

$4 \times 2 = 8$

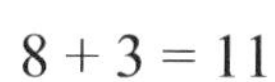

$8 + 3 = 11$

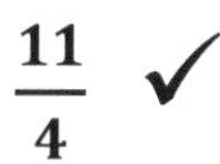

$\frac{11}{4}$ ✓

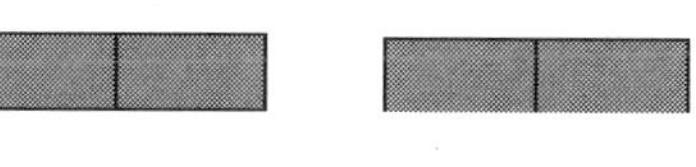

Example 2: Change $\mathbf{5\frac{1}{2}}$ to improper fraction.

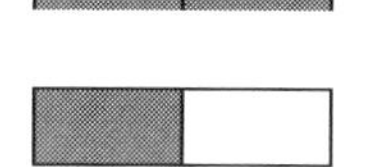

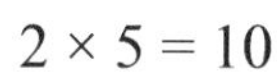

$2 \times 5 = 10$

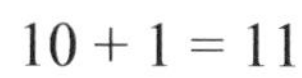

$10 + 1 = 11$

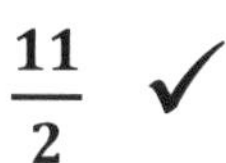

$\frac{11}{2}$ ✓

5.4 CONVERSION FROM IMPROPER FRACTION TO A MIXED NUMBER

Example 1: Convert $\frac{9}{4}$ to a mixed number.

It is a normal division with the remainder on top as the numerator.

Four (4) goes into 9 two (**2**) times, remainder one (**1**) ⟶ $2\frac{1}{4}$ ✓

Example 2: Convert $\frac{29}{8}$ to a mixed number.

8 goes into 29 three (**3**) times remainder five (**5**) ⟶ $3\frac{5}{8}$ ✓

5.5 SHADED FRACTIONS

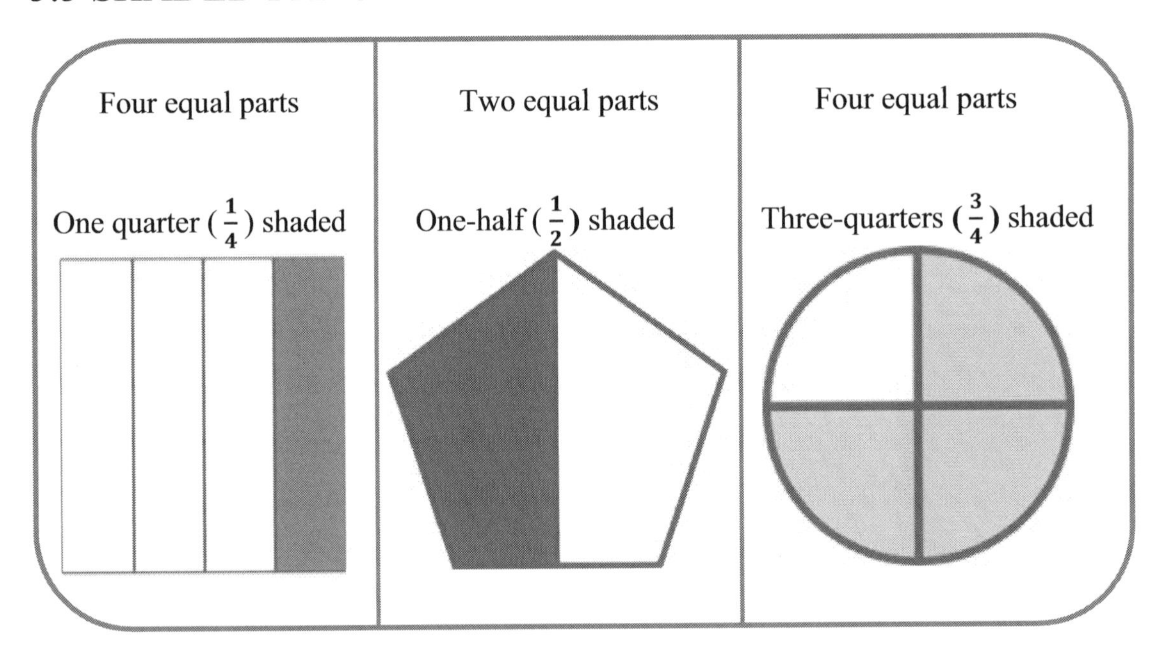

5.6 EQUIVALENT FRACTIONS

These are fractions that have the same value though they may look different.
If **the same number multiplies both the numerators and denominators**, the value of the fraction remains the same provided that the number is not zero (0).

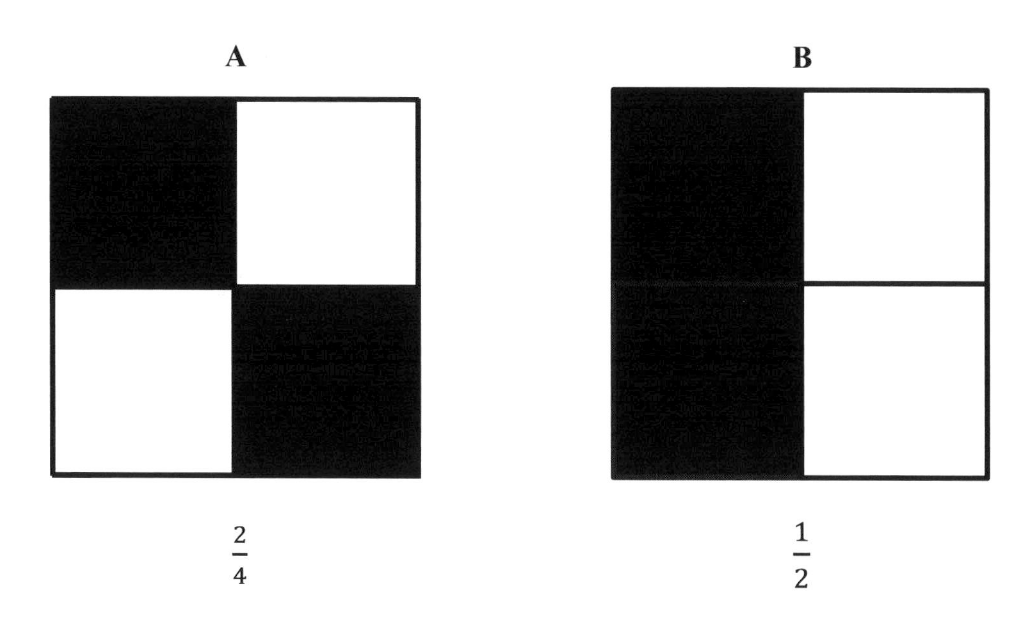

$\frac{2}{4}$ $\frac{1}{2}$

Shapes A and B are the same and are called equivalent fractions. They look different but are the same.

How to form equivalent fractions

$\frac{2}{3}$ is equivalent to $\frac{2}{3} \times \frac{4}{4} = \frac{8}{12}$ **................ Example 1**

$\frac{5}{11}$ is equivalent to $\frac{5}{11} \times \frac{3}{3} = \frac{15}{33}$ **................ Example 2**

Also, equivalent fractions can be obtained by **dividing** the numerator and the denominator by the same number provided that number is not zero (0).

$\frac{16}{20}$ is equivalent to $\frac{16 \div 4}{20 \div 4} = \frac{4}{5}$ .... **Example 3**

Example 4: Draw pictorial diagram to show that $\frac{3}{5} = \frac{9}{15}$

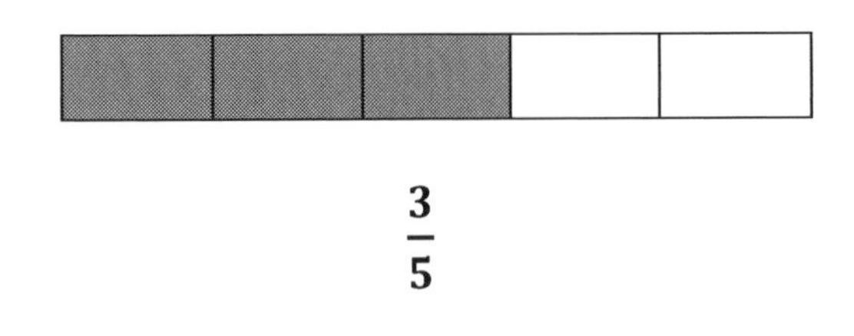

$\frac{3}{5}$

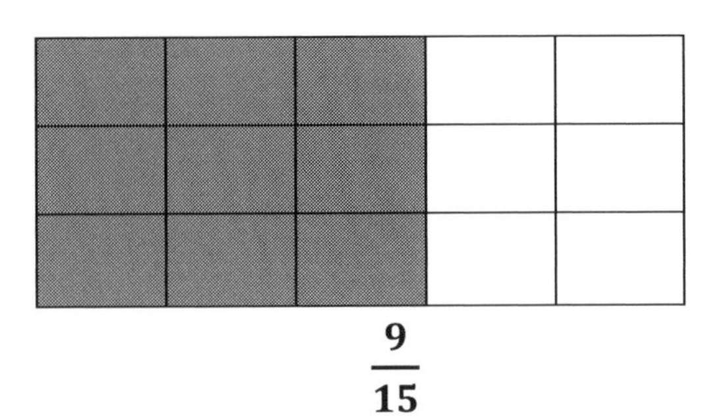

$\frac{9}{15}$

Example 5: Fill in the missing numbers.

a) $\frac{1}{3} = \frac{5}{\square}$ b) $\frac{3}{7} = \frac{\square}{21}$ c) $\frac{12}{18} = \frac{\square}{3}$

Solutions

a) $\frac{1}{3} = \frac{5}{\boxed{15}}$ (× 5 top, × 5 bottom)

b) $\frac{3}{7} = \frac{\boxed{9}}{21}$ (× 3 top, × 3 bottom)

c) $\frac{12}{18} = \frac{\boxed{2}}{3}$ (÷ 6 top, ÷ 6 bottom)

Note: Whatever you do to the top, you must do to the bottom. In equivalent fractions, × or ÷ is used to find any missing number.

Example 6: Arrange the following fractions in order of size, smallest first.

$$\frac{2}{3}, \frac{1}{5}, \frac{5}{6} \text{ and } \frac{7}{10}$$

The easiest way to compare fractions is to make the denominators the same using equivalent fractions. We also use our knowledge of the lowest common multiple.

The LCM of 3, 5 and 6 is 30.
Therefore, we find equivalent fractions with 30 as the denominator.

$$\frac{2 \times \mathbf{10}}{3 \times \mathbf{10}} = \frac{\mathbf{20}}{\mathbf{30}}$$

$$\frac{1 \times \mathbf{6}}{5 \times \mathbf{6}} = \frac{\mathbf{6}}{\mathbf{30}}$$

$$\frac{5 \times \mathbf{5}}{6 \times \mathbf{5}} = \frac{\mathbf{25}}{\mathbf{30}}$$

$$\frac{7 \times \mathbf{3}}{10 \times \mathbf{3}} = \frac{\mathbf{21}}{\mathbf{30}}$$

Since denominators of the equivalent fractions are now 30, the fraction with the lowest numerator is the lowest fraction. Likewise, the fraction with the highest numerator is the highest fraction.

Therefore, the order of size of smallest to biggest is $\frac{1}{5}$, $\frac{2}{3}$, $\frac{7}{10}$ and $\frac{5}{6}$

EXERCISE 5A

In questions **1** to **6**, write down the fractions shaded.

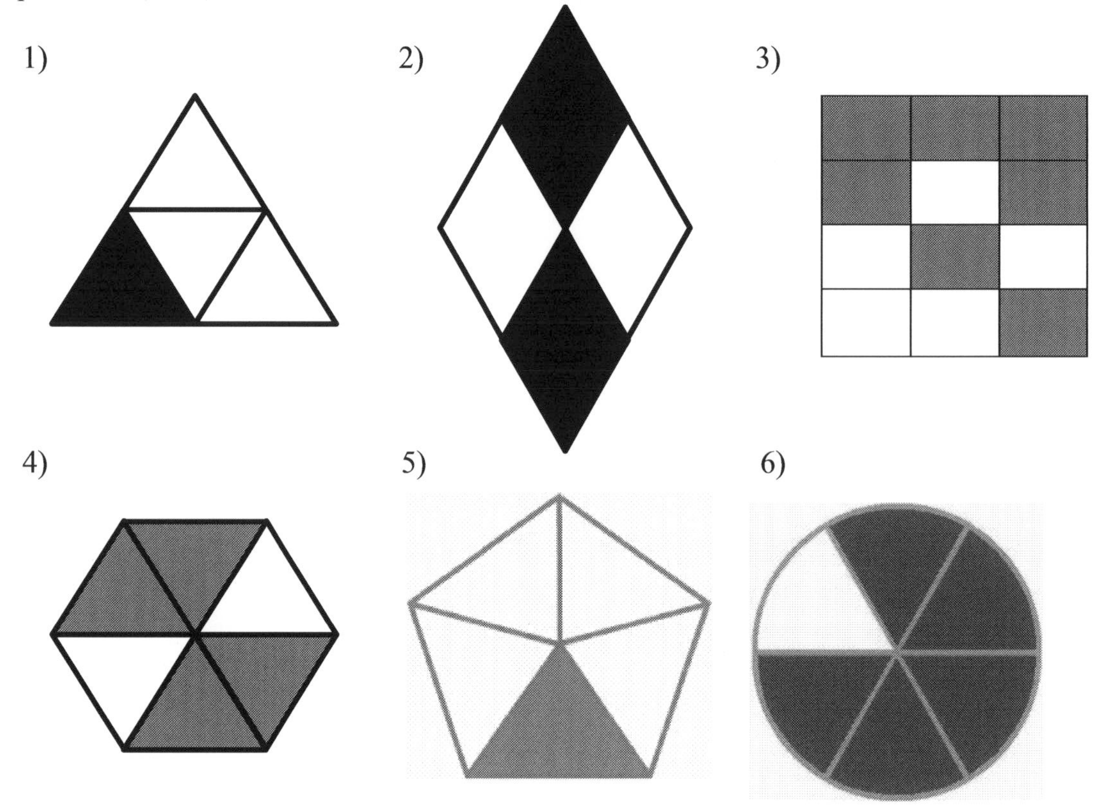

7) Copy and shade $\frac{1}{3}$ of the shape below.

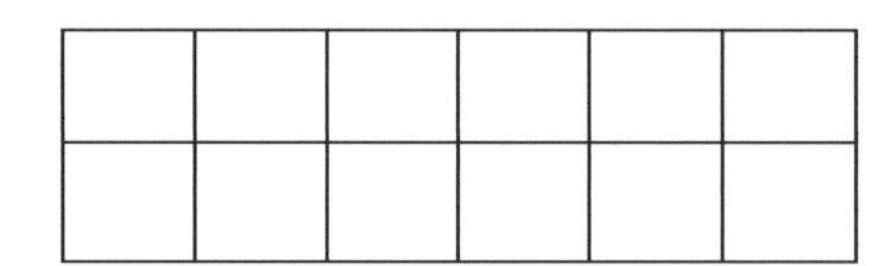

8) Copy and shade $\frac{1}{4}$ of the shape in number 7 above.

9) Copy and shade $\frac{3}{4}$ of the shape in question 7 above.

Change each of these improper fractions to mixed numbers.

10) $\frac{3}{2}$ 11) $\frac{14}{6}$ 12) $\frac{35}{11}$ 13) $\frac{122}{40}$

Change each of these mixed numbers to improper fractions.

14) $1\frac{3}{5}$ 15) $2\frac{1}{3}$ 16) $6\frac{2}{5}$ 17) $9\frac{5}{7}$

In questions 18 to 28, copy and complete the equivalent fractions.

18) $\frac{3}{4} = \frac{9}{?}$

19) $\frac{1}{2} = \frac{7}{?}$

20) $\frac{2}{5} = \frac{?}{15}$

21) $\frac{7}{9} = \frac{?}{18}$

22) $\frac{7}{10} = \frac{?}{30}$

23) $1 = \frac{?}{5}$

24) $\frac{3}{5} = \frac{21}{?}$

25) $\frac{30}{48} = \frac{?}{8}$

26) $\frac{7}{35} = \frac{?}{5}$

27) $\frac{1}{2} = \frac{?}{4} = \frac{?}{8}$

28) $\frac{1}{3} = \frac{?}{9} = \frac{?}{12} = \frac{?}{15}$

5.7 SIMPLIFYING FRACTIONS

When it is ***impossible*** to find a number which is not 1, that will divide exactly into the numerator (top number) and denominator (bottom number), the fraction is said to be in its **simplest form or lowest term.**

The fractions $\frac{2}{3}, \frac{7}{11}, \frac{12}{13}$ are in their simplest forms or lowest terms because no number can divide both the numerator and denominator exactly.

Advice: To simplify a fraction to its lowest form, always divide the fraction by their **highest common factor** (HCF).

Example 1:

Is the fraction $\frac{4}{8}$ in its lowest form?

NO, because 2 or 4 can divide both numbers. In its simplest form,

$$\frac{4 \div 4}{8 \div 4} = \frac{1}{2}$$

$\frac{1}{2}$ is the lowest form of $\frac{4}{8}$

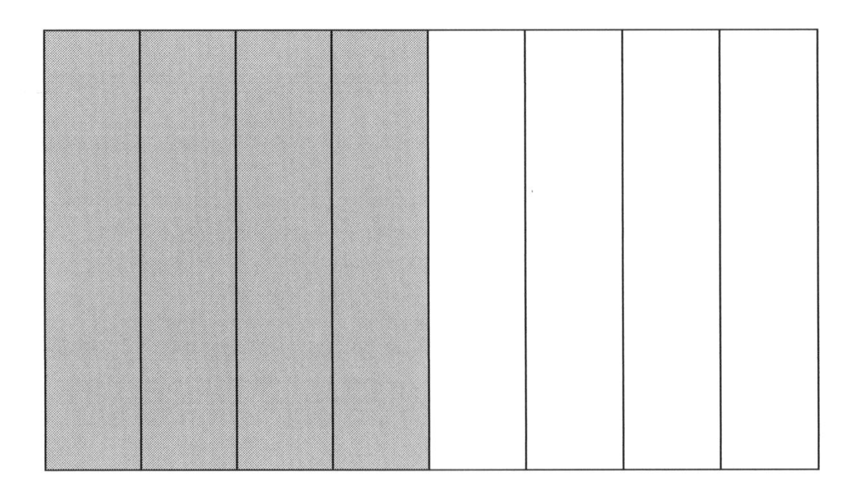

Example 2: Reduce $\frac{25}{30}$ to its lowest form.

Divide both numbers by 5.

$$\frac{25 \div 5}{30 \div 5} = \frac{5}{6}$$

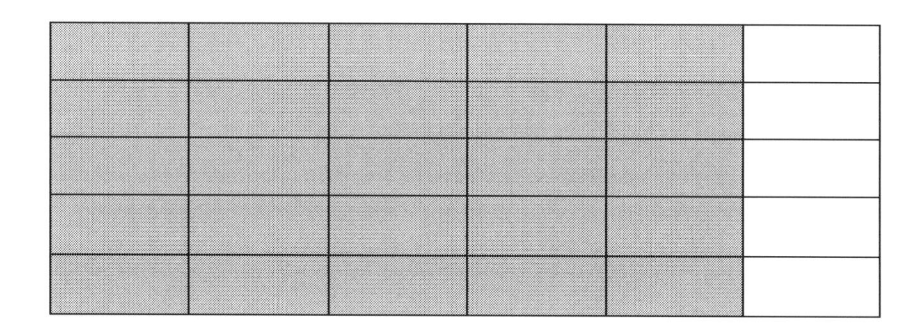

Example 3:

Write $\frac{16}{40}$ in its lowest form.

2, 4 and 8 are the common factors of 16 and 40. We may use any of the common factors to divide.

$$\frac{16 \div 2}{40 \div 2} = \frac{8}{20} \qquad \frac{8 \div 2}{20 \div 2} = \frac{4}{10} \qquad \frac{4 \div 2}{10 \div 2} = \frac{2}{5} \checkmark$$

Not in simplest form   *not in simplest form*   *in simplest form*

It took three calculations to get to the fraction in its lowest form.

However, it is advisable to use the **highest common factor** when dividing both numbers. The answer will be obtained quicker in that way. Dividing by 2 or 4 will still give a correct answer, but more calculations are needed.

The highest common factor, in this case, is 8.

Use 8 to divide both numbers.

$$\frac{16 \div \mathbf{8}}{40 \div \mathbf{8}} = \frac{\mathbf{2}}{\mathbf{5}} \checkmark$$

In one simple calculation, $\frac{16}{40}$ to its lowest form is $\frac{\mathbf{2}}{\mathbf{5}}$

5.8 ADDING AND SUBTRACTING FRACTIONS

Before fractions can be added or subtracted, the **denominators** (bottom numbers) must be the same. If they are the same, simply add the numerators and divide by one of the denominators (since they are the same). Remember, do not add the denominators.

Example 1: $\frac{2}{7} + \frac{4}{7} = \frac{2+4}{7} = \frac{\mathbf{6}}{\mathbf{7}} \checkmark$

(the Same denominator, so add the numerators)

Also when subtracting, $\frac{2}{5} - \frac{1}{5} = \frac{2-1}{5} = \frac{\mathbf{1}}{\mathbf{5}} \checkmark$

(the Same denominator, so subtract the numerators)

WHEN DENOMINATORS ARE NOT THE SAME

When fractions have different denominators, we **must** make them the same by finding a common denominator, preferably, the lowest common multiple (LCM). Then write each fraction as an equivalent fraction and perform the given calculation(s).

See sections 4.1.1 for LCM and 5.6 for equivalent fractions.

Example 2: $\frac{1}{3} + \frac{2}{5}$

Since the denominators 3 and 5 are not the same, we must make them the same before adding. The LCM of 3 and 5 is 15, so we make the denominators equal 15.

Find the equivalent fractions using 15 as the denominator.

$$\frac{1 \times \mathbf{5}}{3 \times \mathbf{5}} + \frac{2 \times \mathbf{3}}{5 \times \mathbf{3}}$$

$$= \frac{5}{15} + \frac{6}{15}$$

$$= \frac{5 + 6}{15}$$

$$= \mathbf{\frac{11}{15}} \checkmark$$

Example 3: $\frac{3}{4} - \frac{3}{5}$

$$\frac{3 \times \mathbf{5}}{4 \times \mathbf{5}} - \frac{3 \times \mathbf{4}}{5 \times \mathbf{4}}$$

$$= \frac{15}{20} - \frac{12}{20}$$

$$= \mathbf{\frac{3}{20}} \checkmark$$

EXERCISE 5B

1) Write each fraction in its simplest form.

a) $\frac{10}{20}$

b) $\frac{14}{21}$

c) $\frac{6}{18}$

d) $\frac{6}{33}$

e) $\frac{10}{40}$

f) $\frac{4}{18}$

g) $\frac{10}{48}$

h) $\frac{55}{88}$

i) $\frac{5}{15}$

j) $\frac{12}{40}$

k) $\frac{20}{75}$

l) $\frac{44}{80}$

m) $\frac{50}{250}$

n) $\frac{200}{500}$

o) $\frac{44}{44}$

2) Make both fractions below, so they have the same denominator.

a) $\frac{4}{5}$ and $\frac{2}{3}$

b) $\frac{1}{6}$ and $\frac{4}{5}$

c) $\frac{3}{4}$ and $\frac{1}{2}$

3) Write each of these fractions as a mixed number.

a) Eight fifths

b) Twenty nine sevenths

c) Fifty-five thirds

d) Eleven sixths

e) $\frac{9}{5}$

f) $\frac{17}{4}$

g) $\frac{142}{13}$

h) $\frac{200}{15}$

4) Work out each of the following calculations in its simplest form.
Write as a mixed number when necessary.

a) $\frac{1}{9} + \frac{3}{9}$

b) $\frac{2}{7} + \frac{1}{7}$

c) $\frac{4}{6} + \frac{1}{6}$

d) $\frac{1}{10} - \frac{1}{10}$

e) $\frac{1}{3} + \frac{1}{4}$

f) $\frac{1}{5} + \frac{1}{9}$

g) $\frac{2}{3} + \frac{1}{5}$

h) $\frac{1}{9} - \frac{1}{9}$

i) $\frac{7}{9} - \frac{1}{5}$

j) $\frac{2}{8} + \frac{3}{5}$

k) $\frac{1}{2} + \frac{1}{7}$

l) $\frac{4}{7} + \frac{1}{10}$

m) $\frac{1}{2} + \frac{1}{3} + \frac{1}{4}$

n) $\frac{2}{3} + \frac{3}{5} - \frac{1}{6}$

o) $\frac{4}{5} - \frac{1}{4} + \frac{3}{5}$

5) Copy and complete

a) $\frac{1}{3} + \square = 1$

b) $\frac{1}{9} + \square = 1$

c) $1 - \frac{2}{3} = \square$

d) $\frac{15}{17} + \square = 1$

e) $\square + \frac{6}{13} = 1$

f) $1 - \square = \frac{15}{19}$

6) Write the fractions below in the order of size, largest first.

a) $\frac{1}{2}, \frac{2}{3}, \frac{5}{6}$

b) $\frac{7}{8}, \frac{1}{4}, \frac{1}{2}$

c) $2\frac{1}{3}, 3\frac{2}{3}, 2\frac{2}{3}$

7) What fraction is shaded

a) red

b) green

c) yellow

d) blue?

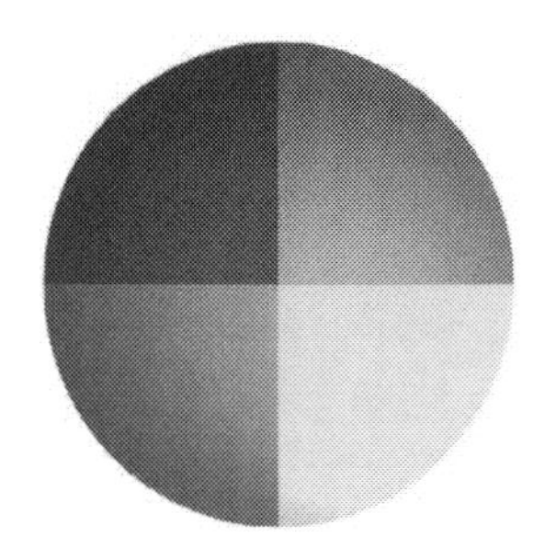

8) Funmi gave $\frac{1}{3}$ of her books to Abubakar and $\frac{1}{5}$ of her books to Chichi. What fraction of Funmi's book does she still have?

9) Five mathematics books are placed side by side and the width shown.

a) What is the total width, ***b metre*** of the five books?

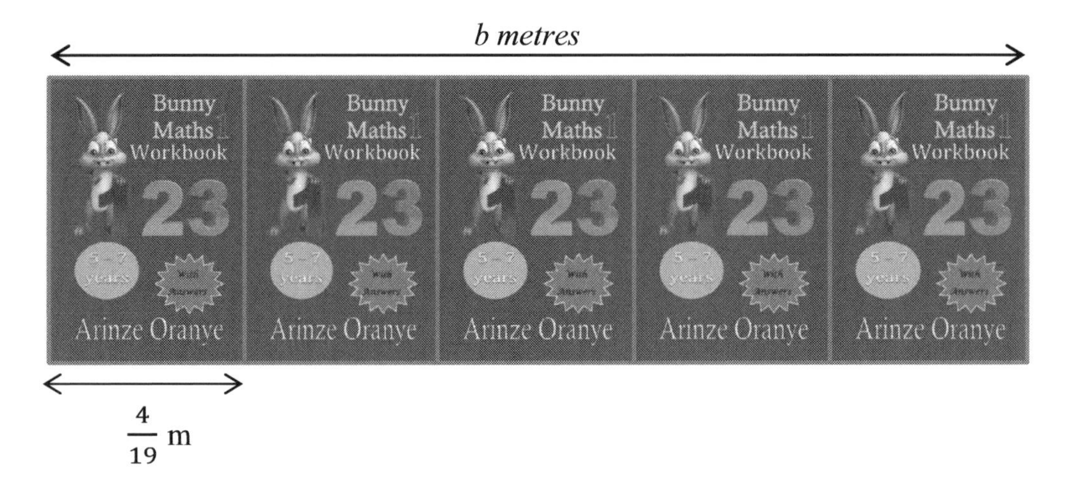

b) A student bought two of the books. What is the combined width of the remaining books?

10) All are equivalent fractions apart from one. Pick the odd one out and give a reason.

$$\frac{4}{15} \qquad \frac{8}{30} \qquad \frac{5}{45} \qquad \frac{20}{75} \qquad \frac{40}{150}$$

11) Complete the missing numbers in the boxes.

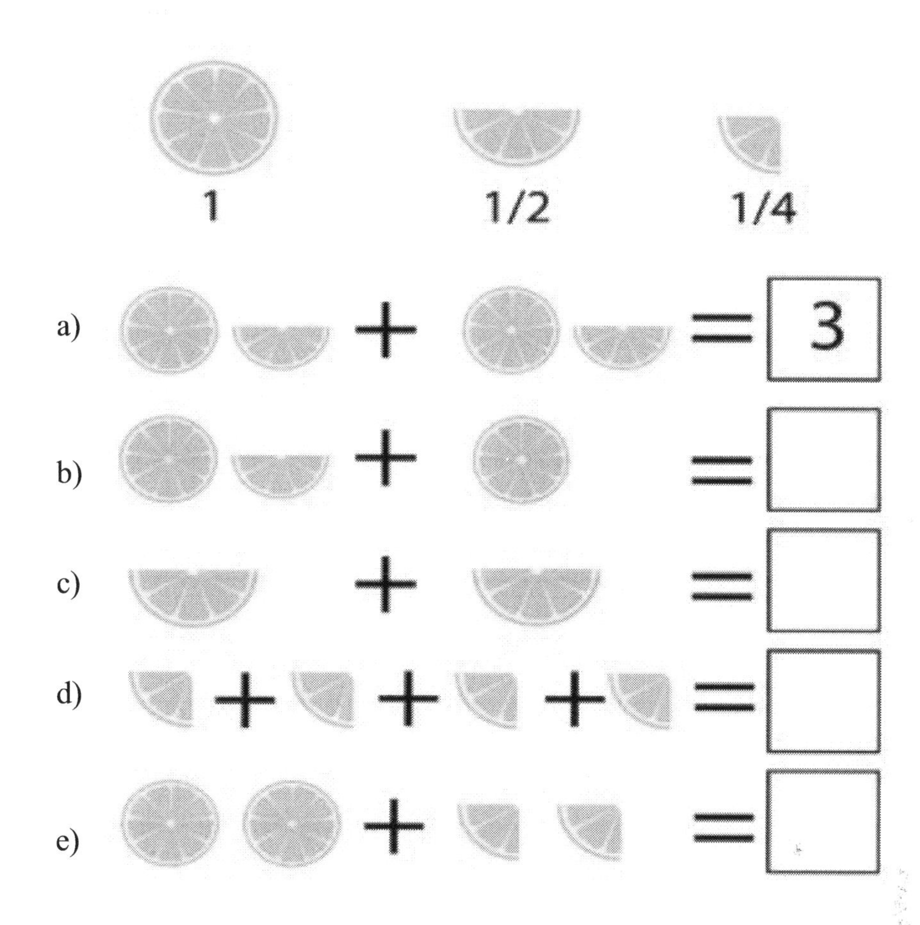

ADDING AND SUBTRACTING MIXED NUMBERS

Example 1: Work out $2\frac{1}{2} + 3\frac{1}{3}$

Solution: Change the mixed numbers to improper fractions (Refer to section 5.3).

For mixed number $2\frac{1}{2}$, change to $\frac{(2 \times 2)+1}{2} = \frac{5}{2}$

For mixed number $3\frac{1}{3}$, change to $\frac{(3 \times 3)+1}{3} = \frac{10}{3}$

Therefore, $\frac{5}{2} + \frac{10}{3}$

The denominators are not the same, so we make them the same.

$$= \frac{5 \times \mathbf{3}}{2 \times \mathbf{3}} + \frac{10 \times \mathbf{2}}{3 \times \mathbf{2}}$$

$$= \frac{15}{6} + \frac{20}{6} = \frac{35}{6}$$

However, the answer is top heavy (improper fraction) and should be written as a mixed number (refer to section 5.4).

Therefore, $\frac{35}{6} = 5\frac{5}{6}$

$2\frac{1}{2} + 3\frac{1}{3} = \mathbf{5\frac{5}{6}}$ ✓

A quicker method is to add up the whole numbers and fractions separately.

$2 + 3 = \mathbf{5}$

$$\frac{1}{2} + \frac{1}{3} = \frac{1 \times \mathbf{3}}{2 \times \mathbf{3}} + \frac{1 \times \mathbf{2}}{3 \times \mathbf{2}} = \frac{3}{6} + \frac{2}{6} = \mathbf{\frac{5}{6}}$$

Therefore, 5 and $\frac{5}{6}$ gives $\mathbf{5\frac{5}{6}}$ ✓ ....same answer.

Example 2: Work out $4\frac{1}{2}$ - $3\frac{1}{3}$

Solution: Change the mixed numbers to improper fractions (Refer to section 5.3).

For mixed number $4\frac{1}{2}$, change to $\frac{(2 \times 4) + 1}{2} = \frac{9}{2}$

For mixed number $3\frac{1}{3}$, change to $\frac{(3 \times 3) + 1}{3} = \frac{10}{3}$

Therefore, $\frac{9}{2}$ - $\frac{10}{3}$

The denominators are not the same, so we make them the same.

$$= \frac{9 \times \mathbf{3}}{2 \times \mathbf{3}} - \frac{10 \times \mathbf{2}}{3 \times \mathbf{2}}$$

$$= \frac{27}{6} - \frac{20}{6}$$

$$= \frac{7}{6}$$

However, the answer is top heavy (improper fraction) and should be written as a mi×ed number (refer to section 5.4).

Therefore, $\frac{7}{6} = 1\frac{1}{6}$

$4\frac{1}{2} - 3\frac{1}{3} = \mathbf{1\frac{1}{6}}$ ✓

EXERCISE 5C

Work out the following and leave your answers in their simplest form.

1) $1\frac{2}{3} + 1\frac{2}{3}$

2) $2\frac{2}{3} + 1\frac{2}{5}$

3) $4\frac{2}{3} + 3\frac{1}{3}$

4) $5\frac{2}{3} - 2\frac{2}{3}$

5) $7\frac{1}{3} - 1\frac{1}{4}$

6) $1\frac{2}{3} + 1\frac{2}{3} + 1\frac{2}{3}$

7) $1\frac{2}{3} - \frac{3}{5} + \frac{2}{7}$

8) $3\frac{2}{7} + \frac{1}{3}$

9) $1\frac{2}{3} - 1\frac{2}{3}$

10) $8\frac{2}{3} - 3\frac{4}{5}$

12) On Tuesday, Edward cycled $9\frac{2}{3}$ kilometres to work from his house.

On his way back, he cycled $4\frac{1}{4}$ km and had a puncture. Edward had a lift back to his house by a friend, Andrew.

How far did Andrew travel to get Edward back to his house?

5.9 MULTIPLYING FRACTIONS

Multiplying fractions is the easiest of all the fraction calculations. To multiply fractions, multiply the numerators (top) numbers and multiply the denominators (bottom numbers). Cancel down when possible.

Example 1: Work out $\frac{1}{4} \times \frac{2}{5}$

Multiply the numerators: $1 \times 2 = 2$

Multiply the denominators: $4 \times 5 = 20$

Therefore, $\frac{1}{4} \times \frac{2}{5} = \mathbf{\frac{2}{20}}$

A more structured approach would be

$$\frac{1}{4} \times \frac{2}{5} = \frac{1 \times 2}{4 \times 5} = \mathbf{\frac{2}{20}}$$

But $\mathbf{\frac{2}{20}}$ is not in its simplest form, so we cancel down

$$\frac{2 \div \mathbf{2}}{20 \div \mathbf{2}} = \mathbf{\frac{1}{10}} \checkmark$$

Example 2: Work out $\frac{3}{7} \times \frac{5}{8}$

$$= \frac{3 \times 5}{7 \times 8}$$

$$= \mathbf{\frac{15}{56}} \checkmark$$

Example 3: The diagram shows a square of side **1 metre** which is divided into Q, R, S, T rectangles.

a) Work out the area of **each** rectangle Q, R, S, and T.

b) Show that the total area of the square is 1m^2

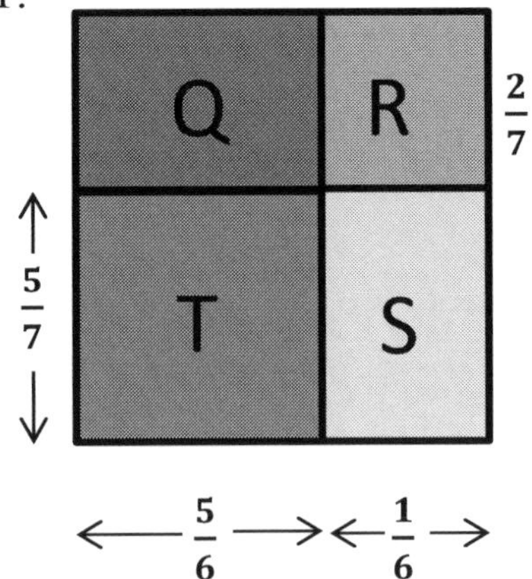

Solution:

Area of a rectangle = length × width

Area of **Q** $= \frac{2}{7} \times \frac{5}{6} = \frac{10}{42} = \mathbf{\frac{5}{21}}\ \text{m}^2$

Area of **R** $= \frac{2}{7} \times \frac{1}{6} = \frac{2}{42} = \mathbf{\frac{1}{21}}\ \text{m}^2$

Area of **S** $= \frac{5}{7} \times \frac{1}{6} = \mathbf{\frac{5}{42}}\ \text{m}^2$

Area of **T** $= \frac{5}{7} \times \frac{5}{6} = \mathbf{\frac{25}{42}}\ \text{m}^2$

For question b, total area of square is the total area of rectangles Q, R, S, and T

$$\frac{5 \times \mathbf{2}}{21 \times \mathbf{2}} + \frac{1 \times \mathbf{2}}{21 \times \mathbf{2}} + \frac{5}{42} + \frac{25}{42}$$

$$= \frac{10}{42} + \frac{2}{42} + \frac{5}{42} + \frac{25}{42}$$

$$= \frac{10 + 2 + 5 + 25}{42}$$

$$= \frac{42}{42}$$

$$= 1$$

Therefore, the total area of the square is $\mathbf{1\ m^2}$ ✓

EXERCISE 5D

1) Work out and simplify your answers where possible.

a) $\frac{1}{5} \times \frac{2}{3}$

b) $\frac{2}{5} \times \frac{2}{7}$

c) $\frac{3}{4} \times \frac{6}{7}$

d) $\frac{5}{6} \times \frac{2}{9}$

e) $\frac{3}{5} \times \frac{3}{5}$

f) $\frac{1}{4} \times \frac{7}{8}$

g) $\frac{5}{7} \times \frac{8}{9}$

h) $\frac{3}{12} \times \frac{1}{8}$

i) $\frac{2}{9} \times \frac{6}{7}$

j) $\frac{12}{15} \times \frac{1}{7}$

k) $\frac{8}{10} \times \frac{3}{4}$

l) $\left(\frac{2}{20}\right)^2$

2) The height of the Nigerian flag is $\frac{6}{7}$ m.
If fourteen flags are placed on top of each other, what would be the total length of all the flags?

3) Work out the area of the rectangular field.

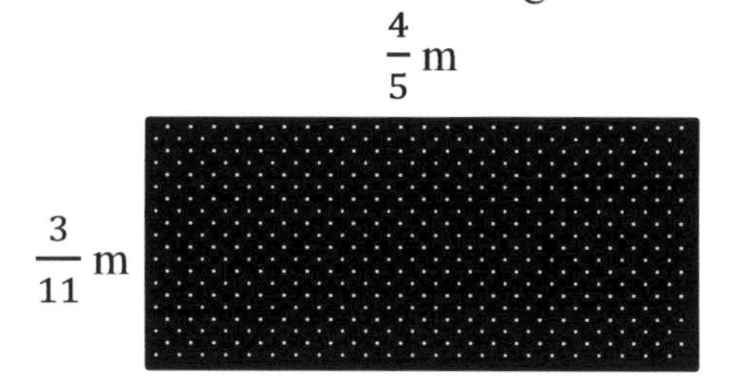

4) Work out the area of the triangle.

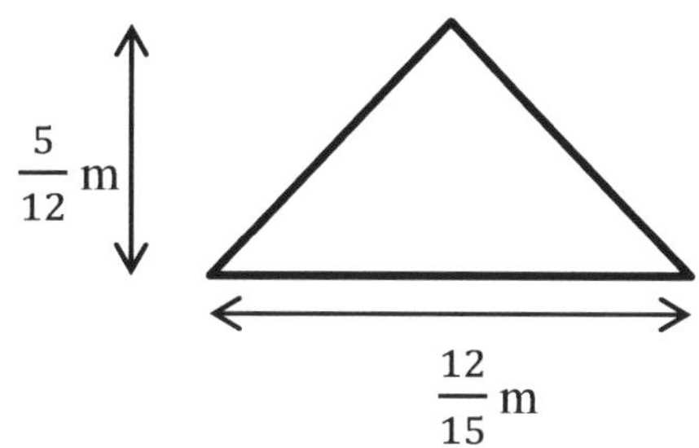

5) A rectangle is divided into four parts as shown below. Work out the area of **each** part of the rectangle. All lengths are in metres.

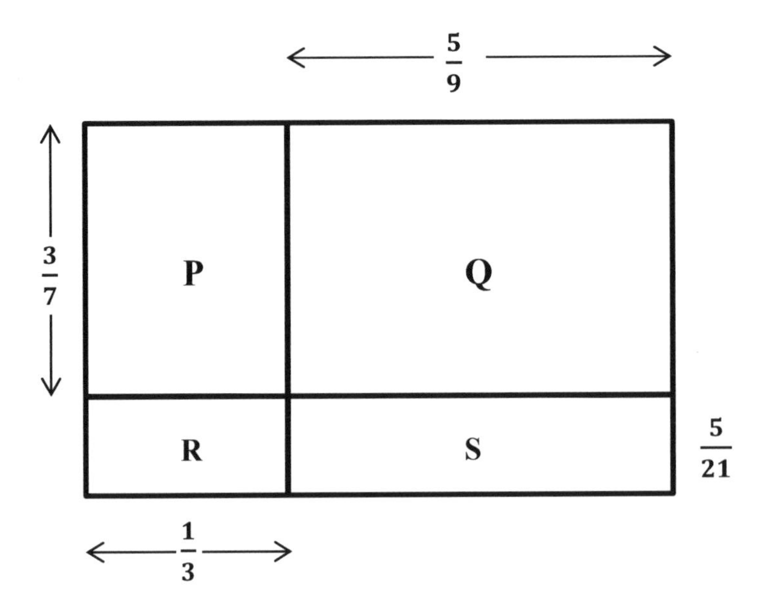

6) Work out the following.

a) $\frac{1}{3} \times \frac{1}{3} \times \frac{1}{3}$

b) $\frac{2}{3} \times \frac{3}{4} \times \frac{1}{5}$

c) $\frac{4}{11} \times \frac{9}{13}$

7) Nnaemeka says "$\frac{5}{6} \times \frac{1}{3} = \frac{6}{9}$"

a) Is he correct? Explain fully

b) If not, what did Nnaemeka do wrong?

FRACTION OF AN AMOUNT

Example 1: Work out $\frac{3}{5} \times 20$

Remember, every whole number can be written as a fraction by dividing by 1.
20 is the same as $\frac{20}{1}$

$$\frac{3}{5} \times \frac{20}{1}$$

$$= \frac{3 \times 20}{5 \times 1}$$

$$= \frac{60}{5} = \mathbf{12} \checkmark$$

Alternatively, you may consider dividing by the denominator and multiplying by the numerator. It is the same thing as cancelling down. 20 ÷ 5 = 4. Then, 4 × 3 = 12.

$$\frac{3}{\cancel{5}_{\mathbf{1}}} \times \frac{\cancel{20}^{\mathbf{4}}}{1} = 3 \times 4 = \mathbf{12}$$

Example 2: Work out $\frac{3}{5}$ of ₦100

In mathematics, '**Of**' in this instance is multiplication (×).

It is the same method of working out fractions as in examples 1 and 2 above.

$$\frac{3}{5} \text{ of ₦}100 = \frac{3}{5} \times \text{₦}100$$

$$= \frac{3}{\cancel{5}_{\mathbf{1}}} \times \cancel{\text{₦}100}^{\mathbf{₦20}}$$

$$= 3 \times \text{₦}20$$

$$= \mathbf{₦60} \checkmark$$

Dividing by the denominator and multiplying by the numerator

Alternatively; $\frac{3}{5}$ of ₦100

$= \frac{3}{5} \times \frac{100}{1}$

$= \frac{3 \times 100}{5 \times 1}$

$= \frac{300}{5}$

= **₦60**

Example 3: $450 \times \frac{3}{9}$

Method 1: Multiply 450 by 3, and then divide by 9

$450 \times 3 = 1350$

$1350 \div 9 =$ **150** ✓

Method 2: Divide 450 by 9, and then multiply by 3

$450 \div 9 = 50$

$50 \times 3 =$ **150** ✓

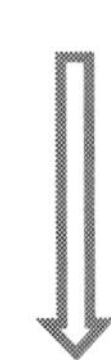

Cancelling down method

$\overset{50}{\cancel{450}} \times \frac{3}{\underset{1}{\cancel{9}}}$

$\frac{50 \times 3}{1} =$ **150** ✓

EXERCISE 5E

1) Work out the following.

a) $\frac{1}{5} \times 5$

b) $\frac{2}{5} \times 10$

c) $\frac{3}{4} \times 12$

d) $\frac{7}{8} \times 24$

e) $40 \times \frac{1}{5}$

f) $45 \times \frac{2}{9}$

g) $120 \times \frac{5}{12}$

h) $300 \times \frac{3}{60}$

i) $\frac{1}{5} \times 200$

j) $275 \times \frac{1}{5}$

k) $2000 \times \frac{7}{40}$

l) $\frac{1}{13} \times 13$

2) Work out the following.

a) $\frac{1}{5}$ of ₦200

b) $\frac{1}{3}$ of ₦150

c) $\frac{1}{8}$ of \$400

d) $\frac{1}{9}$ of 54 kg

e) $\frac{5}{8}$ of 64

f) $\frac{4}{5}$ of ₦1500

g) $\frac{1}{5}$ of £77.50

h) $\frac{17}{35}$ of 2100 kg

3)

Emeka paid $\frac{2}{3}$ of the cost of Obiora's jacket for a similar black jacket.

a) How much did Emeka pay for the jacket?

b) Arinze says "$\frac{4}{5}$ of what Emeka paid is more than half the cost of Obiora's jacket."

Is Arinze correct? Explain fully.

MULTIPLYING WITH MIXED NUMBERS

To multiply mixed numbers, it is advisable to change them to improper fractions (Refer to section 5.3)

Example 1: Work out $3\frac{2}{5} \times 7\frac{1}{2}$

Change to improper fractions

$3\frac{2}{5}$ becomes $\frac{(5 \times 3)+2}{5} = \mathbf{\frac{17}{5}}$ and $7\frac{1}{2}$ becomes $\frac{(2 \times 7)+1}{2} = \mathbf{\frac{15}{2}}$

Multiplying the two improper fractions: $\frac{17}{5} \times \frac{15}{2} = \frac{17 \times 15}{5 \times 2}$

$$= \frac{255}{10} = 25.5$$

$$= \mathbf{25\frac{1}{2}} \checkmark$$

Example 2: Work out $5\frac{1}{4} \times \frac{2}{7}$

$5\frac{1}{4}$ becomes $\frac{(4 \times 5)+1}{4} = \frac{21}{4}$

Multiplying both fractions gives $\frac{21}{4} \times \frac{2}{7}$

$$= \frac{21 \times 2}{4 \times 7}$$

$= \frac{42}{28}$ (Cancel down by dividing both numbers by 2)

$= \frac{21}{14}$ (Cancel down by dividing both numbers by 7)

$= \frac{3}{2}$ (Change back to a mixed number)

$$= \mathbf{1\frac{1}{2}} \checkmark$$

Example 3: Multiply $4 \times 2\frac{3}{5}$

Change $2\frac{3}{5}$ to an improper fraction. $\frac{(5 \times 2)+3}{5} = \frac{13}{5}$

Multiplying gives $4 \times \frac{13}{5}$

$= \frac{4 \times 13}{5}$

$= \frac{52}{5}$

$= \mathbf{10\frac{2}{5}}$ ✓

EXERCISE 5F

1) Write the following mixed numbers to improper fractions.

a) $1\frac{3}{4}$ b) $5\frac{3}{5}$ c) $11\frac{4}{9}$ d) $17\frac{1}{3}$

2) Work out the following but give your answer as a mixed number when applicable.

a) $2\frac{3}{5} \times \frac{3}{5}$ b) $1\frac{3}{5} \times 2\frac{3}{7}$ c) $7\frac{3}{5} \times \frac{3}{5}$ d) $3\frac{1}{5} \times 1\frac{4}{10}$

e) $3\frac{3}{4} \times 1\frac{2}{5}$ f) $2\frac{4}{6} \times 6\frac{3}{7}$ g) $12\frac{3}{5} \times 8\frac{3}{4}$ h) $(2\frac{3}{5})^2$

3) Find the product of these numbers. Write as mixed numbers where possible and simplify.

a) $4\frac{3}{7}$ and $\frac{1}{3}$ b) $10\frac{1}{5}$ and $\frac{3}{5}$ c) 2 and $8\frac{5}{6}$ d) 9 and $6\frac{3}{4}$

4) If it takes $\frac{1}{5}$ of a minute to fill a bucket with cold water, what fraction of a minute will it take to fill $15\frac{1}{2}$ buckets of cold water?

5)

A tin of sweetcorn weighs $\frac{1}{5}$kg.
137 of the tins are packed in a bag.

a) What is the total weight of the sweetcorn?

b) If the bag used in packaging weighs $1\frac{3}{5}$ kg, what is the total weight of the sweet corns and bag?

5.10 DIVIDING FRACTIONS

Our mantra would be "**Keep, Change, Flip**."

Keep the first fraction, **change** the division to multiplication (×) and **flip** the second fraction. Once we have successfully applied the mantra, we then multiply out the fractions (See section 5.9).

Example 1: Work out $\frac{5}{6} \div \frac{3}{4}$

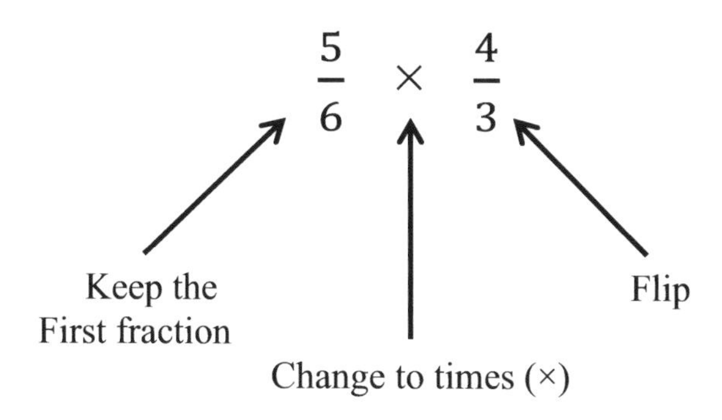

Multiply the two fractions

$$\frac{5 \times 4}{6 \times 3} = \frac{20}{18}$$

Since $\frac{20}{18}$ is not in its simplest form, we cancel down by dividing by 2.

We then have $\frac{20 \div 2}{18 \div 2} = \frac{10}{9}$

Also, the answer is an improper fraction, we then change back to a mixed number (refer to section 5.4)

Therefore, $\frac{10}{9} = \mathbf{1\frac{1}{9}}$ ✓

The same system applies to dividing mixed numbers. However, we must change to improper fractions before applying the *keep, change flip*. See example 2.

Example 2: Work out $3\frac{2}{5} \div 1\frac{3}{4}$

Change the mixed numbers to improper fractions (refer to section 5.3).

$3\frac{2}{5}$ to an improper fraction is $\frac{(5 \times 3) + 2}{5} = \frac{17}{5}$

$1\frac{3}{4}$ to an improper fraction is $\frac{(4 \times 1) + 3}{5} = \frac{7}{4}$

Now, rewrite the fractions using the improper fractions as $\frac{17}{5} \div \frac{7}{4}$

Remember **Keep, change, flip**

$$= \frac{17}{5} \times \frac{4}{7}$$

$$= \frac{17 \times 4}{5 \times 7} = \frac{68}{35} = \mathbf{1\frac{33}{35}} \checkmark$$

Example 3: Work out $10 \div 3\frac{2}{7}$

Change $3\frac{2}{7}$ to improper fraction which will equal $\frac{(7 \times 3 + 2)}{7} = \frac{23}{7}$

Now, rewrite the fractions using the improper fraction as $10 \div \frac{23}{7}$

Remember Keep, change, flip

$$= 10 \times \frac{7}{23}$$

$$= \frac{10 \times 7}{23} = \frac{70}{23} = \mathbf{3\frac{1}{23}} \checkmark$$

Example 4: How many quarters are there in 6?

This means $6 \div \frac{1}{4}$

$$= 6 \times \frac{4}{1}$$

$$= \frac{6 \times 4}{1} = \mathbf{24} \checkmark$$

EXERCISE 5G

1) Work out the fractions and give your answers as a mixed number when possible.

a) $\frac{1}{2} \div \frac{1}{4}$

b) $\frac{4}{5} \div \frac{2}{3}$

c) $\frac{7}{10} \div \frac{3}{5}$

d) $\frac{6}{7} \div \frac{1}{5}$

e) $7 \div \frac{1}{2}$

f) $\frac{1}{3} \div \frac{1}{5}$

g) $10\frac{1}{2} \div \frac{1}{2}$

h) $6\frac{2}{9} \div \frac{5}{6}$

i) $20 \div 1\frac{4}{5}$

j) $2\frac{7}{9} \div 1\frac{7}{10}$

k) $5\frac{2}{3} \div 2\frac{7}{10}$

l) $3\frac{3}{7} \div 1\frac{1}{10}$

2) How many thirds are there in

a) 5 b) 7 c) 9 d) 15?

3) How many tenths are there in

a) 6 b) 8 c) 11 d) 20?

4) The length of the top of a rectangular table tennis table is $2\frac{7}{10}$ m. The area of the top is $4\frac{1}{20}$ m^2.
Work out the width of the top of the table top. Leave your answer as a mixed number.

5) A perimeter fence has seven panels. The width of one panel is $1\frac{8}{10}$ m long. What is the total width of the panels?

6) Work out $15\frac{3}{4} \div 1\frac{3}{4}$, give your answer as a mixed number

7) Work out $12\frac{2}{5} \div 5\frac{7}{8}$, give your answer as a mixed number.

Chapter 5 Review Section
Assessment

1) Look at the picture. What fraction of these students are:

a) males

.... **1 mark**

b) not males

.... **1 mark**

c) wearing white shirts?

.... **1 mark**

2) What fraction of the months of the year starts with the letter J?

.... **1 mark**

3) What fraction of an hour is 20 minutes?

.... **1 mark**

4) Write the following as equivalent fractions with a denominator of 40.

a) $\frac{1}{4}$ b) $\frac{3}{8}$ c) $\frac{4}{5}$ d) $\frac{9}{10}$

.... **4 marks**

5) Change to improper fractions.

a) $2\frac{1}{7}$ b) $1\frac{5}{8}$ c) $12\frac{1}{3}$ d) $20\frac{9}{11}$

.... **4 marks**

6) Add $\frac{4}{7}$ to :

a) $\frac{1}{7}$ b) $\frac{2}{3}$ c) $\frac{5}{6}$ d) $\frac{3}{11}$

.... **8 marks**

7) Simplify each fraction.

a) $\frac{4}{8}$ b) $\frac{9}{27}$ c) $\frac{18}{30}$ d) $\frac{144}{168}$

.... **8 marks**

8) Copy and complete the pairs of equivalent fractions below.

a) $\frac{4}{7} = \frac{?}{21}$ b) $\frac{12}{20} = \frac{?}{100}$ c) $\frac{6}{8} = \frac{54}{?}$ d) $\frac{?}{15} = \frac{24}{30}$

.... **4 marks**

9) Change to mixed numbers.

a) $\frac{7}{3}$ b) $\frac{23}{4}$ c) $\frac{59}{7}$ d) $\frac{204}{20}$

.... **4 marks**

10) Work out and leave your answers as a mixed number where possible. Also, leave your answers in its lowest form where possible.

a) $\frac{4}{17} + \frac{4}{17}$ b) $\frac{8}{9} - \frac{1}{3}$ c) $1\frac{4}{5} + 4\frac{4}{5}$ d) $\frac{2}{3} \div \frac{1}{5}$

e) $3\frac{1}{3} - 1\frac{7}{9}$ f) $3\frac{1}{3} \times \frac{1}{4}$ g) $2\frac{1}{6} \times 2\frac{5}{9}$ h) $8\frac{1}{3} \div 5\frac{2}{5}$

.... **16 marks**

11) Work out

a) $\frac{2}{5}$ of 30 b) $\frac{3}{4}$ of 36 kg c) $\frac{5}{7}$ of ₦420 d) $160 \times \frac{7}{8}$

.... **8 marks**

12) The diagram is divided into rectangles as shown.

a) Work out the area of **each** rectangle.

b) Work out the **total** area of Q, R, S and T

c) What is the mathematical name of the shape formed by Q, R S and T? Give a reason for your answer.

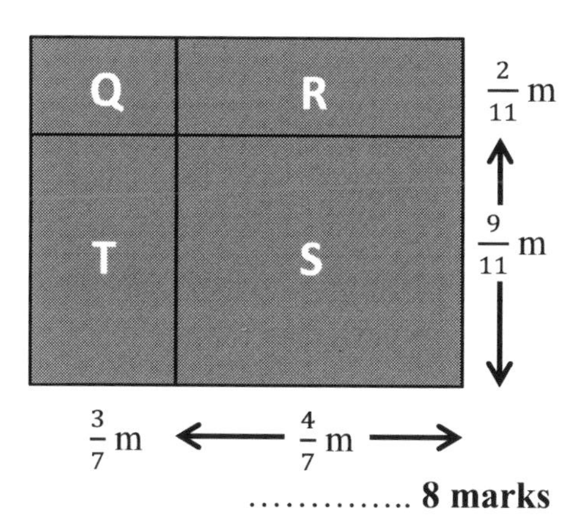

.... **8 marks**

13) Simplify the following.

a) $\frac{16}{20} \div \frac{4}{10}$

b) $\dfrac{\frac{1}{2} \times \frac{6}{7}}{\frac{1}{7}}$

c) $\dfrac{5\frac{3}{5} + \frac{4}{7}}{\frac{2}{3}}$

14)

The weight of each pack of rice is 2kg. Nine packets of rice are packed into a travelling bag of weight $1\frac{5}{6}$ kg.

a) What is the total weight of the bag when loaded with the nine packets of rice?

b) Kola bought 21 of such travelling bags. What is the total weight of the travelling bags?

c) Kola gave three travelling bags away to his friend, Tayo. What is the combined weight of Tayo's bags?

d) If a travelling bag costs ₦5 500, how much did Kola pay for the travelling bags?

6 Percentages 1

This section covers the following topics:

- Fractions, decimals, and percentages
- Percentage of a quantity
- Chapter Review Section

LEARNING OBJECTIVES

By the end of this unit, you should be able to:

a) Understand the word 'percent'
b) Change fractions to decimals
c) Change fractions and decimals into percentages
d) Change percentages into fractions and decimals
e) Write one quantity as a fraction of another

KEYWORDS

- Percentage
- Percent
- Quantity
- Fraction
- Decimal

6.1 UNDERSTANDING PERCENTAGES (%)

The word percent means **'out of a hundred.'** A percentage is a special type of fraction.

1% means 1 part per hundred. This can be written as $\frac{1}{100}$.

1% can also be written as a decimal **0.01**.

Likewise,

2% means $\frac{2}{100}$ or 0.02 as a decimal

10% means $\frac{10}{100}$ or 0.1 as a decimal

47% means $\frac{47}{100}$ or 0.47 as a decimal

100% means $\frac{100}{100}$ which is equal to **one whole** (1)

Percentages are equal to fractions with the denominator (bottom number) equal to 100.

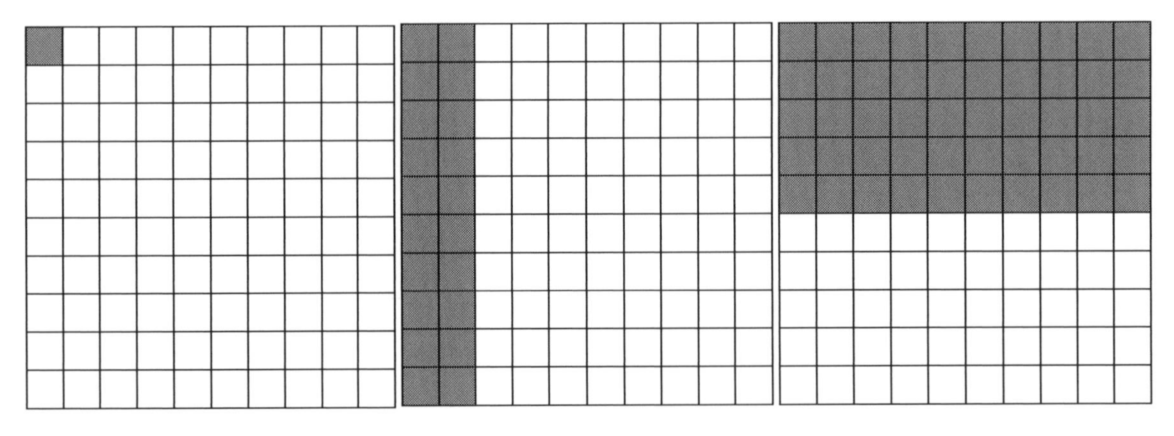

One percent is shaded $\frac{1}{100}$

Twenty percent is shaded $\frac{20}{100}$ or $\frac{1}{5}$ or $\frac{2}{10}$

Fifty percent is shaded $\frac{1}{2}$ or $\frac{50}{100}$ or $\frac{5}{10}$ or $\frac{25}{50}$

Example 1: What percentage of each box has been shaded?

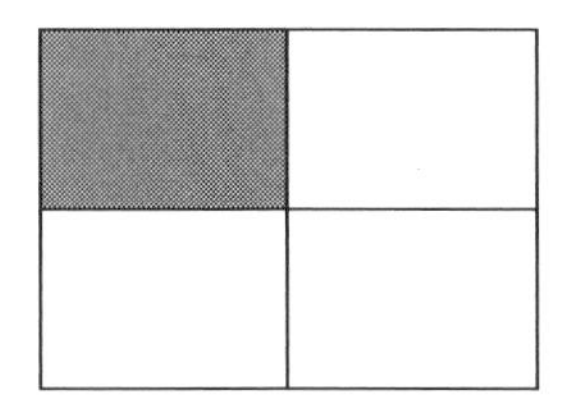

$\frac{1}{4} = 25\%$

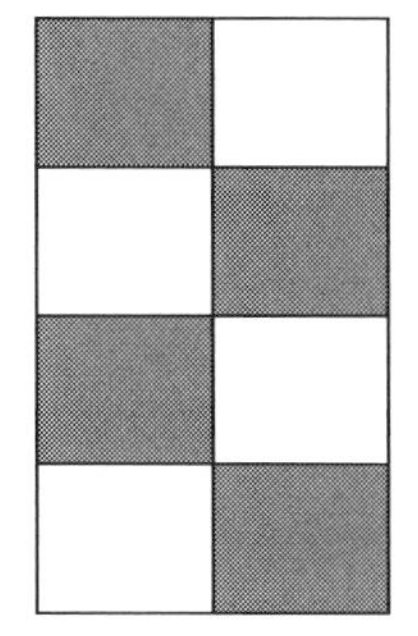

$\frac{4}{8} = \frac{1}{2} = 50\%$

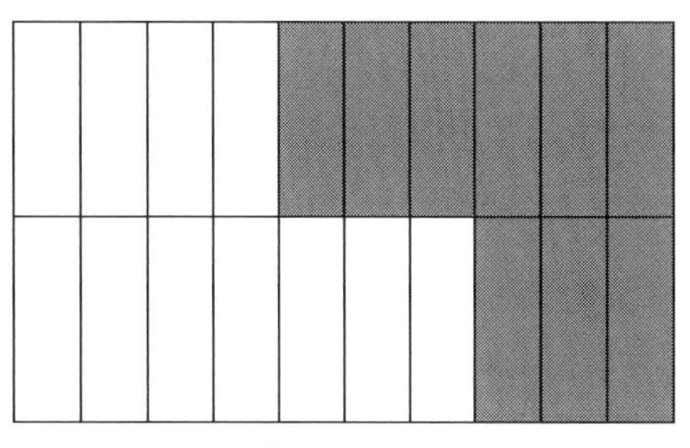

$\frac{9}{20} = 45\%$

Some key percentages to memorise

Fractions	Percentages	
$\frac{1}{4}$	25%	
$\frac{1}{2}$	50%	
$\frac{3}{4}$	75%	
A whole (1)	100%	

Also note:

$\frac{1}{8} = 12\frac{1}{2}\%$

6.2 FRACTIONS TO DECIMALS

A fraction is simply a division. The knowledge in section 3.3 is vital. If you can divide a number, you have successfully converted a fraction to a decimal.

However, the knowledge of equivalent fractions is very important in changing fractions to decimals.

Example 1: Change $\frac{2}{5}$ to a decimal.

Write as an equivalent fraction with a denominator of 100.

$$\frac{2 \times \mathbf{20}}{5 \times \mathbf{20}} = \frac{40}{100} = \mathbf{0.4}$$

Not all numbers will have an equivalent fraction of 100. If that is the case, use other methods like dividing decimals. Alternatively, change the denominator to an equivalent fraction which you may multiply or divide by a number to make 100 or multiples of 10 (if it is easier).You may also cancel down (reducing to its simplest form) and the find an equivalent fraction with a denominator of 100.

Example 2: Convert $\frac{9}{12}$ to a decimal.

As you can see, 100 is not a multiple of 12, and it will not be possible to make an equivalent fraction with a denominator of 100.

Cancelling down gives $\frac{9 \div \mathbf{3}}{12 \div \mathbf{3}} = \frac{3}{4}$

It now possible to make an equivalent fraction with a denominator of 100

$$\frac{3 \times \mathbf{25}}{4 \times \mathbf{25}} = \frac{75}{100} = \mathbf{0.75}$$

Example 3: Change $\frac{13}{20}$ to a decimal.

$$\frac{13 \times \mathbf{5}}{20 \times \mathbf{5}} = \frac{65}{100} = \mathbf{0.65}$$

6.3 PERCENTAGES INTO FRACTIONS

To change a percentage to a fraction, write the percentage as a fraction of a hundred and reduce the fraction to its lowest terms when possible.

Remember, 3% means 3 out of a hundred or $\frac{3}{100}$

4% means 4 out of a hundred or $\frac{4}{100}$

27% means 27 out of a hundred or $\frac{27}{100}$

Example 1: Change 5% to a fraction.

Write as a fraction of 100.

$$5\% = \frac{5}{100}$$

However, $\frac{5}{100}$ is not in its lowest term, so we cancel down.

$$\frac{5 \div 5}{100 \div 5} = \mathbf{\frac{1}{20}}$$

Example 2: Express 38% as a fraction in its lowest form.

Write as a fraction of hundred.

$$38\% = \frac{38}{100}$$

However, $\frac{38}{100}$ is not in its lowest term, so we cancel down.

$$\frac{38 \div 2}{100 \div 2} = \mathbf{\frac{19}{50}}$$

EXERCISE 6A

1) Convert the following fractions to decimals.

a) $\frac{3}{5}$ f) $\frac{1}{8}$ k) $\frac{30}{40}$ p) $\frac{50}{100}$

b) $\frac{4}{5}$ g) $\frac{2}{8}$ l) $\frac{40}{50}$ q) $\frac{2}{20}$

c) $\frac{3}{10}$ h) $\frac{5}{8}$ m) $\frac{7}{20}$ r) $\frac{7}{8}$

d) $\frac{2}{20}$ i) $\frac{3}{50}$ n) $\frac{12}{25}$ s) $\frac{1}{50}$

e) $\frac{7}{25}$ j) $\frac{19}{25}$ o) $\frac{6}{10}$ t) $\frac{16}{20}$

2) In their lowest terms, write the following percentages as fractions.

a) 2% f) 20% k) 50% p) 79%

b) 3% g) 25% l) 55% q) 80%

c) 10% h) 30% m) 60% r) 81%

d) 15% i) 35% n) 70% s) 90%

e) 18% j) 46% o) 75% t) 98%

6.4 FRACTIONS TO PERCENTAGES

Multiplying a fraction or a decimal by **100** changes it to a percentage

Example 1: Express $\frac{1}{4}$ as a percentage.

This means $\frac{1}{4} \times 100$ .... (Refer to section 5.9 on how to multiply fractions and the fraction of an amount.)

$$\frac{1 \times 100}{4} = \frac{100}{4} = 25$$

Therefore, $\frac{1}{4}$ as a percentage is **25%**

Example 2: Write $\frac{2}{20}$ as a percentage.

$$= \frac{2}{20} \times 100$$

$$= \frac{2 \times 100}{20} = \frac{200}{20} = \mathbf{10\%}$$

Example 3: Change $\frac{5}{8}$ to a percentage.

$$= \frac{5}{8} \times 100$$

$$= \frac{5 \times 100}{8} = \frac{500}{8}$$

Refer to section 2.6 on dividing whole numbers

You may also reduce to its simplest form and then change to a mixed number.

$$\frac{500 \div 2}{8 \div 2} = \frac{250}{4} = \frac{250 \div 2}{4 \div 2} = \frac{125}{2} = \mathbf{62\frac{1}{2}\ \%}$$

EXERCISE 6B

1) Write the following fractions as percentages.

a) $\frac{1}{4}$ f) $\frac{3}{20}$ k) $\frac{16}{50}$ p) $\frac{4}{40}$

b) $\frac{1}{5}$ g) $\frac{6}{10}$ l) $\frac{30}{50}$ q) $\frac{7}{14}$

c) $\frac{2}{5}$ h) $\frac{3}{8}$ m) $\frac{6}{25}$ r) $\frac{16}{32}$

d) $\frac{3}{10}$ i) $\frac{7}{20}$ n) $\frac{16}{25}$ s) $\frac{13}{52}$

e) $\frac{4}{5}$ j) $\frac{9}{25}$ o) $\frac{13}{20}$ t) $\frac{16}{320}$

2) Draw each shape below and answer the question for each shape.

a) b) c)

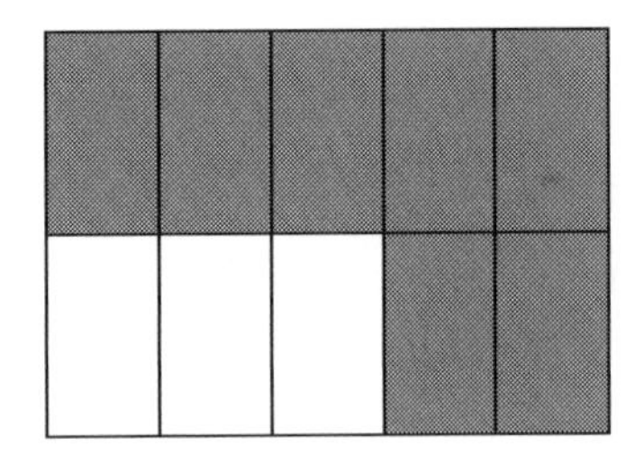

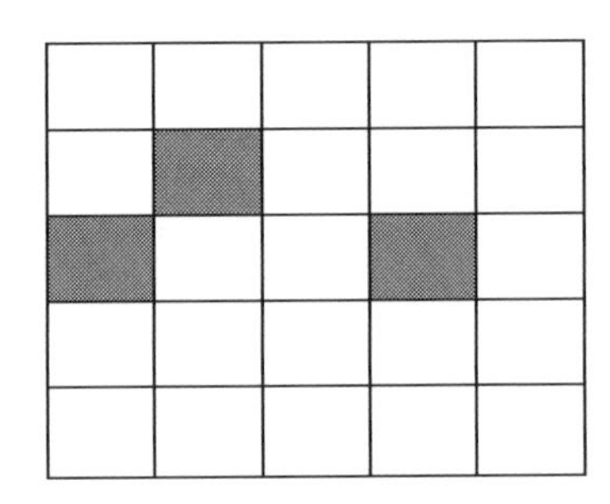

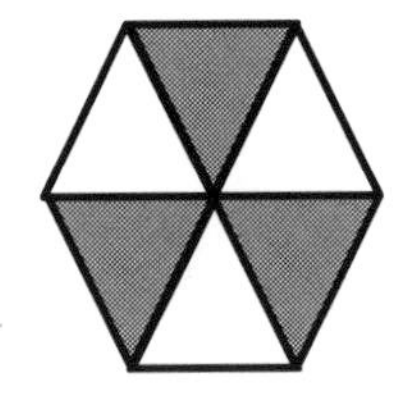

i) What fraction is shaded?
ii) What percentage is shaded?
iii) What fraction is unshaded?
iv) What percentage is unshaded?

3) Copy and complete.

a) $40\% = \frac{?}{10}$ c) $\frac{2}{5} = \square \%$ d) $70\% = \frac{?}{10}$

b) $2\% = \frac{?}{100}$ d) $\frac{7}{20} = \square \%$ e) $60\% = \frac{?}{5}$

6.5 WRITING ONE NUMBER AS A FRACTION OF ANOTHER

Example 1: Express 4 as a fraction of 10.

This is written as $\frac{4}{10}$

Reduce the fraction to its lowest term.

$$\frac{4 \div 2}{10 \div 2} = \frac{2}{5} \checkmark$$

Example 2: Write 3 minutes 20 seconds as a fraction of an hour.

Remember: The quantities must be in the same units before we can successfully work out the calculations.

This is written as $\frac{3 \text{ min } 20 \text{ sec}}{1 \text{ hour}}$

Convert everything to seconds $= \frac{(3 \times 60) \text{ sec} + 20 \text{ sec}}{60 \times 6\ 0}$

$$= \frac{(180 + 20) \text{ sec}}{3600 \text{ sec}}$$

$$= \frac{200 \text{ sec}}{3600 \text{ sec}}$$

> **Remember:**
>
> 60 seconds = 1 minute
>
> 60 minutes = 1 hour
>
> 1 hour = 60 × 60
> = 3600 seconds

In its lowest term, first, divide both numbers by 100

$$= \frac{200 \div \mathbf{100}}{3600 \div \mathbf{100}}$$

$= \frac{2}{36}$ (then divide both numbers by 2)

$$= \frac{\mathbf{1}}{\mathbf{18}} \checkmark$$

EXERCISE 6C

1) Express 5 as a fraction of 7

2) Express 4 as a fraction of 20. Leave your answer in its lowest term.

3) What fraction of 5 cm is 2 m? Leave your answer in its lowest term.

4) What fraction of 30 cm is 3 m? Leave your answer in its lowest term.

5) Write the first number as a fraction of the second number. Leave your answers in their lowest term.

a) 12 week, one year

b) 30 minutes, 2 hours

c) 7 mm, 21 cm

d) $34, $100

e) 6 min 15 seconds, 1 hour

f) 16 cm, 10 m

g) 20m, 3 km

6.6 WRITING ONE QUANTITY AS A PERCENTAGE OF ANOTHER

Two steps to follow:

1) Write the first number as a fraction of the second number (See section 6.5).

2) Multiply the fraction or decimal formed by 100

Example 1: Write 6 as a percentage of 24

Write 6 as a fraction of 24: $\frac{6}{24}$

Multiply by 100 to convert to a percentage

$$= \frac{6}{24} \times 100$$

To make it easier, reduce $\frac{6}{24}$ to its lowest term which is $\frac{6 \div 6}{24 \div 6} = \frac{1}{4}$

Therefore, $\frac{1}{4} \times 100 = \frac{1 \times 100}{4} = \frac{100}{4} = \mathbf{25\%}$ ✓

Alternatively, $\frac{6}{24} \times 100 = \frac{6 \times 100}{24} = \frac{600}{24} = \mathbf{25\%}$

Example 2: Chetanna scored 20 out of 25 in a maths test.
Work out Chetanna's percentage mark.

As a fraction, Chetanna's mark $= \frac{20}{25}$

To convert to percentage, multiply by 100.

$$= \frac{20}{25} \times 100$$

A simpler way is to divide 100 by 25 and multiply by 20.

$$= 100 \div 25 = 4$$

$$4 \times 20 = \mathbf{80\%}$$

Example 3: A kettle is in a sale and reduced to ₦4000. What is the percentage reduction?

Find the reduction:
₦5 000 – ₦4 000 = ₦1 000

The reduction as a fraction is $\frac{1000}{5000}$

To its lowest term the fraction is $\frac{1}{5}$

Percentage reduction $= \frac{1}{5} \times 100$
$= \mathbf{20\%}$

EXERCISE 6D

1) Write

a) 20 as a percentage of 50

b) 5 as a percentage of 20

c) £100 as a percentage of £500

d) 12 as a percentage of 36

2) Express these times as a percentage of an hour.

a) 6 minutes

b) 30 minutes

c) 15 minutes

d) 60 minutes

3) Express 2 hours as a percentage of 5 days.

4) Express 300 ml as a percentage of 2 litres

5)

A polo shirt is reduced from ₦50 to ₦35 in a sale. What is the percentage reduction?

6) The Nigerian Football Association has 80 members as shown in the table below.

Female	Male
30	50

a) What percentages of the members are male?

b) What percentages of the members are female?

7) The results of a science test for three students are shown below.

a) What is Abdul's percentage score?

b) What is Amina's percentage score?

c) Okonkwo says "Uduak scored 71%."
Is Okonkwo correct?
Explain fully.

Names	**Out of 40**
Abdul	8
Uduak	28
Amina	20

Chapter 6 Review Section

Assessment

1) 73% of students walk to school. What percentage of students go to school by other means?

…………..….. **1 mark**

2) Copy and shade in

a) 25%

b) 50%

c) 75%

………….. **3 marks**

3) Write 4cm as a fraction of 2m

…………… **1 mark**

4) Write the following numbers as a percentage of the second.

a) 15 weeks, 1 year

b) 500 g, 2 kg

c) ₦4 000, ₦8 000

…………....**6 marks**

5) Express these fractions as percentages.

a) $\frac{1}{2}$ b) $\frac{15}{25}$ c) $\frac{7}{10}$ d) $\frac{3}{20}$

………….. **8 marks**

6) In summer examinations, Chidimma scored these marks:

Maths: 17 out of 25
Economics: 2 out of 40
English: 30 out of 50.
Physics: 7 out of 20

a) Work out Chidimma's percentage for **each** subject.

…………. **8 marks**

b) Arrange the percentages in question 6a above in order of size, smallest first.

………….. **1 mark**

7) Copy and complete the table below.

Fractions	Percentages	Decimals
$\frac{2}{5}$		
		0.35
	60%	
$\frac{6}{25}$		

…………. **8 marks**

8) Arrange in order of size, highest first.

$\frac{1}{2}$,   0.65, $\frac{3}{5}$,   51%

…………. **2 marks**

9) Which is bigger, 30% or $\frac{8}{25}$?
Explain your decision.

…………. **2 marks**

10) There are 45 women and 15 men in a theatre.

a) What percentage are women? …………. **2 marks**

b) What percentage are men? …………. **2 marks**

7 Algebra 1

This section covers the following topics:

- Understand the meaning of open sentence
- Introduction to basic algebra
- Basic equations

LEARNING OBJECTIVES

By the end of this unit, you should be able to:

a) Find missing numbers in open sentences
b) Use letters for numbers
c) Form simple algebraic expressions
d) Solve basic equations

KEYWORDS

- Open Sentence
- Algebra
- Letters
- Replace
- Numbers
- Equations

7.1 OPEN SENTENCES

An open sentence is a statement that can either be **true** or **false** depending on the values used.

Consider the statement: $n + 5 = 12$

Until we say what the value of "n" is, we do not know whether the statement "$n + 5 = 12$" is true or false.

If we decide that $n = 7$, then the statement "$n + 5 = 12$" is **true** because $7 + 5 = 12$.

If we decide that $n = 3$, then the statement "$n + 5 = 12$" is **false** because $3 + 5$ is 8 and not 12.

If I say that **4** is an **even** number, it is a closed sentence as it is always true.

However, if I say that ***w*** is an **even** number, it is an open sentence (could be true or false depending on the value of ***w***.

Example 1: Find the number that would make the sentence $n + 7 = 10$ true.

For the sentence to be true, n must be $(10 - 7) = \mathbf{3}$

Therefore, **3** is the number that would make the sentence $n + 7 = 10$ true.

Check: $3 + 7 = 10$

Example 2: Find the number which makes the sentence $\square - 5 = 15$ true.

For the sentence to be true, the number must be bigger than 15.

$15 + 5 = 20$

Therefore, the missing number is **20**.

Check: $20 - 5 = 15$.

EXERCISE 7A

Find the numbers that make the open sentences true.

1) $\square + 6 = 8$

2) $\square + 4 = 13$

3) $\square + 11 = 19$

4) $\square - 2 = 3$

5) $\square - 23 = 45$

6) $4 + \square = 11$

7) $8 + \square = 30$

8) $11 + \square = 98$

9) $32 - \square = 26$

10) $5 - \square = 2$

11) $40 \div \square = 4$

12) $57 = 17 + \square$

13) $300 = \square - 56$

14) $123 \times 0 = \square$

15) $27 \div \square = 3$

If empty boxes have the same value in each question, find the number that makes the sentences true.

16) $\square + \square = 2$

17) $\square + \square = 8$

18) $40 = \square + \square$

19) $20 - \square = \square$

20) $81 = \square \times \square$

21) $\square + \square + \square = 27$

22) $\square = 8 - \square$

23) $\square + \square - \square = 10$

24) $\square \div \square = 1$

25) $45 = \square + \square + \square$

26) $144 = \square \times \square$

Simple substitution

Depending on what number goes into the box, the values may change.

Example 3: If 3 goes in the box, work out the value of $\square + 11$.

The answer will have **only one** value as we have specified what the unknown is.

3 + 11 = **14**

EXERCISE 7B

1) If **9** goes into the boxes below, work out the values of each sentence.

a) $\square + 2$

b) $\square + 6$

c) $5 + \square$

d) $13 - \square$

e) $90 \div \square$

f) $\square \div 9$

g) $144 - \square$

h) $500 - \square$

i) $\square \times \square$

2) If **15** goes in the boxes, work out the value of each sentence.

a) $\square \times 2$

b) $60 \div \square$

c) $\square + \square - \square$

d) $3 \times \square + 20$

e) $\square - \square + 50$

f) $34 - \square$

g) $\square \div 5$

h) $\square - \square$

7.2 USING LETTERS TO REPRESENT NUMBERS

Most times in mathematics, we have to use a letter to represent an unknown number. Using letters to represent numbers in mathematics is called ***algebra***.

Using boxes is obsolete though some teachers still use them.

Instead of writing $\square + 3$, we replace the box with any letter of the alphabet using lower case.

$n + 3$ is more appropriate. As stated above, n stands for any number.

$\boldsymbol{n + 3}$ is called an **expression** in terms of n.

The value of the expression $n + 3$ depends on the value of n.
When $n = 5$, $n + 3$ has the value 8.
When $n = 20$, $n + 3 = 23$…..and so on.

Example 1: Using algebra, write an expression for 2 more than n.
Answer: $\boldsymbol{n + 2}$

Example 2: Write an expression for c add p
Answer: $\boldsymbol{c + p}$

Example 3: Write an expression for $x + x + x$
Answer: $\boldsymbol{3x}$

Example 4: Write an expression for r take away 9
Answer: $\boldsymbol{r - 9}$

Example 5: Write an expression for 7 less than n.
Answer: $\boldsymbol{n - 7}$

Be careful here as this example is different from Example 4 above.

Example 6: Six people share w apples equally.
How many apples does each person receive?

Answer: $w \div 6 = \frac{w}{6}$

Example 7: The letter f represents the number of football games that Dagogo owns.

a) Write the number of football games owned by Abiye, who owns 7 football games less than Dagogo

Answer: Abiye owns $\boldsymbol{f}$ **– 7** games

<u>Points to note:</u>

In algebra, one letter means one of such letters though we do not write the number "1" beside the letter. We know it is there.

m is the same as $1m$
n is the same as $1n$.

As a convention, we **do not** write the number **1** beside a letter.
m + m + m + m = 4m
8m - m………… This is: 8m - 1m = 7m

EXERCISE 7C

1) Write an algebraic expression for the statements below.

a) 7 more than a
b) 20 more than b
c) 9 less than c
d) d take away 2
e) 17 more than e
f) f add 6
g) g subtract 4
h) h take away 13

2) Use the letter ***k*** to represent Kenechukwu's age in years and write expressions for the ages of:

a) Chuka, who is 5 years younger than Kenechukwu
b) Okoro, who is 10 years older than Kenechukwu

3) a. Write an expression for the combined ages of Chuka and Okoro in question 2.
b. If Kenechukwu is 20 years old, what are the ages of Chuka and Okoro?

4) Write expressions for the following and use the letter y as the starting number.
a) Starting number, add 20
b) Starting number, take away 2
c) Starting number, multiply it by 6, then add 23
d) Starting number, treble it

5) Adamu packed table tennis balls in 24 cartons.
a) Write a sentence (expression) for the total number of tennis balls in w cartons.
b) Tennis balls could also be packed in cartons of 72.
Write an expression for the total number of table tennis balls in w cartons of 24 and n cartons of 72

6) Three students buy a box of chocolate to share. Emma ate ***b*** number of chocolates. Write an expression for the number of chocolates eaten by:

a) Uchechi, who ate 7 more chocolates than Emma
b) Nduka, who ate 4 fewer chocolates than Emma

EXERCISE 7D

Find the number that each letter stands for if each sentence is **true**.

1) $n + 5 = 8$
2) $v + 3 = 7$
3) $m + m = 24$
4) $b - 4 = 9$
5) $k = 54 - 50$
6) $x = 16 - 5$
7) $7 + 11 = y$
8) $3 + y = 8$
9) $7 + d = 15$
10) $15 + c = 50$
11) $24 = d + 6$
12) $43 = c + 9$
13) $100 - x = 60$
14) $40 \div 4 = w$
15) $40 \div d = 8$
16) $k \times 7 = 42$
17) $g = 5 \times 11$
18) $w \div 4 = 24$
19) $5 \times 0 = t$

20 $134 - w = 29$

When $c = 3$, work out the value of the following expressions.

21) $c + c$
22) $c + 7$
23) $c - 1$
24) $c \times c$
25) $18 \div c$
26) $0 + c$
27) $c + c - c$
28) $78 - c$
29) $16 \times c - 3$
30) $30 \div c + 100$

31) In 1990, Chinedu was y years old. In 1995, Chinedu was 24 years old. What is the value of y?

32)

w cm

The Bunny maths book is placed on top of the dictionary. If w is 19 cm, what is the total height of the dictionary and Bunny maths book?

Chapter 7 Review Section

Assessment

Find the numbers that make the open sentences true.

1) $\square + 20 = 48$ 4) $14 + \square = 11$ 7) $40 \div \square = 4$

2) $\square + 14 = 53$ 5) $28 + \square = 50$ 8) $57 = 17 + \square$

3) $\square + 13 = 19$ 6) $91 + \square = 98$ 9) $30 = \square - 56$

.... **9 marks**

If empty boxes have the same value in each question, find the number that makes the sentences true.

10) $\square + \square = 2$ 13) $\square + \square - \square = 20$

11) $\square + \square = 8$ 14) $\square \div \square = 1$

12) $\square + \square + \square = 24$ 15) $\square \times \square = 289$

.... **6 marks**

16) Write an algebraic expression for the statements below.

a) 5 more than a b) 4 less than c
c) 12 more than b d) h subtract 4

.... **4 marks**

17) Use the letter m to represent Kola's age in years and write expressions for the ages of:

a) Taiwo, who is 3 years younger than Kola
b) Olu, who is 7 years older than Kola

………… **2 marks**

18) **a.** Write an expression for the combined ages of Taiwo and Olu in question 17.
b. If Kola is 9 years old, what are the ages of Taiwo and Olu?

………… **2 marks**

19) Write expressions for the following and use the letter y as the starting number.
a) Starting number, add 3
b) Starting number, take away 8
c) Starting number, multiply it by 7, then add 54
d) Starting number, square it

………… **4 marks**

20) Three students bought a box of chocolate to share. Ada ate g number of chocolates. Write an expression for the number of chocolates eaten by:

a) Ifeoma, who ate 7 more chocolates than Ada
b) Chieme, who ate 4 fewer chocolates than Ada

…………**2 marks**

Find the number that each letter stands for if each sentence is **true**.

21) $n + 9 = 18$
22) $v + 13 = 17$
23) $m + m = 128$
24) $b - 4 = 34$
25) $k = 34 – 5$
26) $64 = d + 16$
27) $40 = c + 3$
28) $200 - x = 120$
29) $90 \div 3 = w$
30) $140 \div d = 7$

………… **10 marks**

When $f = 8$, work out the value of the following expressions.

31) $f + f + f + f$
32) $f + 17$
33) $f - 1$
34) $f \times f$

………… **4 marks**

8 3d Objects

This section covers the following topics:

- Three dimensional-shapes
- Sketching and recognising 3-D objects
- Nets of 3-D shapes
- Properties of 3-D shapes

LEARNING OBJECTIVES

By the end of this unit, you should be able to:

a) Identify the properties of 3-D objects including cubes, cuboids, prisms, cones, pyramids cylinders, and spheres
b) Draw the net of 3-D objects
c) Draw skeletal views of 3-D shapes
d) Name vertices, edges, and faces of a 3-D object

KEYWORDS

- 3-Dimensional
- Solids
- Prisms
- Nets
- Faces, Edges, and Vertices

8.1 THREE – DIMENSIONAL SHAPES

When two-dimensional shapes are extended to three dimensions, they form three-dimensional shapes.

A square (2-D shape) extends to form a cube (3- D shape)

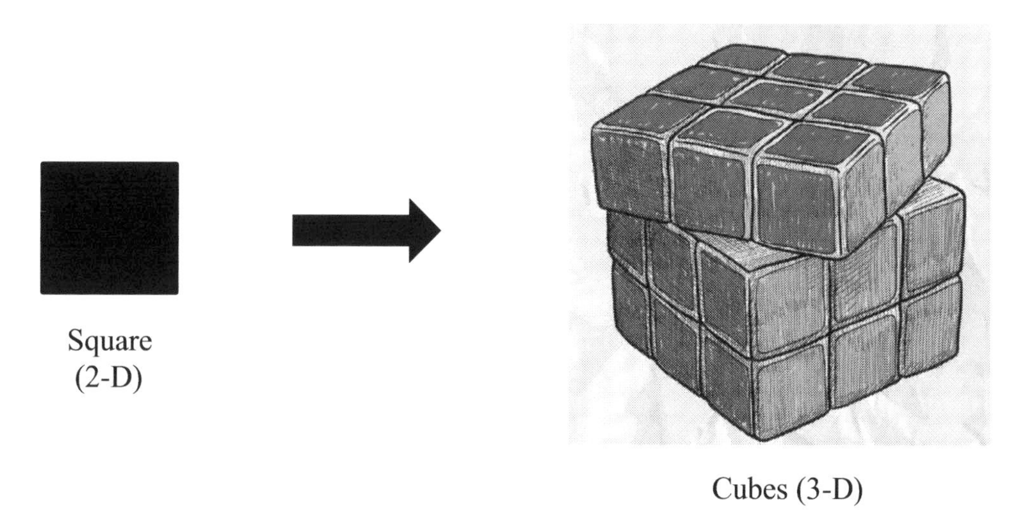

Cubes (3-D)

3 – Dimensional shapes are solid shapes with 3 dimensions: **Length**, **width,** and **height** as shown with the cuboid below.

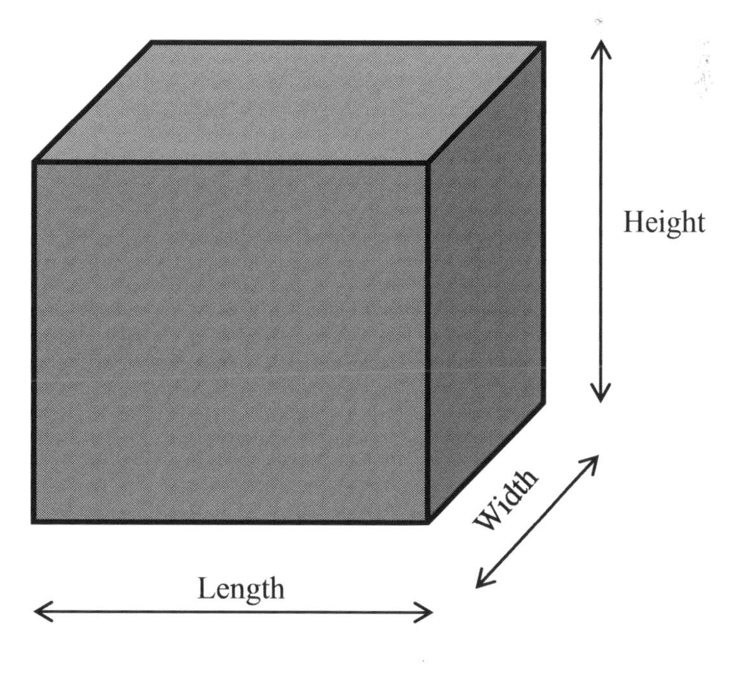

Some 3-D objects in real life have irregular shapes while some are human-made.

Most of the human-made 3-D shapes have regular shapes. They are called **geometrical shapes**. Some examples are shown below.

PROPERTIES OF SOME 3-D SOLIDS

Most 3-D shapes have edges, faces, and vertices.

- **Cuboid**

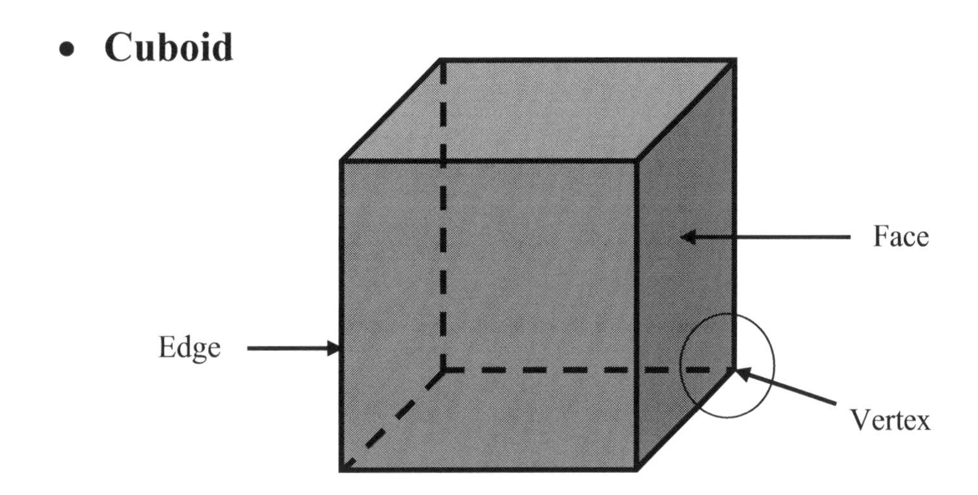

A cuboid has 6 faces, 12 edges, and 8 vertices.

Remember:
An edge is where two faces meet
A face is a flat or curved surface
A vertex (plural: vertices) is a corner where edges meet

- **Cylinder**

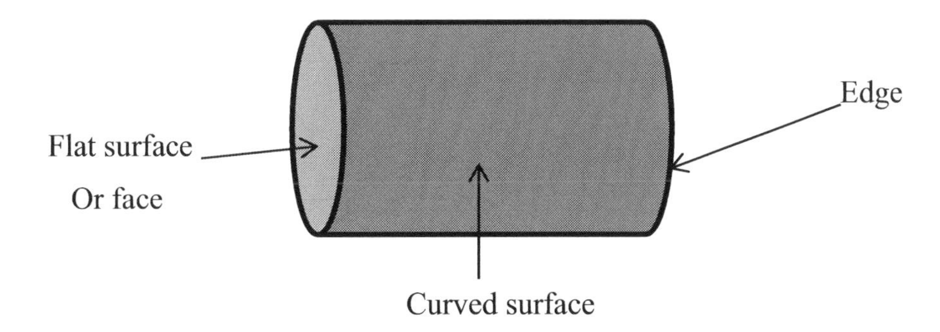

A cylinder has 3 faces, 2 edges and zero (0) vertex

- **Triangular prism**

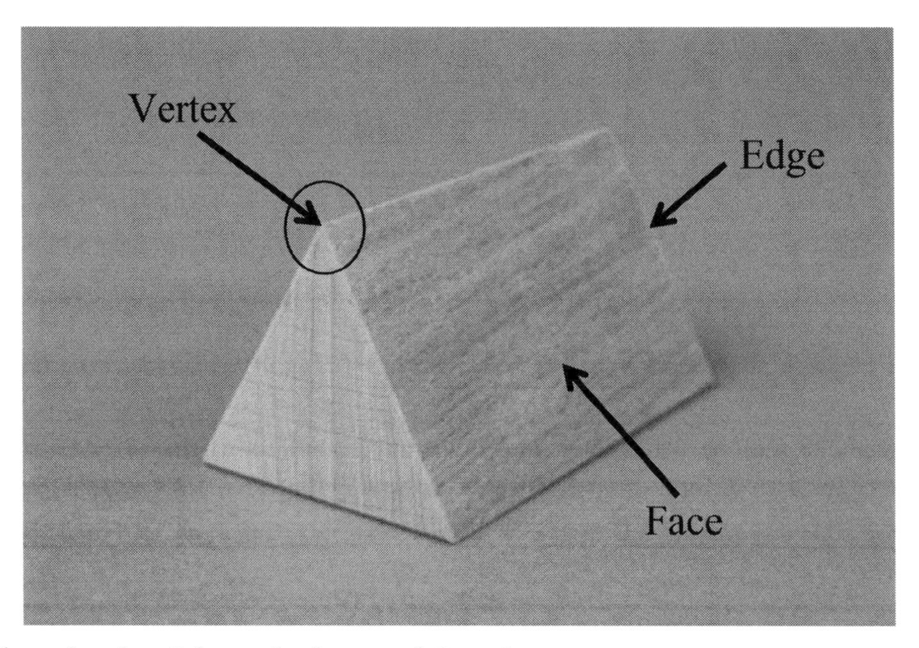

A triangular prism has 5 faces, 9 edges, and 6 vertices.

- **Square based Pyramid**

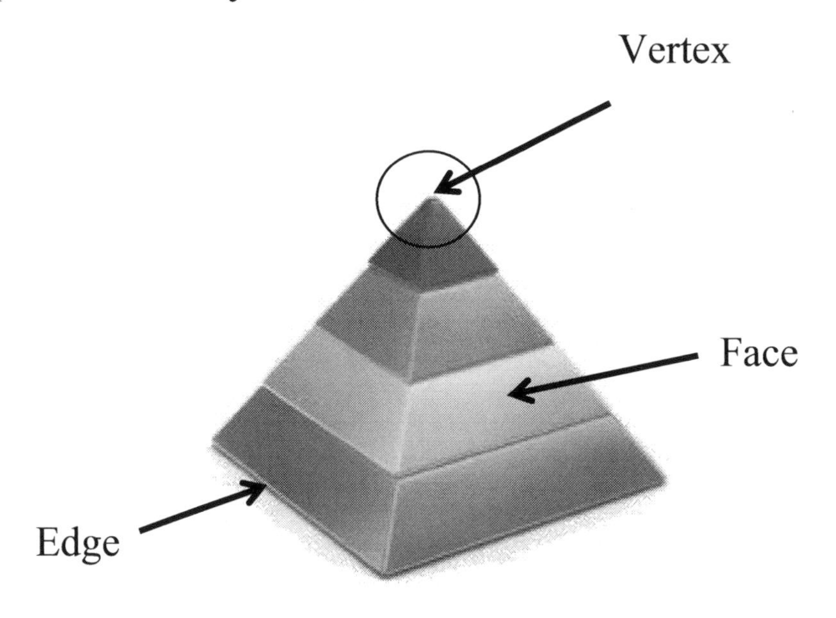

A square based pyramid has 8 edges, 5 faces and 5 vertices.

8.2 DRAWING CUBOIDS – SKELETAL VIEW

Cuboids and cubes are widely used in the day-to-day teaching of mathematics, and as a consequence, we must learn how to draw them. A cube is a cuboid having all the faces equal (squares).

Skeletal views are often used to represent cuboids and cubes. Practice how to draw them and speak to your art teacher for guidance.

However, you may follow the rules below to draw a rectangle.

- Draw two rectangles of the same size

- Place one above the other but **not** in line with the corresponding corners

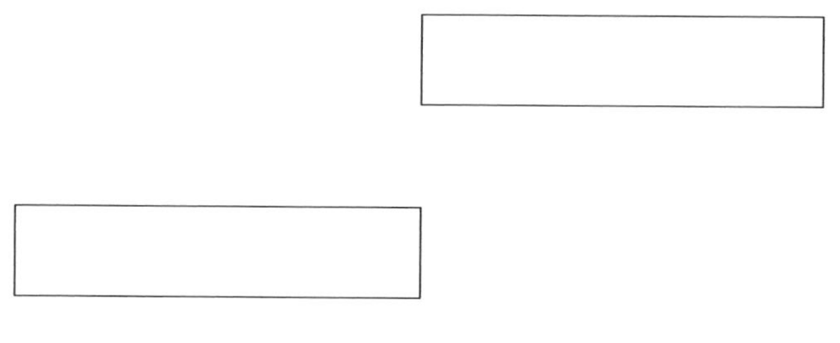

- Join the corresponding corners and use BROKEN LINES for the edges not seen by the naked eye.

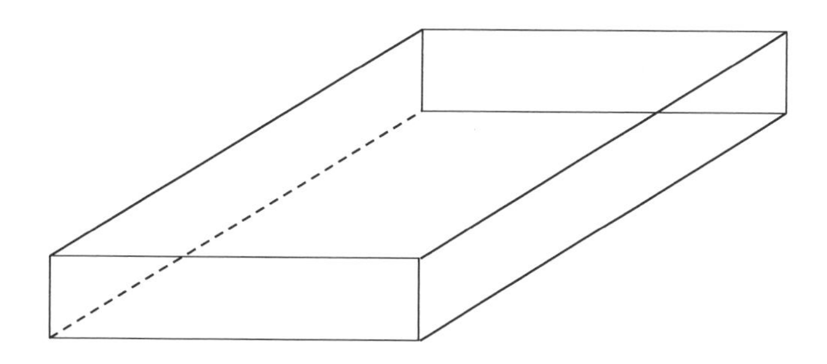

There are obviously other methods of drawing cuboids and cubes. Use any method of your choice, but practice makes perfect.

8.3 PRISMS

Prisms are 3-dimensional solid that must satisfy the three criteria below.

Three points for identifying a prism:

1) They must have identical ends
2) They must have flat faces
3) They must have the same **cross section** all through the length.

The end or front face of a prism is called the cross section. In geometry, the cross section is the shape made when a solid is cut through parallel to the base.

To understand cross sections, take a look at the diagram below of slices of bread.

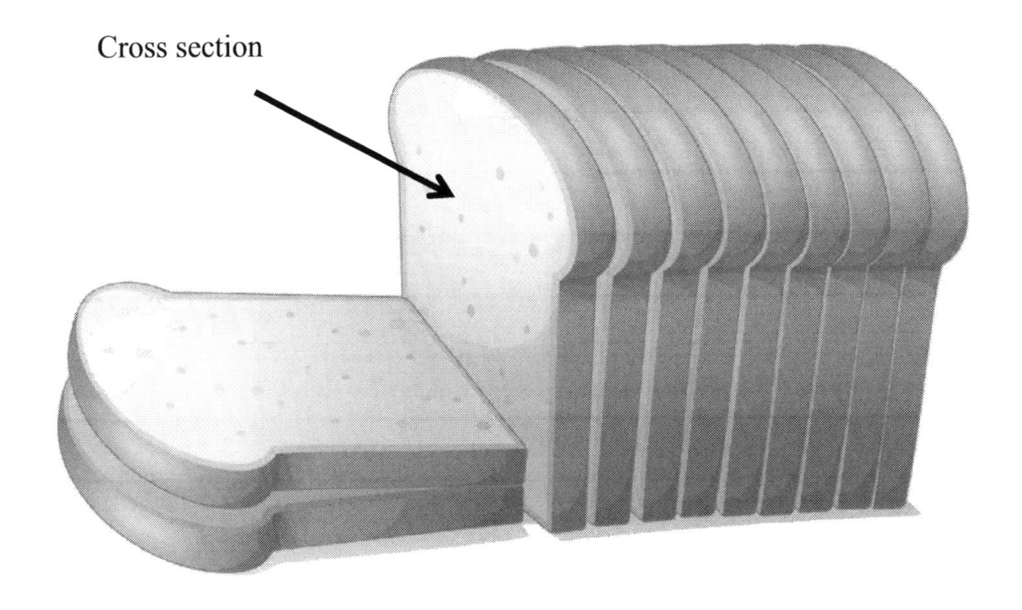

The cross section of the bread is uniform throughout the length of the bread.

However, the bread is **not** a prism by definition given, since it has curved sides.

Most prisms have their names from the cross-section. Consider the following prisms:

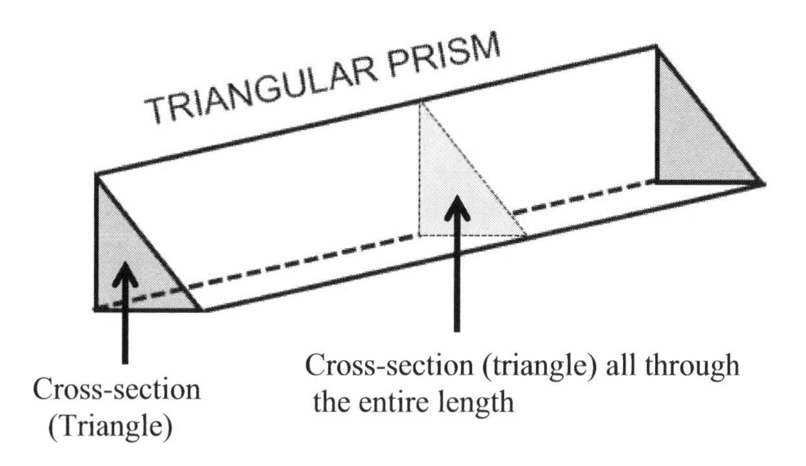

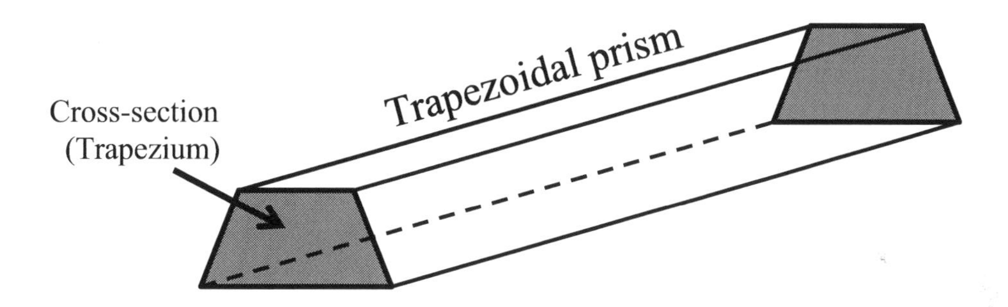

Examples of prisms

- Triangular Prism
- Cubes
- Cuboids
- Pentagonal Prism
- Hexagonal Prism
- Trapezoidal Prism

Please note: A cylinder is **not a prism** because it has curved surfaces (sides).

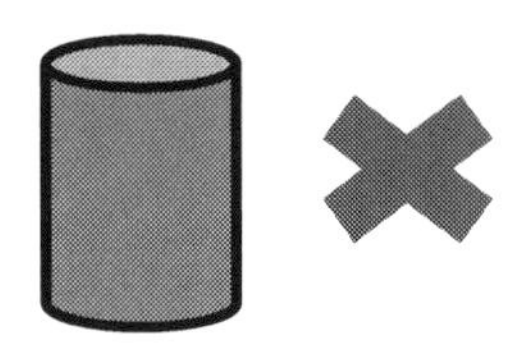

EXERCISE 8A

1) Copy and complete the table below.

SHAPES	MATHEMATICAL NAMES	EDGES	VERTICES	FACES

2) In real life, write down an object which is a:

a) cylinder
b) cube
c) cuboid
d) sphere

3) In real life, write down an object that is:

a) cone
b) pyramid
c) prism

4) Identify the solids that are **prisms** from the diagrams below.

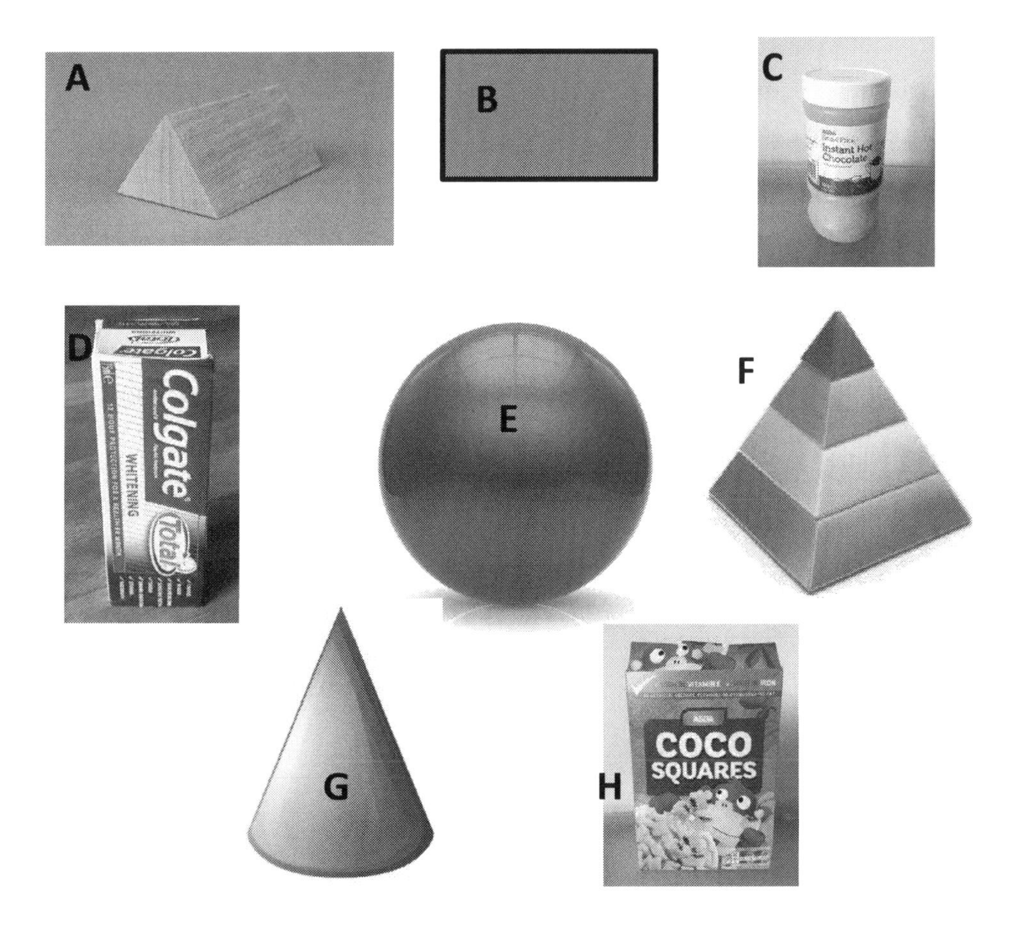

5) Write the mathematical names of the shapes that **are not** prisms.
6) From your choices in question 5, explain why you made those decisions.

8.4 NETS OF 3-DIMENSIONAL (3D) SOLIDS

The net of a 3-dimensional object is a **flat or two-dimensional shape** that can be folded to make the 3-dimensional object. The net of a 3-D object must have the same number of faces. For example, a cuboid has six faces, and the net must have six faces as shown below. It is easier to start drawing the net from **the base** of the object.

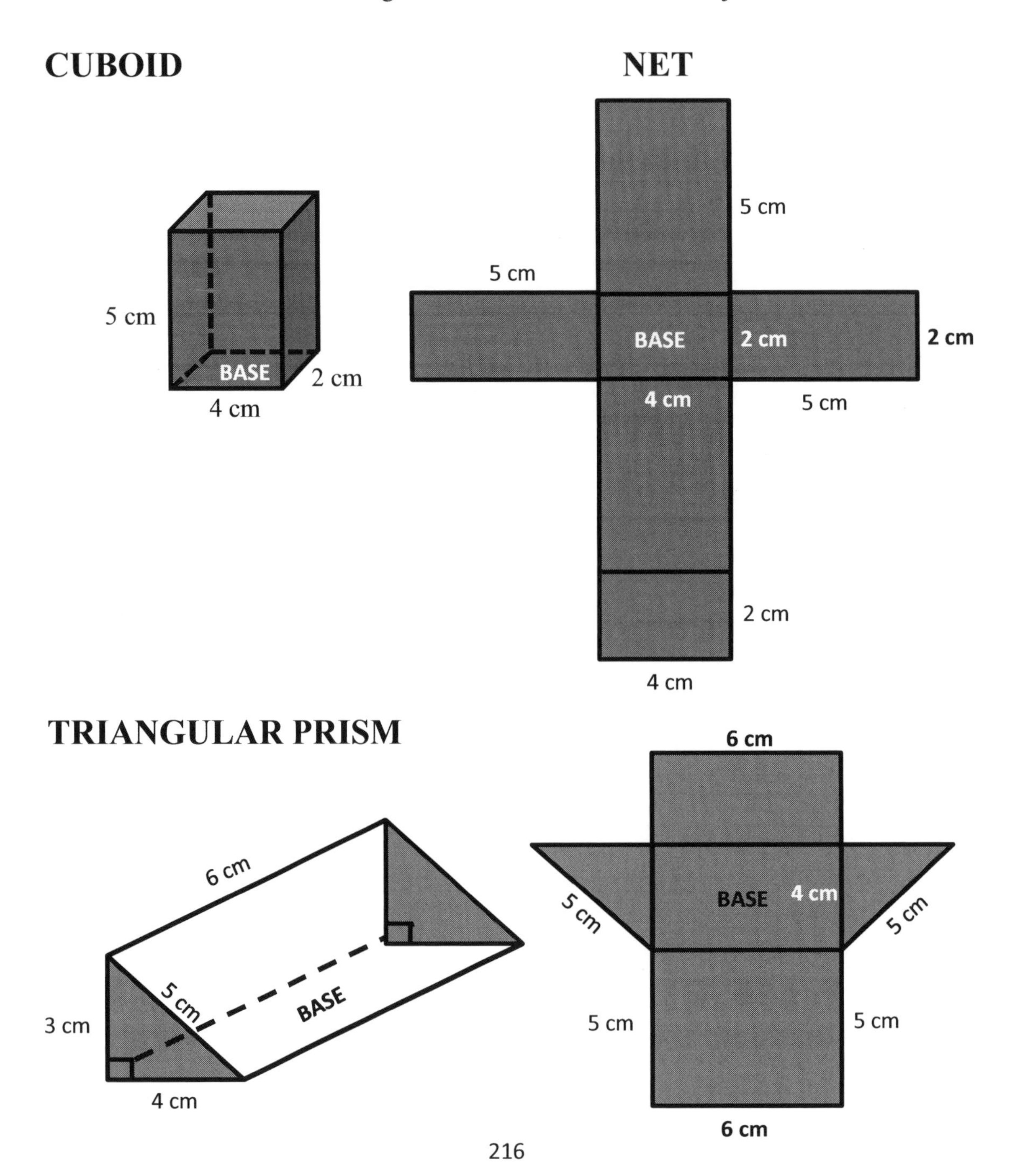

CYLINDER

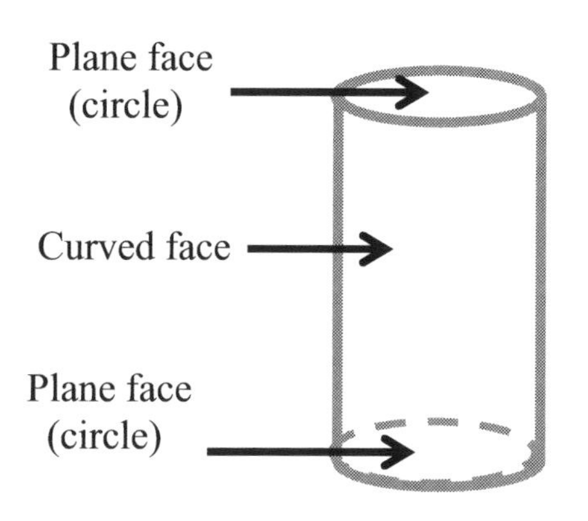

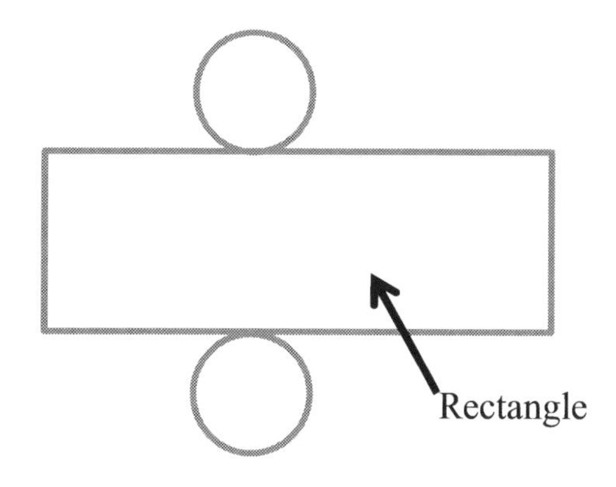

Some cylindrical Objects in real life:

PYRAMID

Most times, the name of pyramids comes from the **base**. You may have a square-based pyramid, a triangular-based pyramid which is often called a **tetrahedron**, pentagonal-based pyramid, hexagonal-based pyramid and the list goes on.

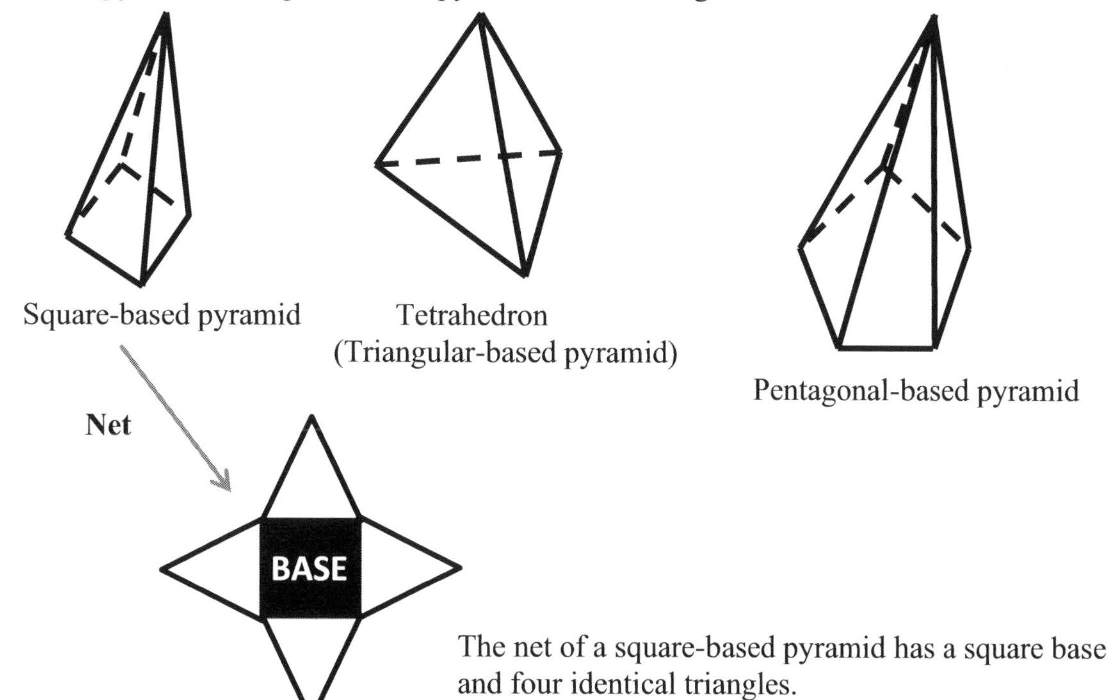

The net of a square-based pyramid has a square base and four identical triangles.

CONE

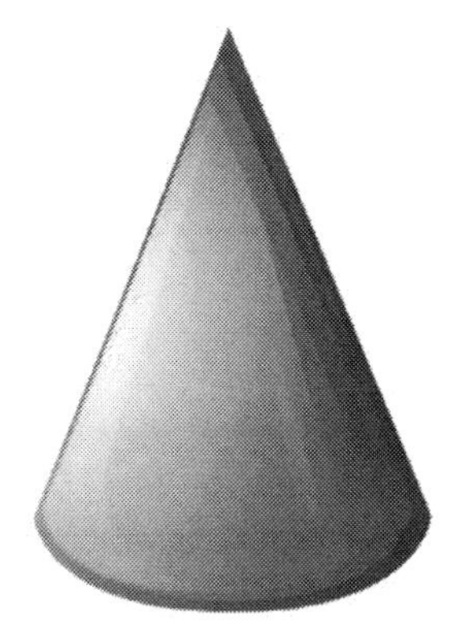

Ice cream cones are examples of real life application of cones. Cones appear everywhere and usually used to support other objects.

Net of a cone

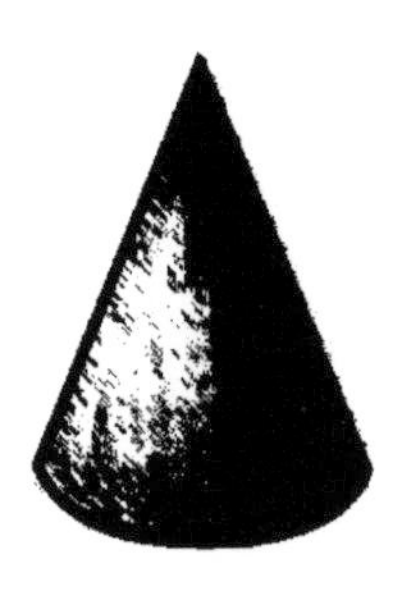

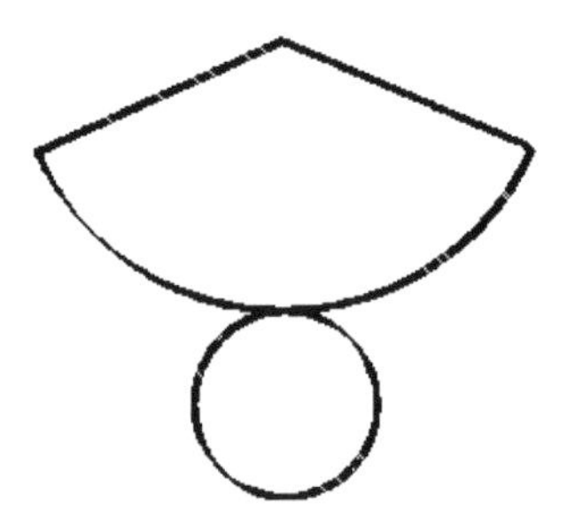

HEMISPHERE

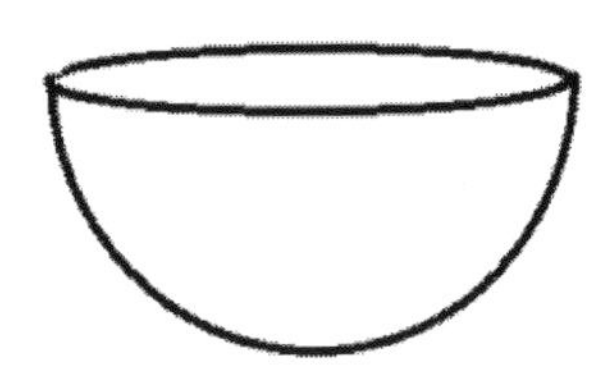

A hemisphere is half of a sphere

8.5 LABELLING IN 3-D SHAPES

Example 1: Look at the triangular based pyramid (tetrahedron) below.

The edges can be labelled as AB, BC, CD AD, AC, and BD.
The vertices are labelled with capital letters.
The right-hand face is labelled BCD.
There are vertices A, B, C and D.

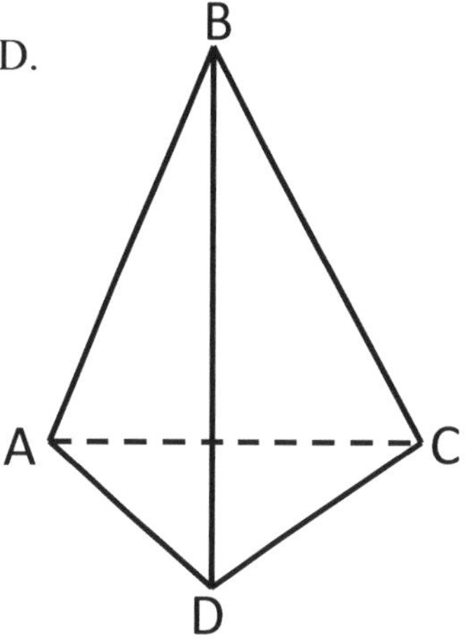

Example 2: Look at the cuboid below.

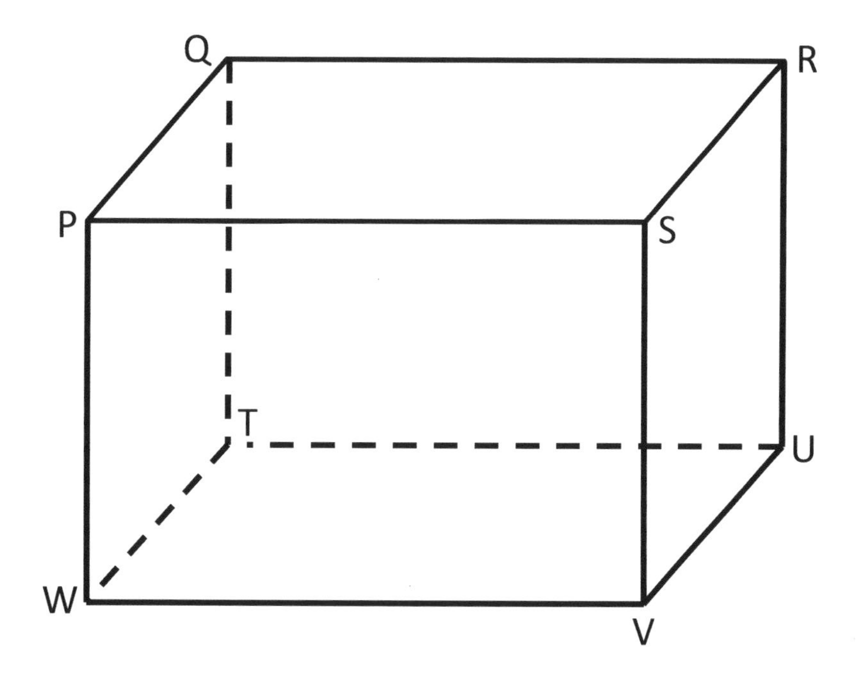

The bottom face is WTUV. The right-hand face is VSRU. The bottom face WTUV meets at edge VW. There are other edges like WP, PQ, QR, RU, SV, TU.
We have vertices at P, Q, R, S, T, U, V, and W.

EXERCISE 8B

1) Draw the accurate net of the cuboid.

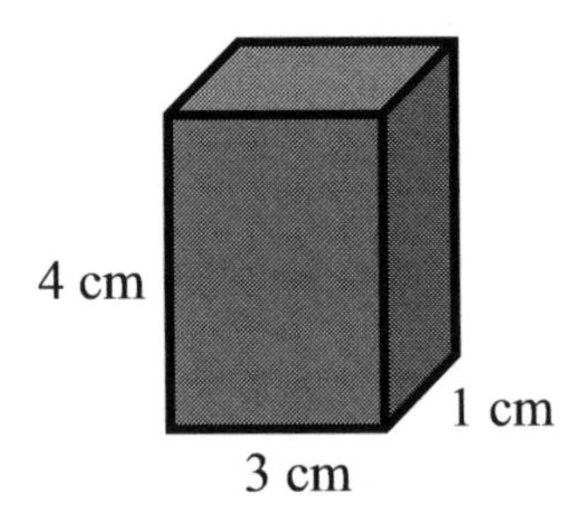

2) Draw the accurate net of the triangular prism.

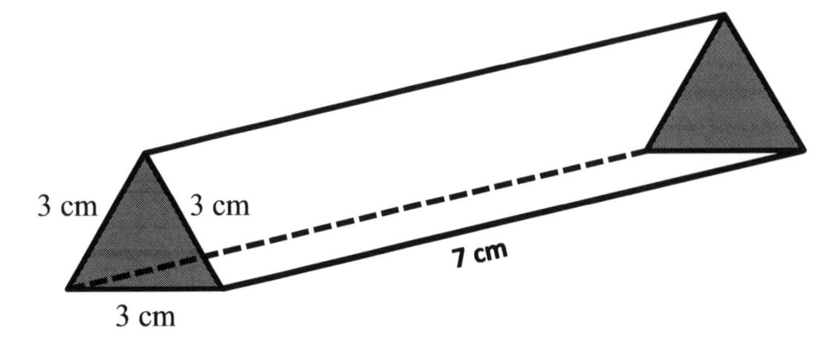

3)

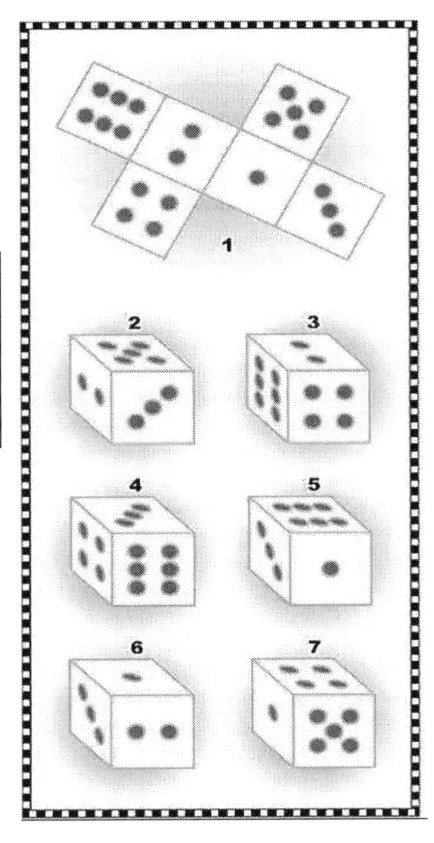

The net of a cube is shown as number **1**.

When 1 is folded to form a cube, which of the 2 – 7 solids can be produced?

4) In question 2 above, how many
   a) Rectangles, b) triangles, c) faces, d) edges and e) vertices are there in the net of the triangular prism?

5)

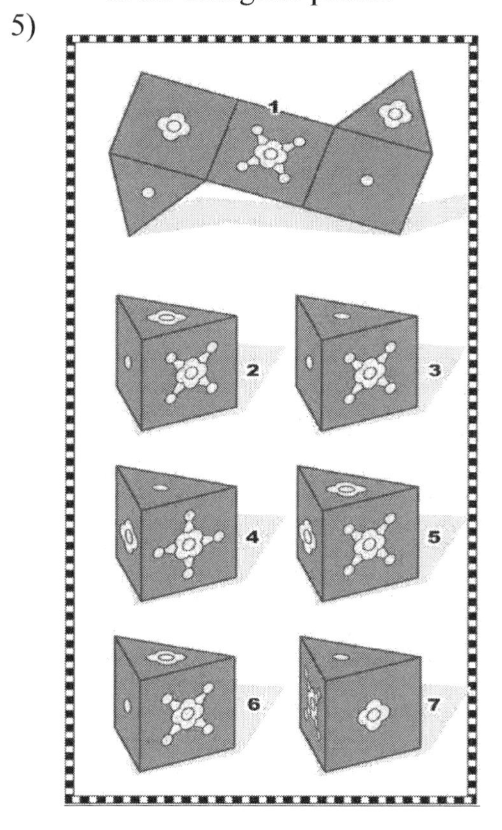

When diagram **1** is folded to form a triangular prism,

a) which of the 2 – 7 objects **cannot** be produced?

b) How many vertices?

c) How many edges?

d) How many faces?

6) i) Name the 3D solids represented by their nets below.
   ii) Draw the nets of A and B accurately on squared paper, adding glue flaps where necessary.
   iii) With a pair of scissors, cut out the nets and glue them together
   iv) How many vertices and edges do the solid have?

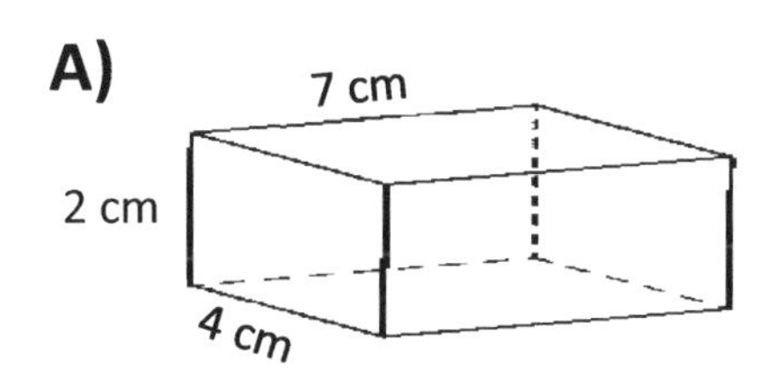

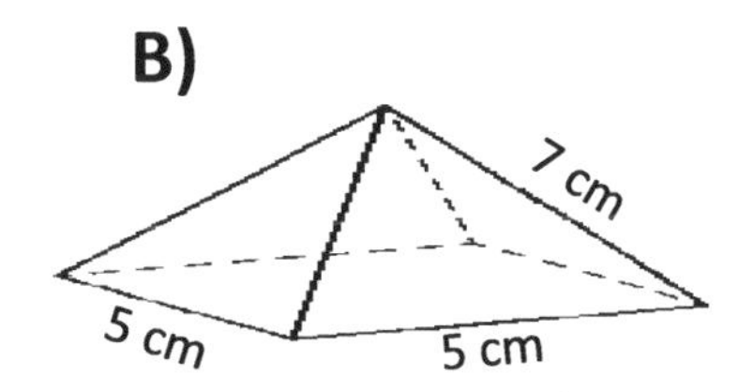

Chapter 8 Review Section
Assessment

1) The diagrams below show some three-dimensional objects.

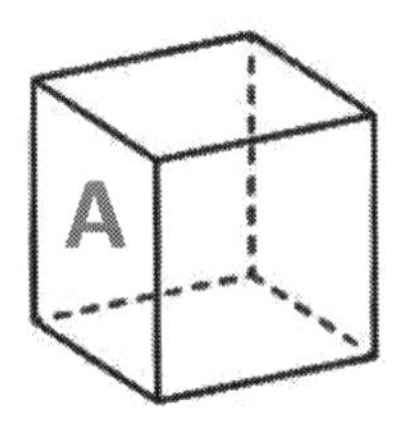

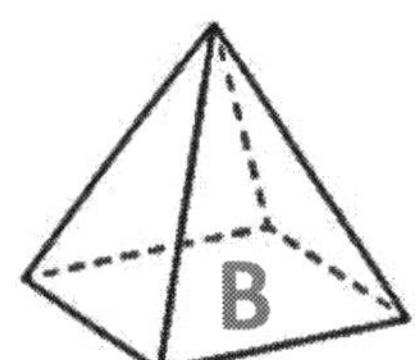

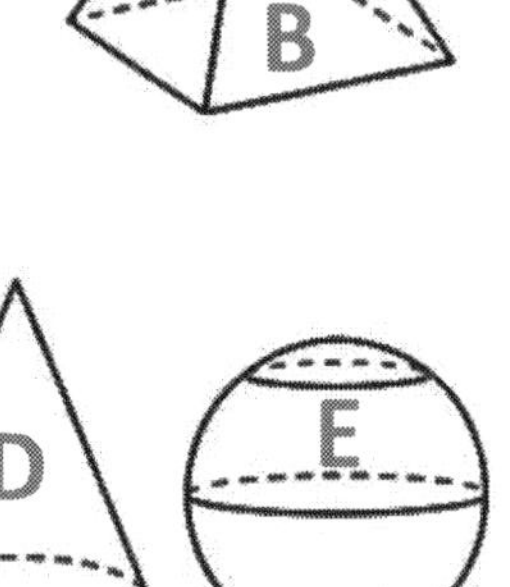

a) What is the mathematical name for objects A, B, C, D and E? ……………5 marks

b) How many faces does object B have? ……………..1 mark

c) How many vertices does object C have? ……………. 1 mark

d) Is object D a prism? Explain fully. …………... 2 marks

e) Sketch the net of object B …………... 2 marks

2)

a) What is the mathematical name for this object? ……………. 1 mark

b) Write down the number of vertices and faces. …………… 2 marks

3)

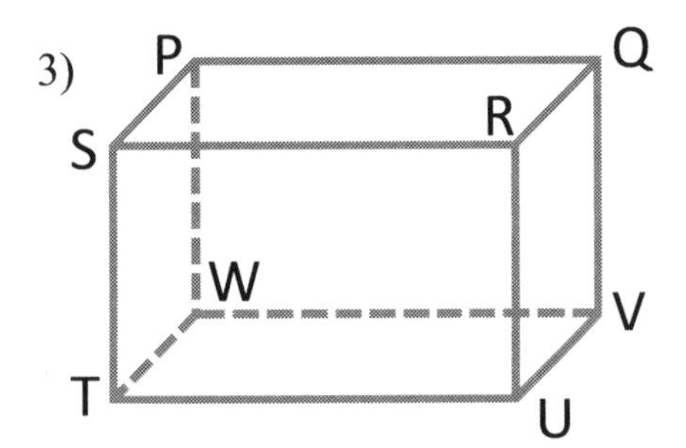

a) What is the mathematical name for this solid?
b) What is the shape of face PQRS?
c) Which edges meet at the vertex V?
d) What is the shape of face URST?
e) How many edges meet at vertex W?
………… 5 marks

4) Write the names of

a) One shape with 5 flat faces
b) One shape with 9 edges
c) One shape with 8 vertices
d) One shape with 6 faces
e) One shape with twelve edges

.... **5 marks**

5)

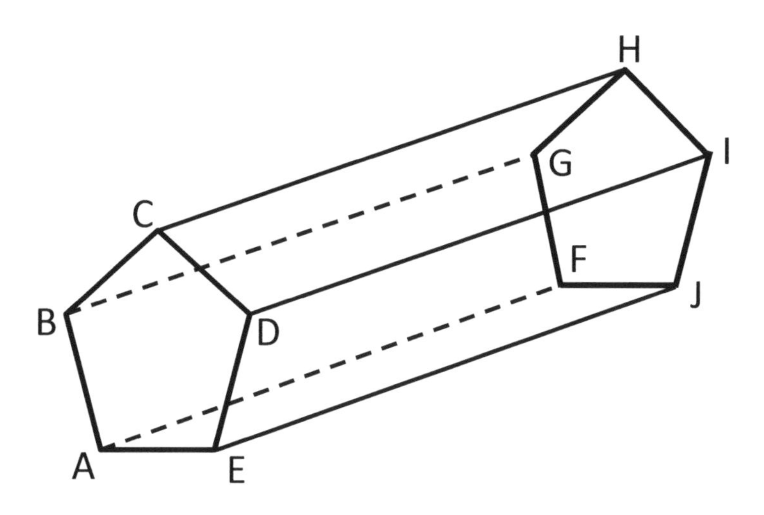

a) What is the shape of face FGHIJ? .... **1 mark**

b) How many edges meet the face ABCDE? .... **1 mark**

c) Faces EDIJ and DCHI meet at what edge? .... **1 mark**

d) How many edges meet the vertex H? .... **1 mark**

6) Copy and complete the table.

NAME	VERTICES	FLAT FACE	EDGES
Triangular Prism			
	8	6	12
Square-based Pyramid			
	0	2	2
Hexagonal-based pyramid			

.... **9 marks**

9 Algebra 2

This section covers the following topics:

- Letters and Symbols
- Introducing formulae
- Collecting Like terms
- Coefficients

LEARNING OBJECTIVES

By the end of this unit, you should be able to:

a) Use letters and symbols in algebra
b) Use formulae
c) Collect like terms
d) Use letters for numbers

KEYWORDS

- Algebra
- Letters
- Replace
- Expressions
- Like terms
- Coefficients

9.1 INTRODUCING FORMULA

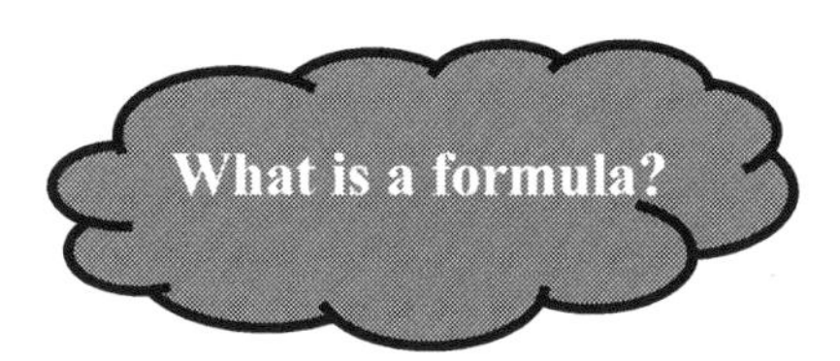

Note: The plural of formula is formulae

Formulae are written using letters to represent the quantities. It may also contain numbers. In mathematics, a formula is a **rule** for working out something. A formula must have an **equal (=)** sign.

Example 1
D = ST is a formula because when the two variables S and T are known, the value of D can be calculated.

Formula connects two or more **expressions** containing variables. The value of one variable depends on the others. The part after the equal sign in a formula is an algebraic expression. Expressions **do not** have an equal sign.

Example 2
Angela is five years older than Daniel. When Daniel is 15 years old, Angela is 20 years old. When Daniel is 45 years old, Angela is 50 years old.

The general rule for the above will be written as
Angela's age = Daniel's age + 5.

If **a** represents Angela's age and **d** represents Daniel's age, the formula would be written as **a = d + 5**

Example 3
A plumber calculates the cost of a repair like this:

Cost (c) = ₦1 000 × number of hours worked (h) + ₦50

a) Write a formula for the cost of repairs.
Answer: **c = 1 000h + 50**

b) Calculate the cost if the repair takes 5 hours.
Answer: from the formula c = 1 000h + 50,
= (1 000 × 5) + 50
= 5 000 + 50
= **₦5 050**

9.2 FORMING EXPRESSIONS

In example 2, the right part of the formula is an algebraic expression.
That is **d + 5**.

Example 1
Suppose I start with the number x and add 10, the expression would be $\boldsymbol{x + 10}$.
Notice: there is no equal sign.

Example 2
I start with a number $\boldsymbol{c}$, add 7 and then divide the result by $\boldsymbol{y}$.
The expression would be $\frac{c+7}{y}$

Example 3
Suppose a price of **₦p** is shared equally between $\boldsymbol{c}$ people. Write an expression for what each person would receive.

The expression is ₦$\frac{p}{c}$

Example 4

Start with $\boldsymbol{a}$, subtract 2 and then multiply the result by 5.

That gives: (a - 2) × 5
The correct form of the expression is **5(a - 2)**

Note: a-2 was put in the bracket because five is multiplying the whole expression. Also as a convention, the number comes first before letters or expressions. Hence, the number 5 is written before the expression and without the multiplication sign.

Remember: 2a means 2 × a, 4(c + 6) means 4 × (c + 6) or (4 × c) + (4 × 6),

5ab means 5 × a × b and acd means a × c × d

EXERCISE 9A

1) Write down the expression you would get following the instructions below.

a) Start with p, add 4
b) Start with c, take away 3
c) Start with n, double it
d) Start with k, treble it
e) Start with p, double it and add 8
f) Start with q, treble it and add 2
g) Start with f, add 7 and multiply the result by 6
h) Start with y, subtract c, add p and then divide the result by n
i) Start with b, add 10 and then divide the result by h
j) Start with r, subtract t and add w

2) Musa (m) is 10 years older than his brother Ibrahim (i).
a) What formula would you write to work out Musa's age if we are given Ibrahim's age?
b) If Ibrahim is 37 years, work out Musa's age.

3) Chuba works in a hotel. He uses this formula to calculate his wages in Naira.

Wages (w) = hours worked (h) × ₦100

a) Write a formula to calculate Chuba's wages using the letters in brackets.
b) Calculate Chuba's wages if he works:
   i) 3 hours　　　iii) 1.5 hours
   ii) 10 hours　　　iv) 5.5 hours

4) Write an expression for the total cost, in Naira of the following statements:

   a) p chocolates at y Naira each
   b) 7 chocolates at w Naira each
   c) m shirts at ₦2500 each
   d) b shirts at ₦3000 each and k chocolates at ₦150 each
   e) 13 chocolates at y Naira each and 25 shirts at f Naira each.

5) A shop lady uses crates which hold 24 bottles each. How many bottles can be placed in
   a) d crates　　　b) m crates　　　d) 5 crates?

6) There are 15 doses in a bottle of medicine. How many doses are there in
   a) y bottles
   b) d bottles
   c) 3 bottles?

7) A doctor hires a bus to take nurses to a cinema. The doctor uses this formula to calculate the cost per nurse

   Cost per nurse (c) = ₦3000 ÷ number of nurses (n) + 5

   a) Write a formula using the letters in brackets to calculate the cost per nurse
   b) Calculate the cost per nurse if the number of nurses travelling is
   i) 15 ii) 20

8) The amount in Naira, of an advertisement in "This Day" newspaper, is calculated using the formula

   Amount (a) = 60 + 140 × number of words (w)

   a) Write a formula using the letters in brackets for the amount.
   b) Calculate the amount of money to be paid when advertising with
   i) 40 words
   ii) 100 words.

9) You have ***w*** apples. You lose half of the apples, but your friend Kenneth gave you 11 of their apples.

   Write an expression for the number of marbles you have now.

10) Ubaka owns ***n*** shirts. Chika owns three times as many shirts as Ubaka.

    a) How many shirts does Chika own?
    b) How many shirts do Ubaka and Chika own together?

11) Edwin has ***m*** shoes, which is five more than Henry's.
    How many shoes does Henry have?

12) Ekpeyong has ***f*** cats. Offiong gives him three more cats. How many cats does Ekpeyong have now?

9.3 COEFFICIENTS AND EXPRESSIONS

An expression is a collection of letters and numbers. For example, 5x + 2y + 3 is an expression. 5x, 2y and 3 are the **terms** of the expression.

Terms with the same letter part are called **like terms.**
8x, 13x and x are all like terms because they contain the letter $\boldsymbol{x}$ only. Simplifying an expression could mean collecting the like terms.

Focusing on the expression $8x$, the 8 is referred to as the **coefficient** of $\boldsymbol{x}$. It means that eight $\boldsymbol{x}$'s have been added $(x + x + x + x + x + x + x + x)$ together.

Other examples of coefficients:

$5n$ → **5** is the coefficient of $\boldsymbol{n}$. It means that five $\boldsymbol{n}$'s have been added together.
$9w$ → **9** is the coefficient of $\boldsymbol{w}$. Nine $\boldsymbol{w}$'s have been added together.
y → **1** is the coefficient of y or any other letter standing alone. There is only one y and it is the same as writing $1y$. However, it is unconventional to include the number 1.

$\frac{1}{4}n$ → $\frac{1}{4}$ is the coefficient of n

$\frac{2}{5}w$ → $\frac{2}{5}$ is the coefficient of w

$\frac{2w}{5}$ → $\frac{2}{5}$ is still the coefficient of w

$\frac{x}{2}$ → $\frac{1}{2}$ is the coefficient of x

EXERCISE 9B

1) Write down the coefficient of the letter in each expression.

a) 4c	e) 5t	i) $\frac{1}{7}x$	m) $\frac{2x}{7}$
b) g	f) 7p	j) $\frac{2}{3}e$	n) $\frac{7x}{11}$
c) 6y	g) 24w	k) $\frac{3}{7}x$	o) $\frac{n}{2}$
d) 3b	h) 8x	l) $\frac{4}{9}x$	p) $\frac{1}{8}a$

9.4 SIMPLIFYING EXPRESSIONS (COLLECTING LIKE TERMS)

Example 1: Simplify $x + x$. This means $1x + 1x =$ $\mathbf{2x}$
Another way to look at it is $2 \times x =$ $\mathbf{2x}$ because in simple arithmetic, 2×3 is the same as writing 3 + 3. Both answers will give 6.

Example 2: Simplify c + c + c + c + c. This means 1c + 1c + 1c + 1c + 1c = **5c**

Example 3: Simplify 5x + 2x.
Since they contain the same letter x, they are **like** terms. We then add up as if adding ordinary numbers.

Therefore, 5x + 2x = **7x**

Example 4: Simplify 10y + y
Again, since they are like terms, we add up as you would ordinary numbers. Remember that a letter on its own has a coefficient of 1. For example, y is equal to 1y although we do not have to write the number one.

Therefore, 10y + y = **11y**

Example 5: Simplify 7c – 3c
Again, this equals **4c**

Example 6: Simplify 6x - 2x + x – 3x

Starting from the right, put a circle round the terms including the signs (+ or -).
This routine helps in identifying the terms to add or subtract.

(6x) (-2x) (+ x) (– 3x)

The positive and negative terms are very clear now. Grouping the positive and negative numbers together gives

6x + x - 2x - 3x
= 6x + x = 7x
= 7x - 2x = 5x
= 5x - 3x
= **2x**

EXERCISE 9C

Simplify these expressions by collecting like terms.

1) $w + w$
2) $6x + 3x$
3) $5w - 2w$
4) $7p - 2p$
5) $3w + 6w - 4w$
6) $11x + 5x + 3x - 2x$
7) $20y - y$
8) $y + y + y + y + y$
9) $8e - 8e$
10) $7g - 3g + 4g + 3g$

Simplify these expressions by collecting like terms.

11) $7n + 6n - 4n$
12) $2e + 9e + 3e - 4e$
13) $8y + 7y - 4y$
14) $3p + 7p + 11p$
15) $7u + 3u - 2u - u$
16) $8p - 3p - 3p$
17) $18u + 4u - 7u - 9u$
18) $5a + a - 3a + 4a$
19) $5c + 3c + 2c - 7c$
20) $16t + 19t - 20t - 3t$

Simplify each expression by collecting like terms.

21) $1.2m + 5.3m$
22) $3.7x + 5.4x - 2.5x$
23) $10v - 3.4v$
24) $5z - 2z + 6.6z$
25) $20.3f + 12.4f - 2.9f$
26) $5g + 7g - 2g - 3g + 4g$
27) $6w - w - w - w$
28) $4k + 8k - 7k + 3k - k$
29) $12c - c - 3c + c - 2c + c + c$
30) $100u - 30u - 2u + u + 7u$

31) What is the total width of the cylinders?

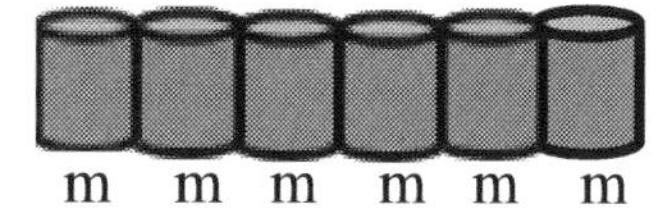

32)

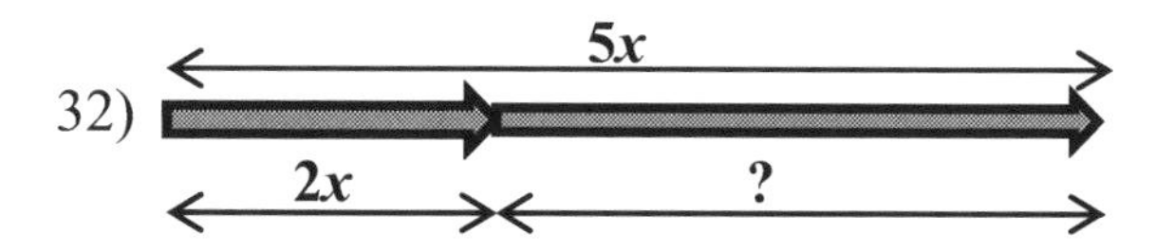

Two bars are joined together. What is the length of the longer bar?

9.5 SIMPLIFYING WITH BOTH LIKE AND UNLIKE TERMS

Expressions such as $2y + 5c$ cannot be written as a single term or simplified. They are **not** like terms and are in their simplest form already.

Example 1: Simplify $5x + 6y + 2x + 4y$
Identify the like terms by looking at the letters.

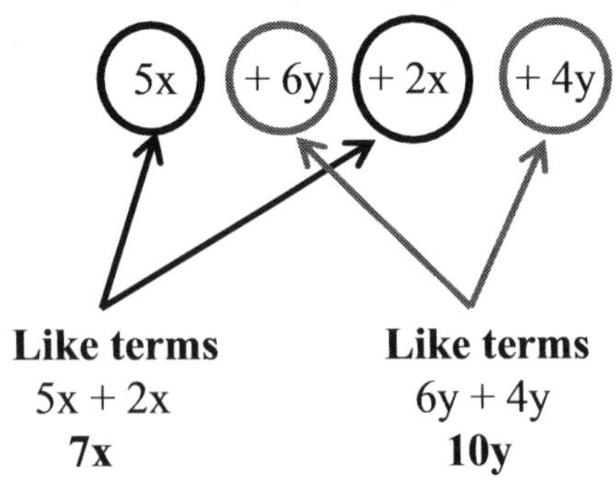

The answer is **<u>7x + 10y</u>**

Example 2: Simplify $8w - 3w - 7$

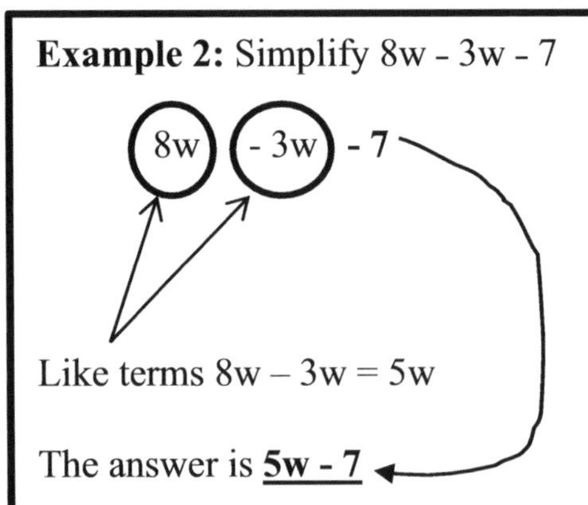

Like terms $8w - 3w = 5w$

The answer is **<u>5w - 7</u>**

Note: - **7** must be included in the final expression.

Example 3: Simplify $5c + 3n - 2c + n - 9$

5c | + 3n | - 2c | + n | - 9

Collect like terms by first putting ‘circles’ around the terms including the signs.

$5c - 2c = 3c$ and **$3n + n = 4n$**

Therefore, it simplifies to **$3c + 4n - 9$** ✓
Note: -9 must be included in the simplified expression.

Example 3: Simplify $9w + 5 + 6w - 3 - 4w$ by collecting like terms

9w | + 5 | + 6w | - 3 | - 4w

$9w + 6w - 4w = 11w$ **and** **$5 - 3 = 2$**

= **$11w + 2$** ✓

EXERCISE 9D

Simplify by collecting like terms where possible.

1) $d + d + e + e + e$
2) $5s + 3 + s + 9$
3) $6t + 9t - 6$
4) $7y + 7t$
5) $6x - 3x - t$
6) $4d + 6 + 3x + 3 + 3d$
7) $6x - x + 3y - y$
8) $5g + 4x + 2g + 5$
9) $p + 4e + p + 7e$
10) $c + 8 + 6c$
11) $8x + 7y - 4x$
12) $5a + b - 3a - 4b$
13) $3a + 6 - a - 7$
14) $8p + 3 - 4p + 13$
15) $7b - 5c + 8c - 3b$
16) $13a + 3 - 4a - 1$
17) $7s - s + 4v - 3v$
18) $6y - 6y - 6y$
19) $8t - 3u + 2v - 3t + 6v - 3$
20) $12b + 6c - 5 - 3b + 2$

EXERCISE 9E

1) The expression in each block is worked out by adding the two blocks below it.
   Complete the blank blocks.

a)

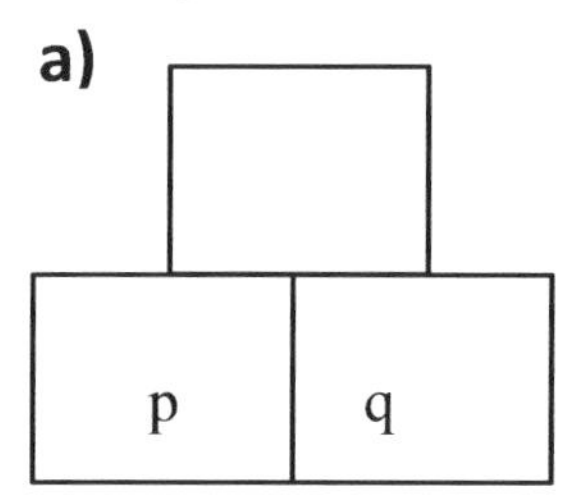

b)

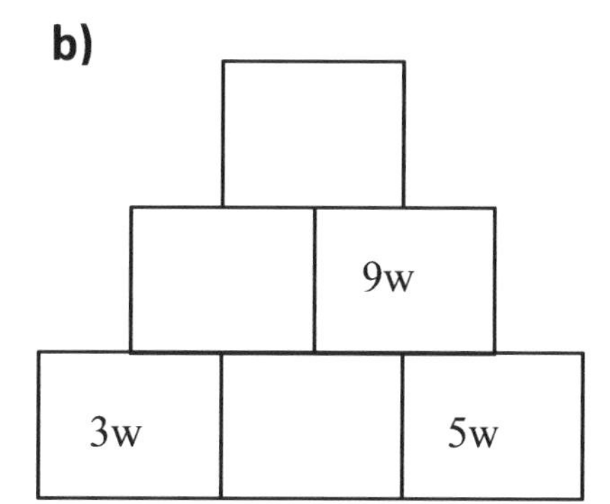

c)

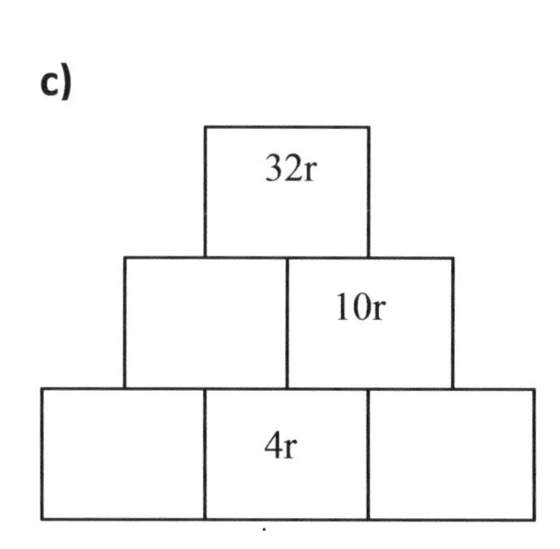

d)

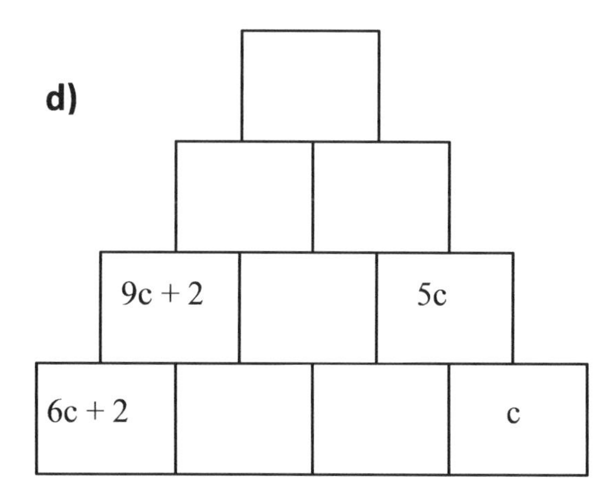

9.6 APPLYING SIMPLIFICATION CONCEPTS

The knowledge of simplifying expressions will be useful in this section.

Example 1: Write an expression for the perimeter of rectangle **A** below.

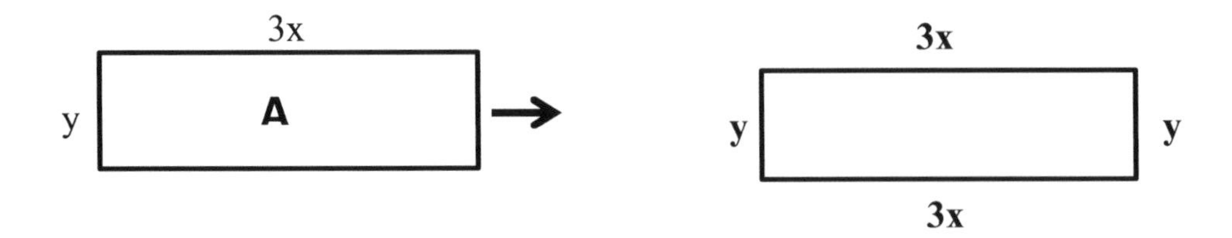

Perimeter is the distance round the shape, so add up all the sides. $3x + 3x + y + y = 6x + 2y$. Therefore, $\mathbf{6x + 2y}$ is the expression for the perimeter of the rectangle.

Example 2: Write an expression for the perimeter of the shape below.

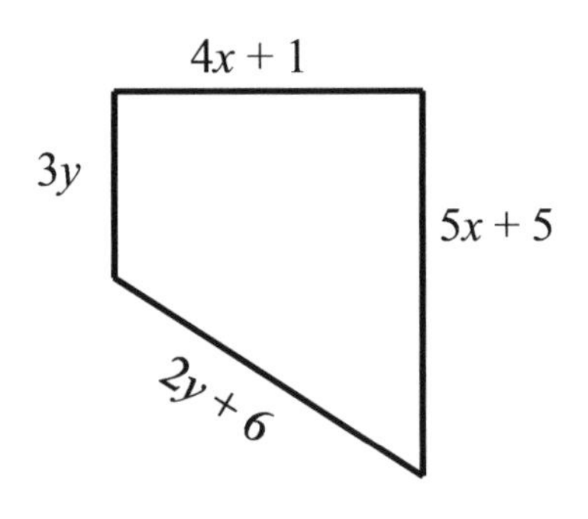

Solution:

Add up all the lengths and collect like terms.

$4x + 1 + 5x + 5 + 2y + 6 + 3y$

$= 4x + 5x + 2y + 3y + 1 + 5 + 6$

$= \mathbf{9x + 5y + 12}$ ✓

Example 3: Write an expression for the perimeter of the triangle below.

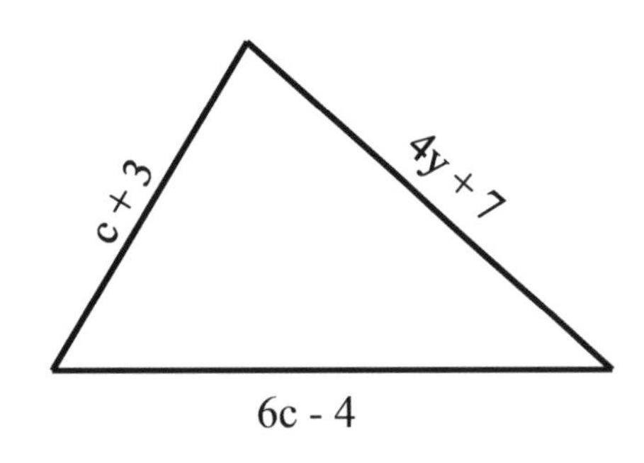

Solution:

Add up all the lengths and collect like terms.

$c + 3 + 4y + 7 + 6c - 4$

$= c + 6c + 4y + 3 + 7 - 4$

$= \mathbf{7c + 4y + 6}$ ✓

EXERCISE 9F

1) Find an expression for the perimeter of the shapes below.

a)

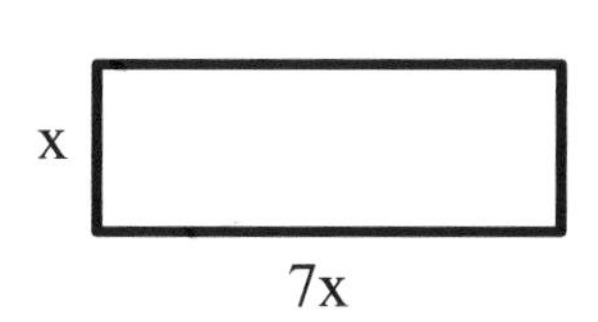

b)

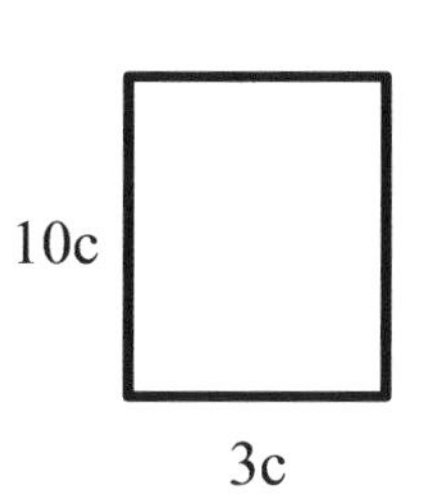

c)

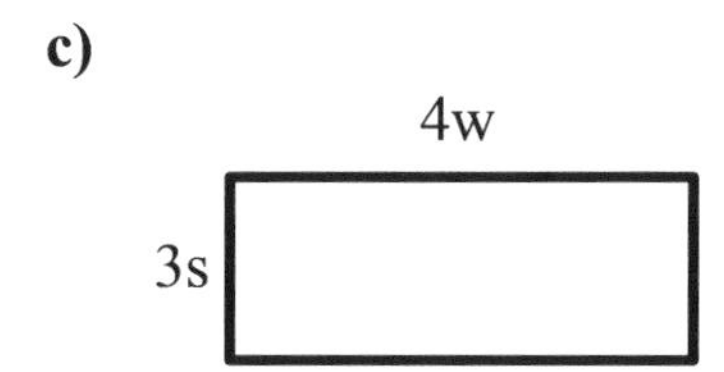

d)

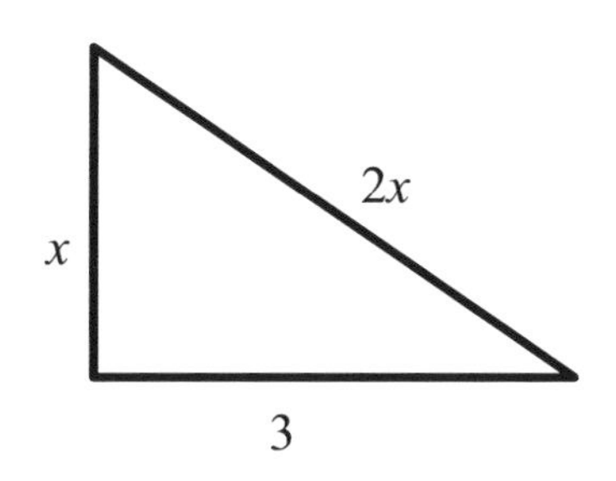

e)

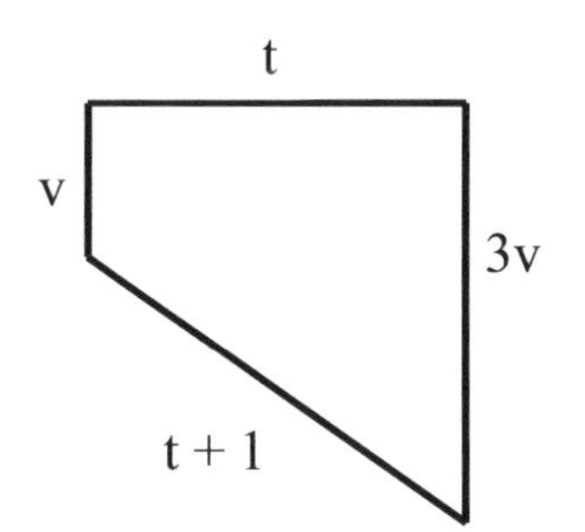

f)

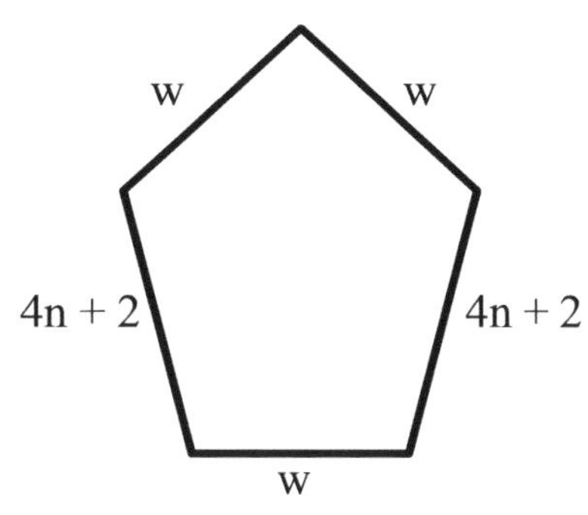

g)

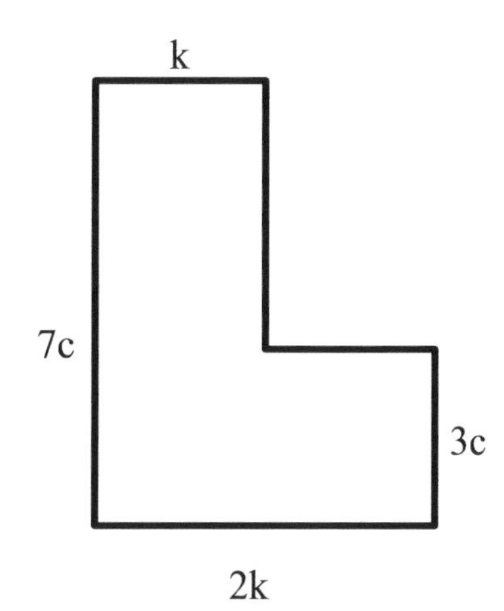

h)

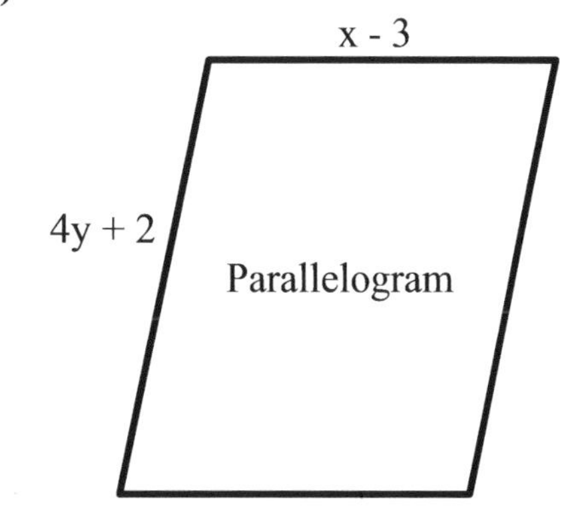

Chapter 9 Review Section

Assessment

1) Write an expression for these statements.

a) 8 more than w
b) x more than 4
c) 5 less than t
d) 2 times c
e) Product of c and 4
f) x divided by 5
g) $\frac{1}{3}$ of m
h) p divided by x
i) a times c
j) 30 more than n

………..…. **10 marks**

2) Simplify the following by collecting like terms.

a) $a + a + a + a + a + a + a$
b) $t + t + t$
c) $2x + 4x$
d) $5a - 3a$
e) $7n + 4m + 3n + 2m$
f) $2v - v + v$
g) $8m + 3 + 5m - 1$
h) $3s + 7j - 2s - 4j$
i) $10c - c + 4e - e + 3$
j) $4 + y + 2 + 5y$

………… **10 marks**

3) The rule to fine p is "treble c and add 7."

a) Write this rule in symbols
b) Calculate the value of p when $c = 5$
c) Calculate the value of c when $p = 28$

………….. **3 marks**

4) $P = 5t + 2u$ $Q = 3t + 10u$ $R = 15t + 3u$

Simplify the following and collect like terms where possible.

a) $P + Q$
b) $P - Q$
c) $R + P$
d) $P + 2Q + 3R$

………….. **8 marks**

5) Find the perimeter of the shapes below. Write your answer as a simplified expression.

a)

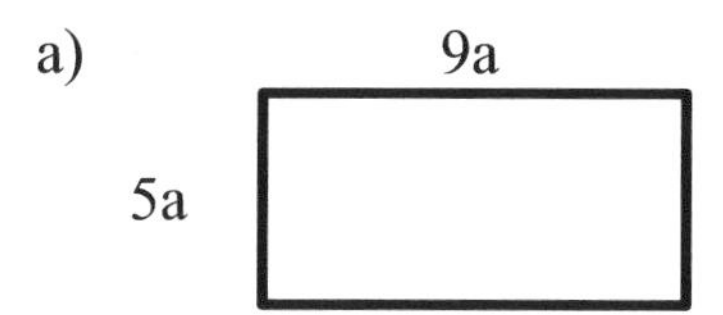

..............2 marks

b)

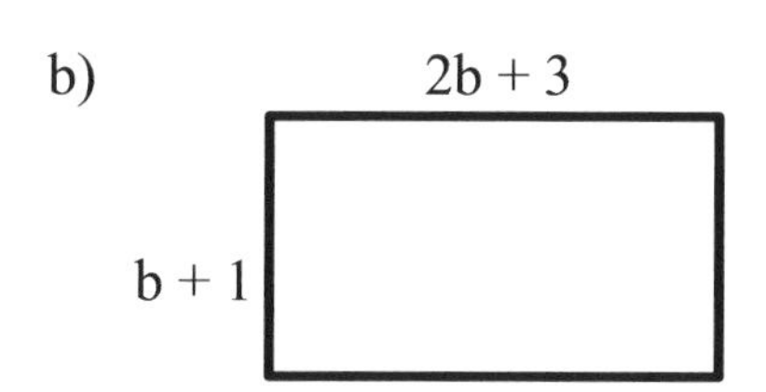

.... 2 marks

c)

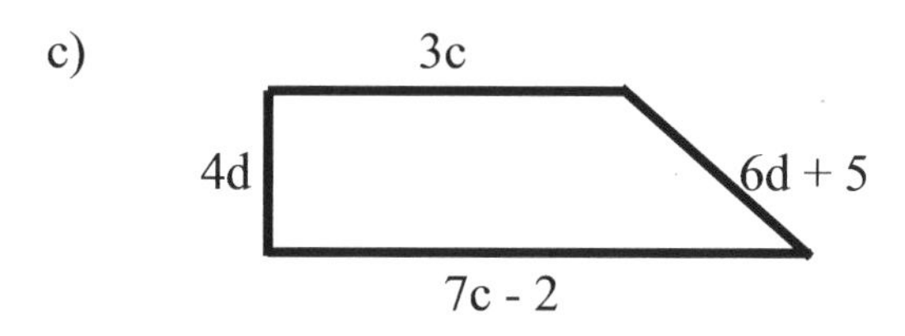

.... 3 marks

6) If x = 8 and y = 4, find the value of

a) $3x$ .... **1 mark**
b) $x + y$ .... **1 mark**
c) $4y – 2$ .... **1 mark**
d) x^2 .... **1 mark**

7) Alice performed a mathematical calculation, and the answer was 10w.
Write down **four** questions that could have this answer. .... **4 marks**

8) Anthony says "$4x + 5y – 3 + 2x – 6y$ will simplify to $6x + 11y – 3$"
Is Anthony correct?
Explain fully .... **2 marks**

9) Assuming the rule for changing temperatures in degrees Celsius (C) into Fahrenheit (F) is:

Temperature (F) = 3 × temperature (C) + 30

a) Write a formula using the letters in the brackets. .... **1 mark**
b) Change these temperatures to Fahrenheit:
i) 20°C ii) 30°C

.... **2 marks**

10 Angles 1

This section covers the following topics:

- Angle as a measure of rotation
- Name angles
- Measure angles
- Use of a protractor

LEARNING OBJECTIVES

By the end of this unit, you should be able to:

a) Estimate angles
b) Understand and interpret angles as a measure of rotation
c) Understand the differences between acute, obtuse and reflex angles
d) Construct and measure angles using a protractor

KEYWORDS

- Angle
- Measurement
- Acute
- Obtuse
- Reflex
- Rotation
- Degree
- Protractor

10.1 ROTATION THROUGH AN ANGLE

A given angle is a measure of **turn** or **rotation**. One complete turn measures 360 degrees. Angles are measured in degrees (^{o}). A full circle is 360°.
One revolution is a complete turn.

Therefore, 360 degrees = 1 complete revolution

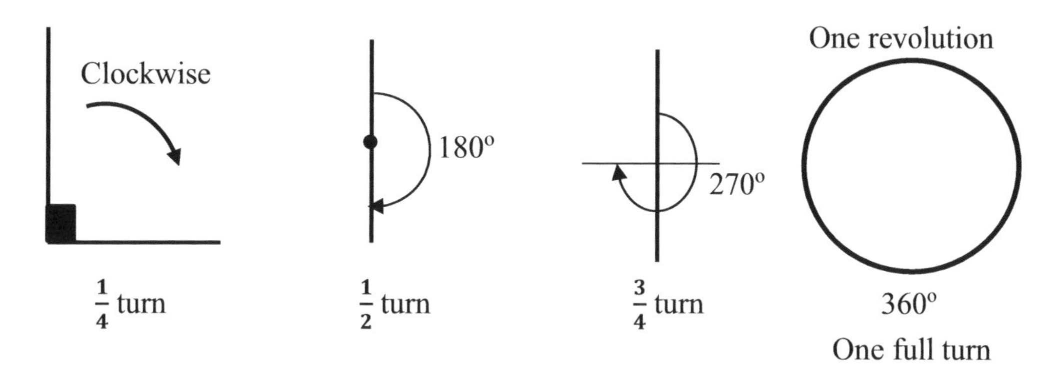

Example 1: Below is a compass showing directions.

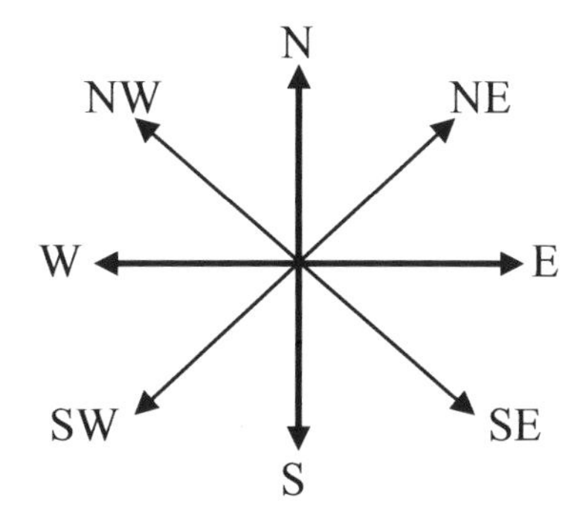

From North (N), clockwise to North East (NE), the angle (turn) is $\frac{1}{8}$ of a revolution (full turn) or 360°.

Therefore, the angle between N and NE in the clockwise direction is $\frac{1}{8} \times 360 = 45°$.

Example 2: The angle between N and SE in the clockwise direction is $\frac{3}{8} \times 360 = 135°$

Example 3: Tolu leaves her house at 4 pm and arrives at the hospital at 5 pm.

The minute (long) hand has completed a full turn (revolution) = 360°.

The hour (short) hand has completed $\frac{1}{12}$ of a revolution = $\frac{1}{12} \times 360 = 30°$.

10.2 LABELLING ANGLES

Generally, angles can be labelled in the following ways:

1) Three letters of lines that make the angle

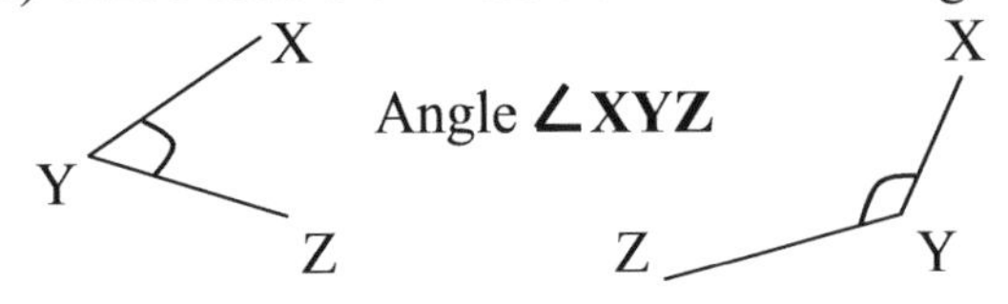

The middle letter denotes the angle.

2) Three letters with an arrow to the middle letter.

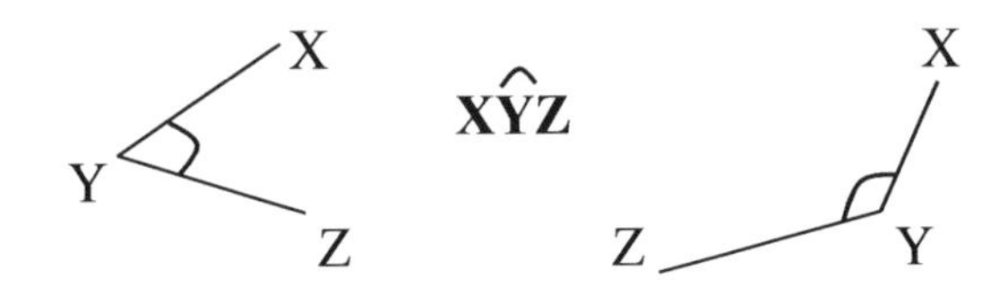

EXERCISE 10A

Describe the following **clockwise turns** in degrees.

1) North West to North
2) South West to North
3) South East to East
4) South East to North West
5) North East to West
6) South to West
7) North West to West
8) North East to North East
9) North to West
10) West to South East

Describe the following **anti-clockwise** turns in degrees.

11) East to North
12) South to West
13) North East to South East

14) The time is $\frac{1}{4}$ past 1.
a) What direction is the minute hand?

b) What is the smaller angle between the minute and the hour hand of the clock?

15) The time is $\frac{1}{2}$ past 4.
a) What is the direction of the minute hand?

b) What is the larger angle between the hour and minute hand?

16) Work out the following:
a) $\frac{1}{2}$ of 90°

b) $\frac{1}{4}$ of 60°

c) $\frac{3}{4}$ of 20°

d) $\frac{1}{4}$ of 270°

17) Work out the angles between the hour hand and the minute hand of a clock at the following times shown.

a) 1 o'clock

b) 2 o'clock

c) 10 o'clock

18) Chude faces west. He turns to face north. Describe the turn he made in degrees.

10.3 ANGLES AND THEIR NAMES

1) Angles between 0° and 90° are called **acute angles**. They are angles less than 90°. Examples of acute angles are 7°, 50°, 89°...

2) An angle that is **exactly** 90° is called a **right angle**. They are recognised with a small square.

3) Angles between 90° and 180° are called **obtuse angles**. They are angles greater than 90° but less than 180°. Examples of obtuse angles are 91°, 130°, 179°....

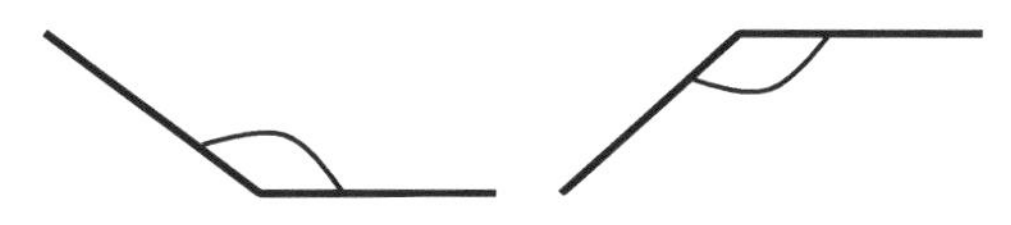

4) A **straight angle** is 180°. It changes the direction to point the opposite way.

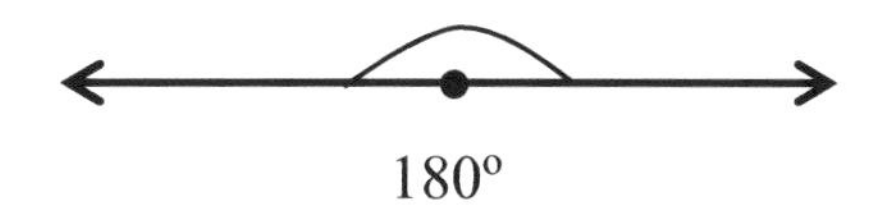

5) Angles between 180° and 360° are called **reflex angles**. They are angles greater than 180° but less than 360°. Examples of reflex angles are181°, 270°, 350°....

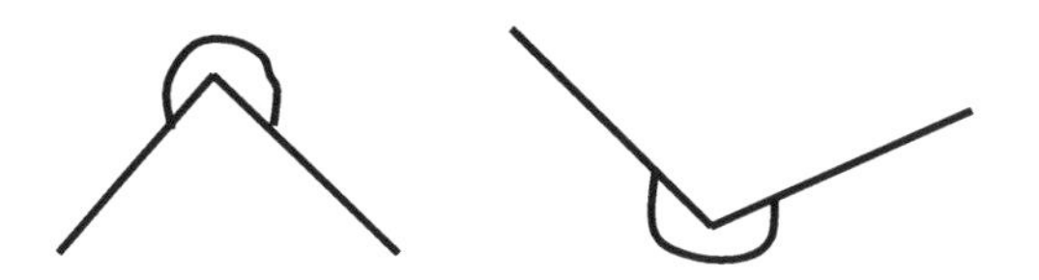

EXERCISE 10B

1) The diagram below shows an angle.

a) What type of angle is this?
b) Using the letters on the diagram, describe the angle.

2) Copy each angle onto your exercise book. What type of angle is each of these?

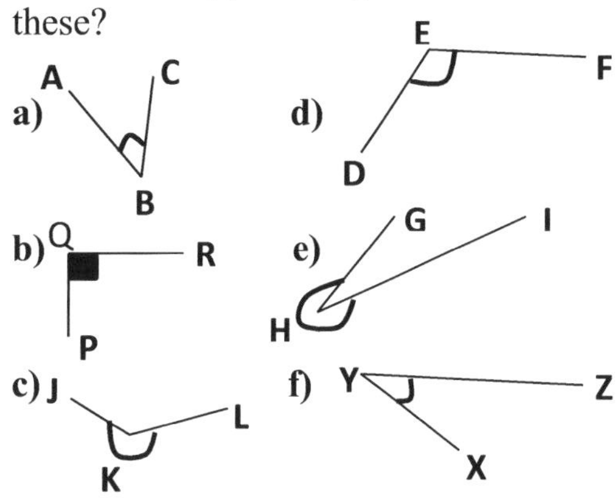

3) Describe each of the angles in question 2 using the letters in the diagram.

4) Write down each of these angles and indicate if it is acute, right angle, obtuse, straight line or reflex.

a) 13°	c) 120°	e) 89°
b) 270°	d) 90°	f) 180°

5) What type of angle is each of these?

a) $P\hat{Q}R$
b) $R\hat{S}T$
c) $Q\hat{P}T$
d) $Q\hat{R}S$
e) $P\hat{T}S$

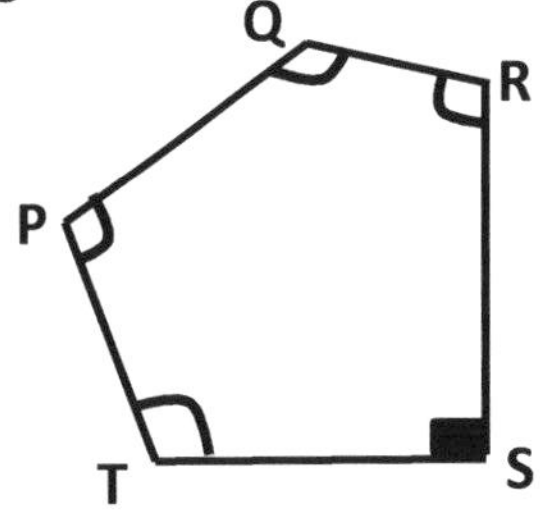

10.4 CONSTRUCTING ANGLES USING A PROTRACTOR

Angles are constructed and measured using a protractor. Most commonly used protractors are

- 180° (semi-circular) protractor
- 360° (circular) protractor

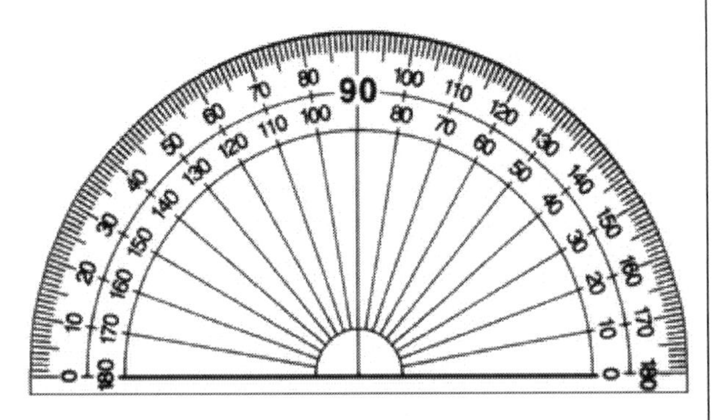

180° protractor
(Most commonly used)

Example 1: Draw (construct) angle 30°

It is always advisable to draw a straight line and mark a point for the centre of the angle.

Carefully place the protractor on the line and the marked point and make a mark at the 30° point on the protractor or paper.

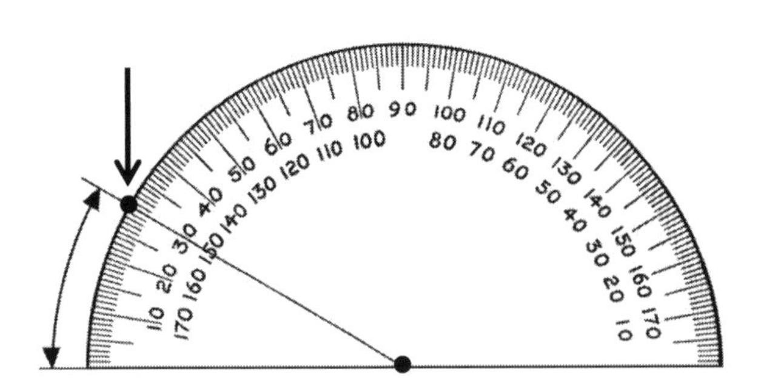

The required angle is then counted from the base line. Notice that there are two starting points on the protractor with zeros. You may measure angles clockwise or anticlockwise. Always check that you have drawn the right angle by noting whether it is acute or obtuse.

There are two readings on the protractor, 30° and 150°. In this case, the angle is acute (30°) and we are drawing clockwise.

Make a mark on the paper at the 30° point. Remove the protractor and join the mark to the centre.

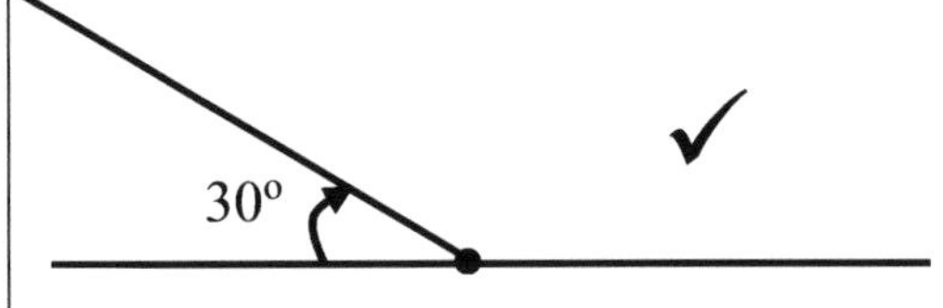

Note: Always start from zero to measure or read angles when using a protractor.

EXERCISE 10 C

1) Construct and label these angles using a protractor.

a) 15° b) 60° c) 70° d) 95°
e) 100° f) 130° g) 173° h) 210°

2)

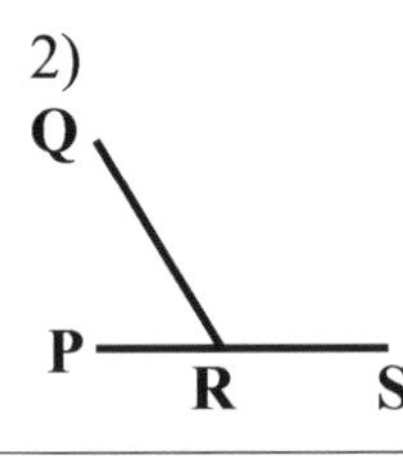

Copy the diagram and make $\widehat{SRQ}$ 120°. Measure the opposite angle. Comment on your angles.

10.5 MEASURING ANGLES

Example 1: Measure angle $A\hat{C}B$.

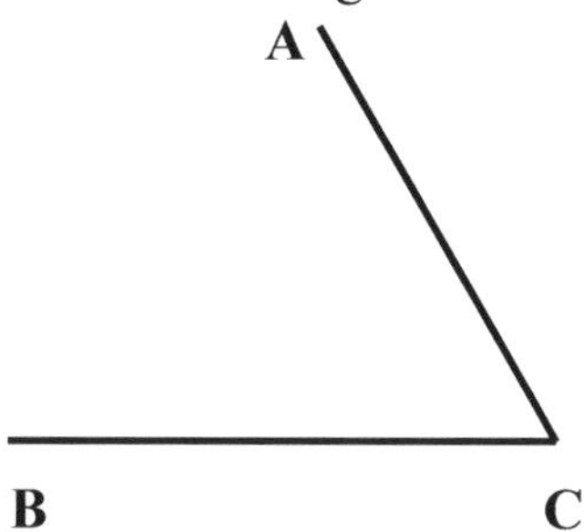

Solution: Carefully place the protractor on the line BC and make sure the centre of the protractor is at point C.

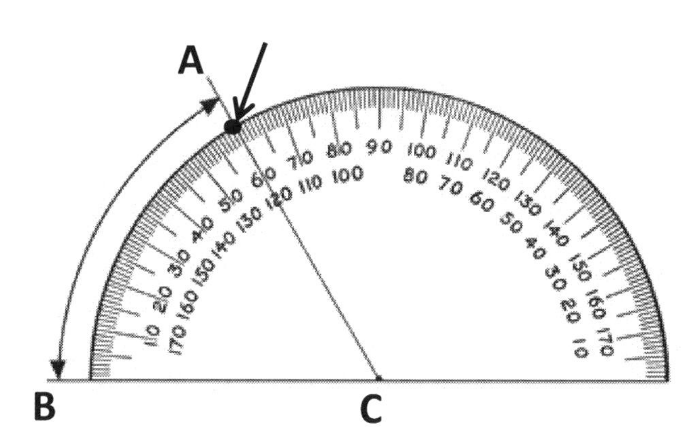

From zero, read the size of the angle. There are two possible angles: 60° and 120°.
Since the diagram shows an acute angle, the angle shown must be **60°**.

Measuring a reflex angle
Since the common 180° protractor used do not have angles more than 180°, measure the acute or obtuse angle and then subtract the result from 360°. That would give the required reflex angle.

See example 2.

Example 2: Measure the reflex angle shown.

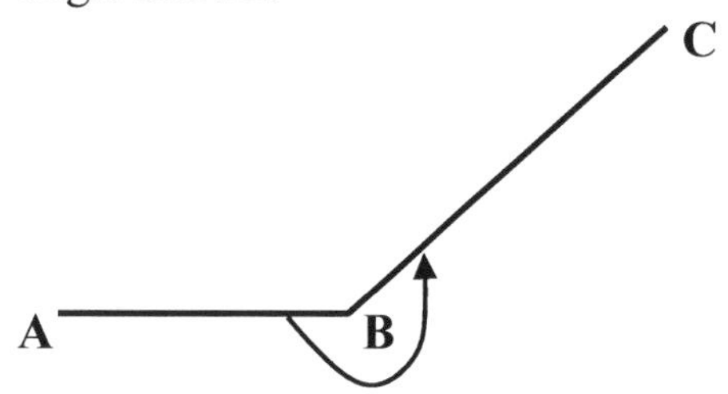

Carefully position the protractor and measure the obtuse angle ABC.

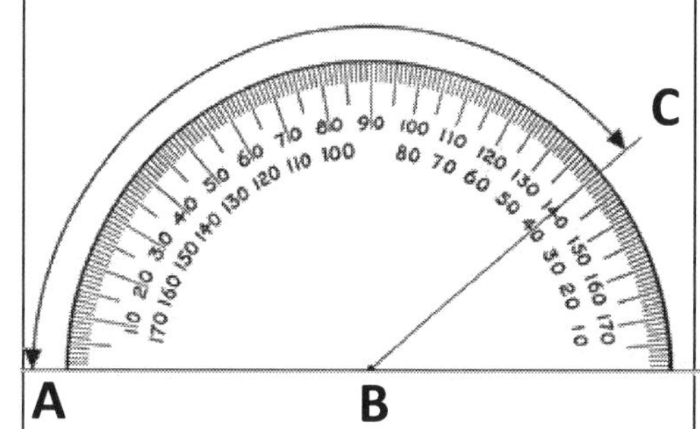

From the diagram, the obtuse angle is 140°. Therefore, the reflex angle is 360 – 140 = **220°**.

EXERCISE 10D

1) Write down any ten obtuse angles.
2) Write down any four acute angles.
3) From the diagram below, **name** the
   a) acute angle
   b) obtuse angles
   c) right angles.

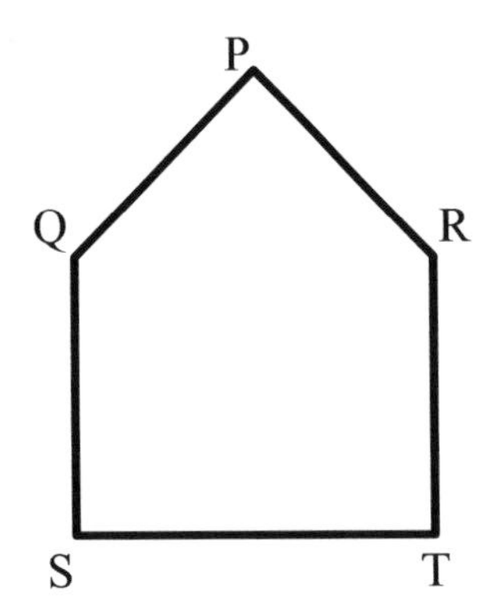

4) Read the sizes of the marked angles.

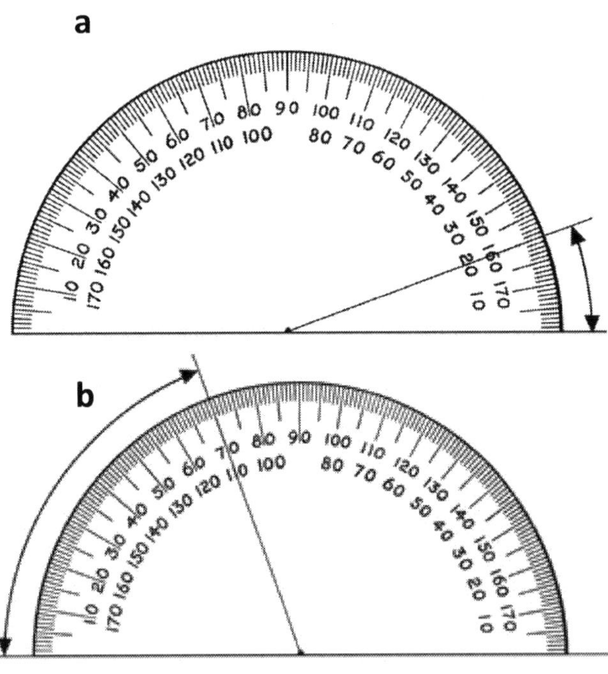

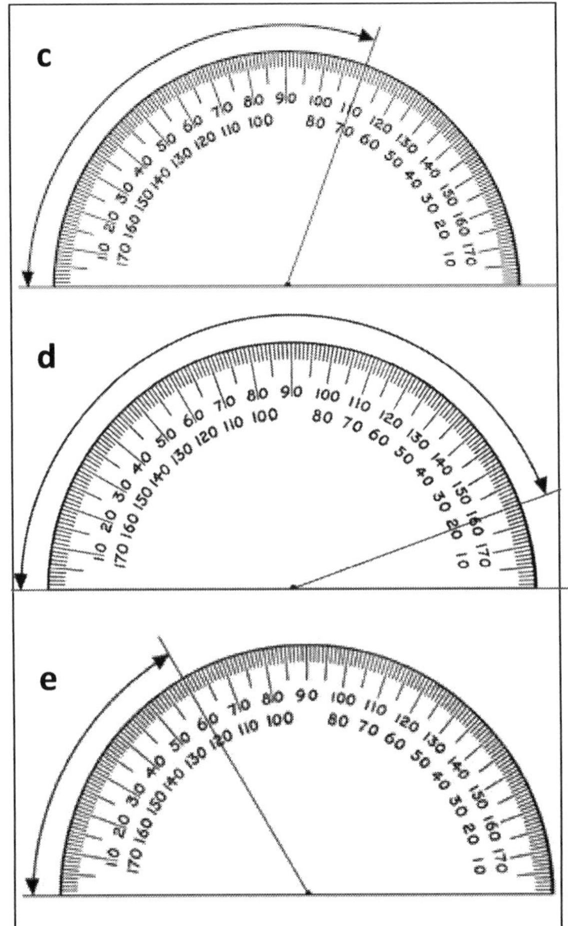

5) The angles in the diagram below have been marked with letters. Identify and write down whether each is an obtuse, acute, right angle or reflex angle.

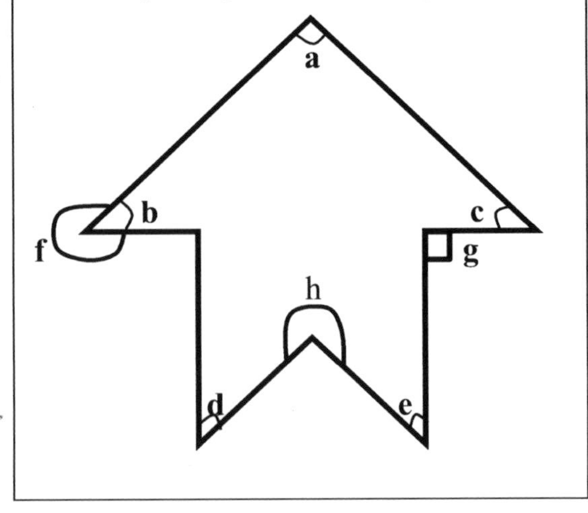

6) Write down whether each of the angles below is acute, right angle, obtuse or reflex angle.

a) 300° b) 34° c) 124° d) 90° e) 2° f) 278° g) 179°

h) 78° i) 189° j) 250° k) 12° l) 80° m) 270° n) 99°

7) You will need a tracing paper, protractor, pencil, and ruler for this exercise. Put the tracing paper on top of the angles and trace the angles on to your exercise book. Next, measure the angles.

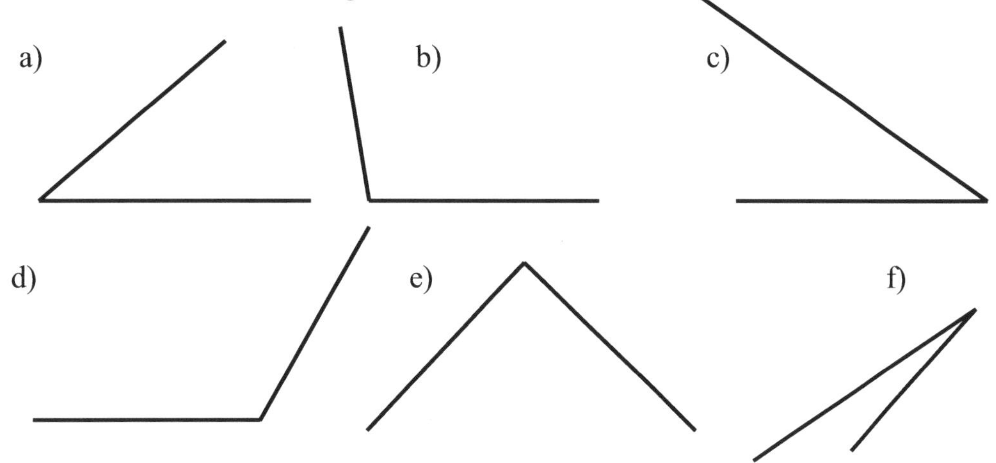

8) Try and sketch the angles below without using a protractor.

a) 30° b) 70° c) 180° d) 270° e) 350°

9) Draw the following angles using a protractor.

a) 30° b) 70° c) 180° d) 270° e) 350°

10) In question 7 above, state the type of angles you measured.

11) The diagram below shows a five-sided shape.

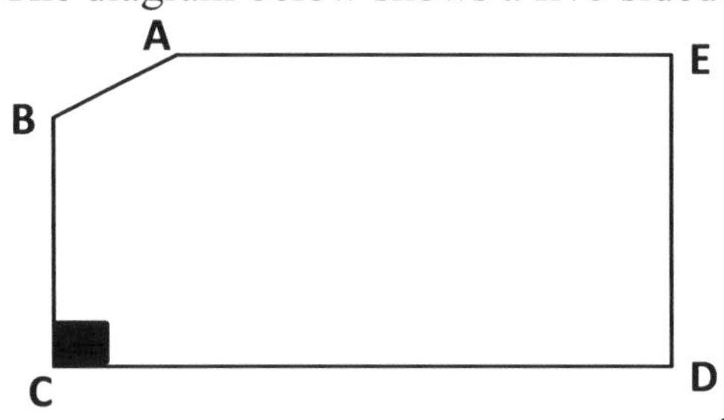

What type of angle is each of these?

a) $A\hat{B}C$
b) $B\hat{C}D$
c) $E\hat{A}B$

Chapter 10 Review Section
Assessment

1) Describe the anti-clockwise turn from South West to North East in degrees.
.... **1 mark**

2) Work out $\frac{3}{4}$ of 40°
.... **1 mark**

3) Work out the angle between the hour hand and the minute hand of a clock at 10 o'clock. .... **1 mark**

4) Estimate the size of these angles.

a) b) c)

.... **3 marks**

5) By using a protractor, construct the following angles.

a) 40° b) 65° c) 137° d) 200° e) 300°
.... **5 marks**

6) Copy the diagram below onto your exercise book.

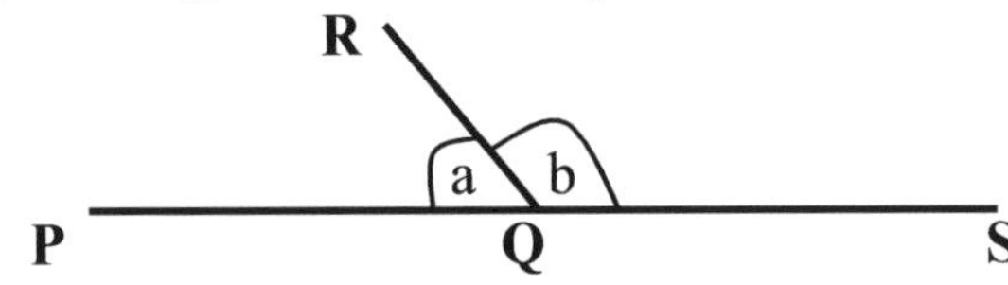

a) Measure angles **a** and **b**. .... **2 marks**
b) Explain why line PQS is a straight line. ....**1 mark**

7)

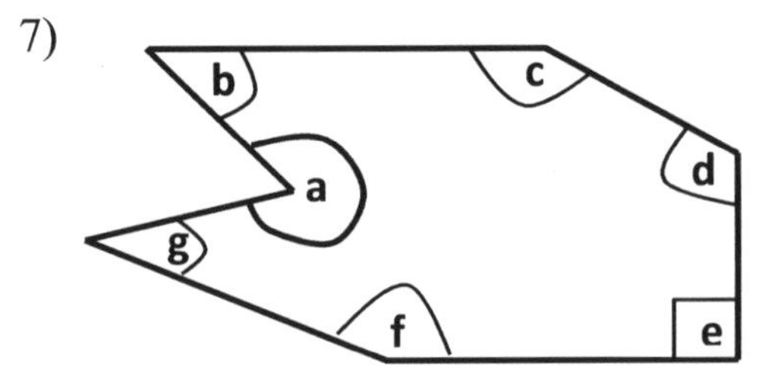

Write down the different types of angles marked with letters.
..............**7 marks**

11 Directed Numbers

This section covers the following topics:

- Directed numbers
- Understanding number lines
- Addition and subtraction of negative numbers

LEARNING OBJECTIVES

By the end of this unit, you should be able to:

a) Understand the meaning of directed numbers
b) Use a number line
c) Add and subtract negative numbers
d) Multiply and divide negative numbers
e) Apply negative numbers to real life situations
f) Understand the concept of integers

KEYWORDS

- Directed numbers
- Add
- Subtract
- Negative numbers
- Number line
- Positive numbers
- Integers
- Multiply

11.1 UNDERSTANDING NEGATIVE NUMBERS

Before we can successfully understand negative numbers, we must familiarise ourselves with directed numbers and integers.

Directed numbers are numbers which have a direction and a size. Positive and negative numbers fall into this category and as such, directed. The positive (+) or negative (–) sign indicates which direction to go from zero (0) until the position of the number is reached.

The numbers below are directed number:

-30, -20.4, - 10, - 5.6, - 5 4, 7, 20

From the directed numbers above, only -30, -10, -5, 4, 7 and 20 are **integers**. Integers are positive or negative **whole** numbers **including zero**.

As a summary, if a directed number is a whole number, then it is called an integer.

All numbers above zero or to the right of zero are positive numbers.
All numbers below zero or to the left of zero are negative numbers
Zero (0) is neither positive nor negative. It is a **neutral** number.

-5 is read as a negative five. It is 5 less than zero. 0 - 5 = **-5**

We can show positive and negative integers on a **number line** below.

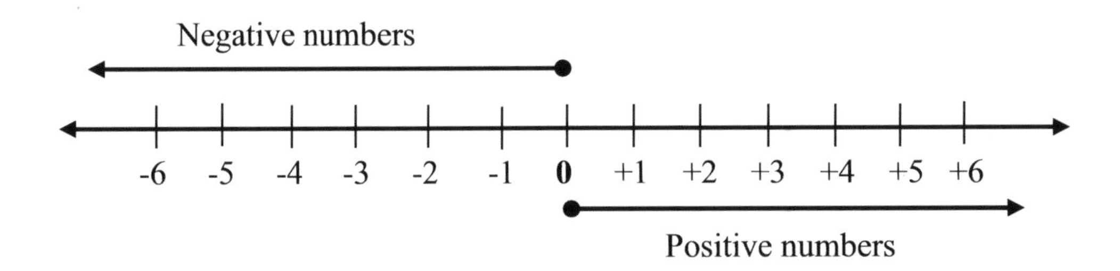

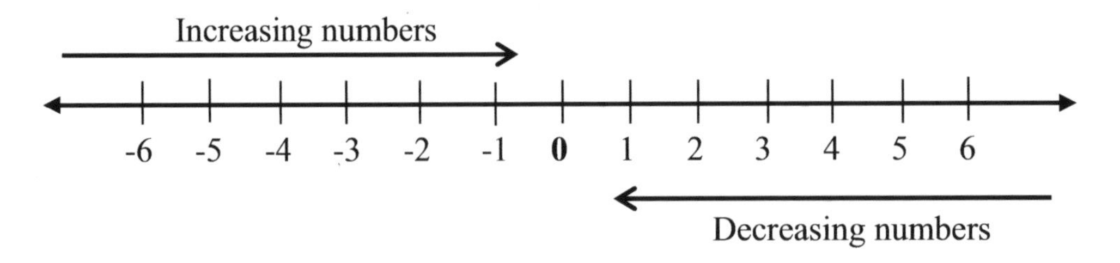

Do not forget:

-1 is greater than -2, -6 is greater than -7, +2 is greater than -2, and +6 is greater than -6

We often use negative numbers with temperatures. During winter in some countries, the weather is very cold, and the temperature can go as low as -10°C. It means that the temperature is 10 below zero.

Water freezes at 0°C approximately. Temperatures are given in degrees Celsius (°C).

Example 1: Suppose the temperature of a room at 3 am was -3°C and it rises **to** 5°C at 9 am. What is the difference in temperature between 3 am and 9 am?

Solution: we may use a number line to help us.

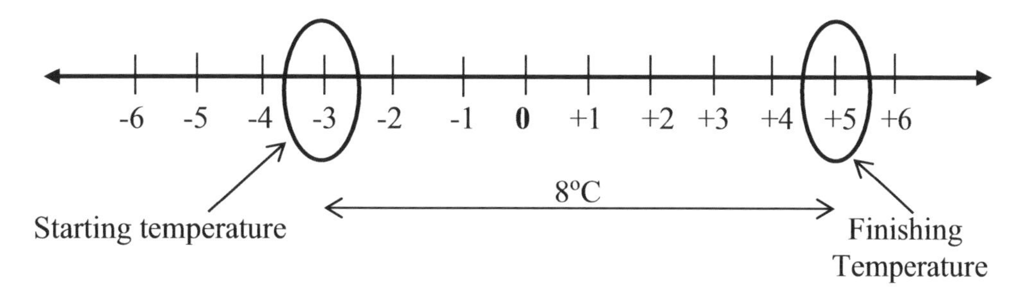

The difference in temperature is clearly **8°C** because +5 - (-3) = 5 + 3 = 8 ✓

Example 2: Suppose the temperature is -4°C and it rises **by** 7°C. What is the new temperature?

Solution: Use the number line to help. Start from -4 and count 7 to the right.

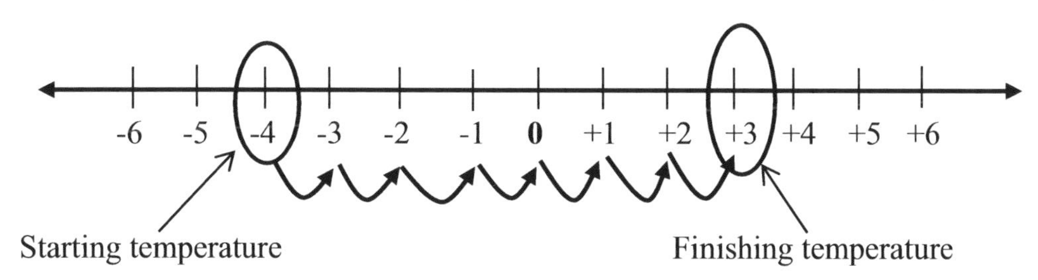

The new temperature is **3°C**. ✓

Example 3: Suppose the temperature is 4°C and it falls by 6°. What is the new temperature?

Solution: Start from 4°C and count six units to the left. The new temperature is **-2°C.**

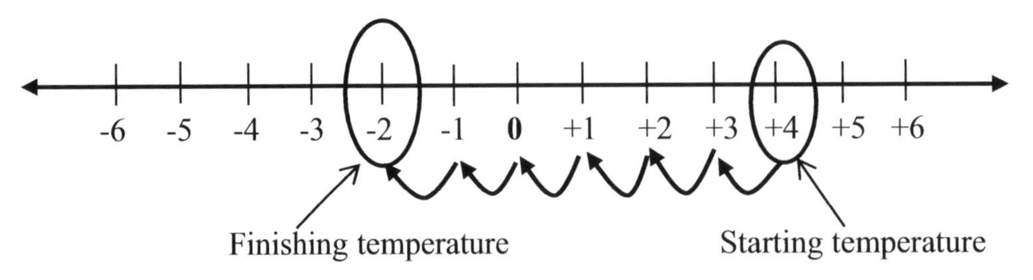

Alternatively, (+4) - (+6) = +4 - 6 = **-2**

Example 4: The diagram shows a thermometer with two arrows.

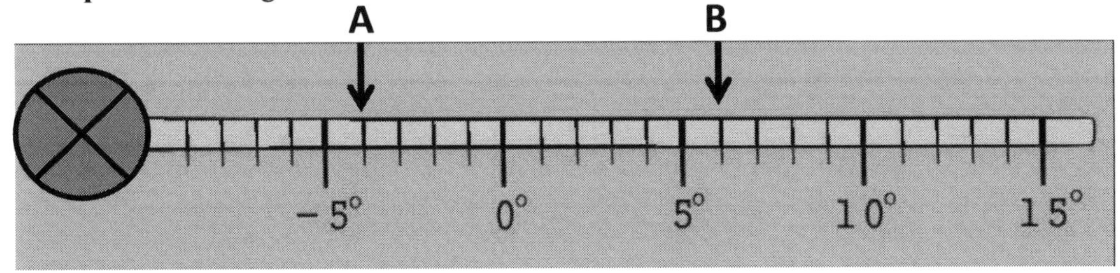

a) Which temperature is cooler?
b) What is the difference between the two temperatures?

Solution:

a) **A (-4°C)** is cooler as it is lower than B (6°C)
b) 6 - (-4) = 6 + 4 = **10°**................The difference is 10°.
(Alternatively, you may count the units between A and B.

Example 5: Use a number line to work out the value of -2 + 5.

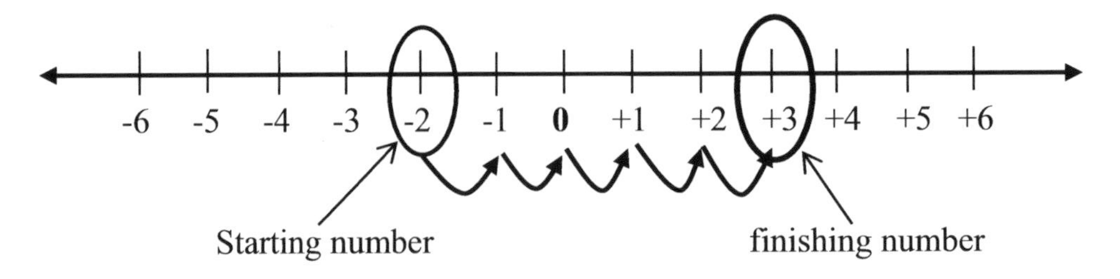

Start from -2 and go to the right five times. The answer is **3**.

EXERCISE 11A

1) Write these temperatures from hottest to coldest. All number in °C.

a) -10, 4, 5, -8, -3, 9
b) 10, -7, -15, 9, 40, -2
c) 42, -20, -10, -15, 17
d) -1, -7, 3, 5, -5,
e) -8, 8, -9, 9, -4, 4
f) -16, 14, -12, 10, -18, 30

2) What temperature in Celcius is shown by the alphabets below?

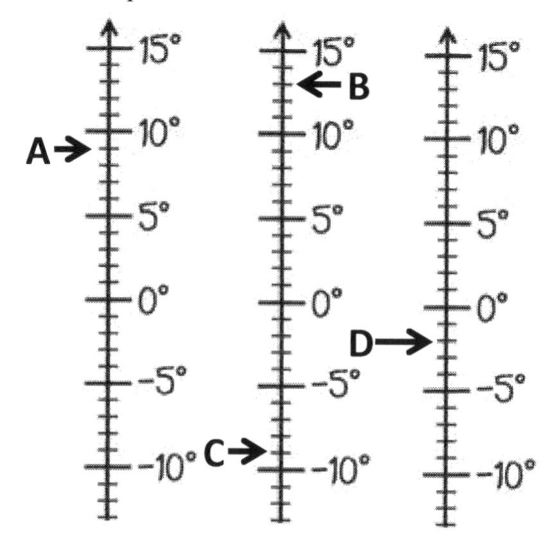

3) What would the temperatures in question 2 above be if it was
a) at C and fell 3°?
b) at A and rose 5°?
c) at B and fell 9°?
d) at D and rose 11°?

4) In the statement below, write <, > or = in the box.

a) 4.5 ☐ 2
b) -3 ☐ - 4
c) -10 ☐ 40
d) -6 ☐ - 6

5) Find the differences between each pair of temperatures in °C.

a) -3, 5
b) -13, 3
c) -7, -10
d) -3, -8
e) 0, -3
f) -2, -5
g) 10, -3
h) 2, -4

5°C

Which of the temperatures above are
a) below freezing point
b) above freezing point

Which of the temperatures above is
c) the hottest
d) the coldest?
e) What is the difference between the coldest and hottest temperatures?

7) For each question below, state whether the temperature has fallen or risen and by how many degrees.

a) It was -4°C and now -10°C
b) It was -5°C and now 10°C
c) It was 7°C and now -3°C
d) It was 2°C and now 25°C

8) Use a number line to perform the following calculations.

a) 2 + 3
b) -2 + 3
c) -4 + 2
d) -1 - 3
e) - 4 - 5
f) 7 - 3

EXERCISE 11 B

1) Write the next two numbers in the sequences below.

a) -4, -6, -8, _ , _

b) 7, 4, 1, _, _

c) 10, 5, 0, _, _

d) 20, 14, 8, 2, _, _

e) -7, -5, -3, _, _

f) -3, -5, -7, _, _

2) Put each set of numbers in order, lowest first.

a) -4, 5, -9, 0, -3

b) -50, 55, -30, 40, 20,

c) -7, -3, 1, -2, -1, 0, 4

3) Copy and complete the table below.

Temperature °C	Change °C	New Temp. °C
-3	+4	
	-3	-6
-7	-3	
	+6	12
-9	-9	

4) Find the range of these temperatures in °C.

a) -3 and -5

b) 2 and -7

c) 13 and -15

5) The table shows different temperatures in different towns on one day.

Town	**Temperature °C**
Lagos	+25
Onitsha	+27
Essex	-3
Manchester	-1
Kano	+32
Abuja	+30

a) Which town is the warmest?
b) Which town is the coldest?
c) Write down the names of towns in order, from coldest to warmest
d) At 3 pm on the same day, the temperatures in each town are 4°C higher. Write down the new temperatures for Essex and Onitsha.

11.2 ADDING AND SUBTRACTING DIRECTED NUMBERS

Number lines are useful in adding or subtracting directed number.

Example 1: -3 + 4

This means start from -3 and move to the **right,** four times.

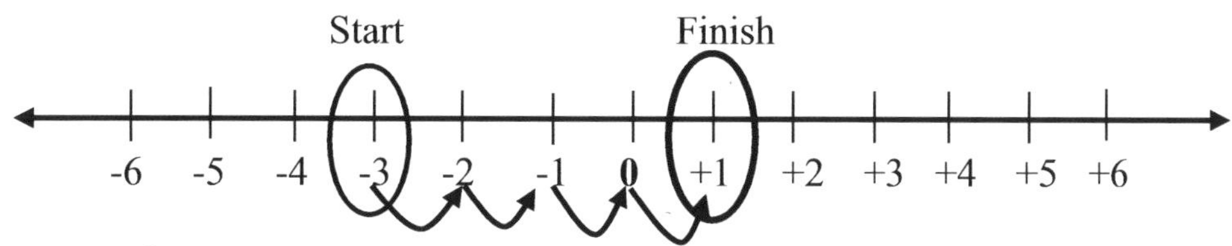

The answer is **+1** or **1**

Example 2: 5 - 7

It means start from +5 and go **left** seven times. The answer is -2

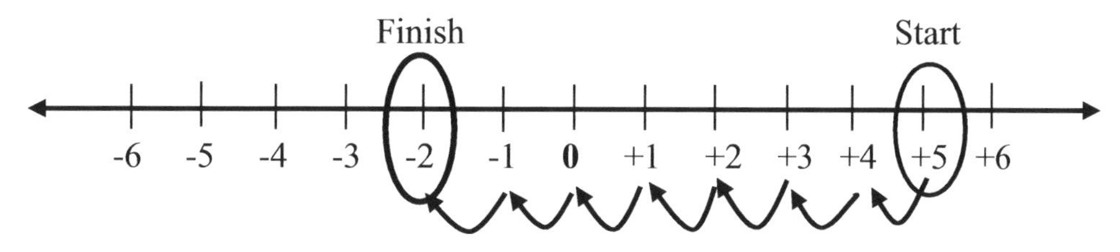

Example 3: -2 + (-3)

To add a negative number, go left from the starting number. The answer is -5.

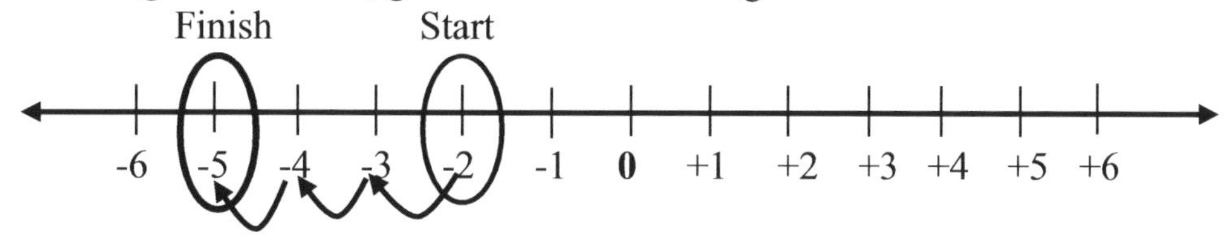

When you have two sighs together, use this rule:

+ + = +
- - = +
} **Same signs**

+ - = -
- + = -
} **Different signs**

Example 4: 4 - (-2)

Replace - - with (+) It becomes 4 + 2 = **6**

Example 5: -5 + (-2)

Replace + - with (-) It becomes -5 - 2 = **-7**

* You may use a number line to help by starting from -5 and go left 2 times.

Example 6: -3 - (+2)

Replace - + with (-) It becomes -3 - 2 = **-5**

Example 7: 5 + (-6)

Replace + - with (-) It becomes 5 - 6 = **-1**

Example 8: -6 - (-4)

Replace - - with (+) It becomes -6 + 4 = **-2**

EXERCISE 11C

Work out without using a calculator. You may use a number line to help.

1) a) 8 + 3 b) 8 – 8 c) 4 – 5 d) 6 – 10 e) 5 – 9 f) 8 + (-5)
g) 7 + (-2) h) 3 + (-7) i) 9 + (-6) j) 2 + (-4) k) 3 + (-8) l) 9 + (-2)
m) 6 + (-3) n) 7 + (-5) o) 10 + (-8) p) 1 + (-1) q) 7 + (-7) r) 3 + (-2)

2) a) -3 + 5 b) -8 + 10 c) -6 + 3 d) -5 + 4 e) -2 + 2 f) -8 + 8
g) -10 + 3 h) -8 + 5 i) -6 + 5 j) -12 + 5 k) -9 + 1 l) -3 + 7
m) -10 + 5 n) -13 + 20 o) -9 + 3 p) -7 +1 q) -3 + 3 r) -6 + 9

3) a) -1 -5 b) -6 -10 c) -7 -3 d) - 4 - 2 e) -2 - 1 f) -8 - 8
g) -10 -3 h) -8 - 5 i) -6 - 5 j) -12 - 5 k) -7 - 4 l) -3 -7
m) -10 -5 n) -1 - 8 o) -2 - 3 p) -5 - 7 q) - 4 - 6 r) -20 - 3

4) a) -3 + (-8) b) -6 + (-9) c) -2 + (-5) d) -4 + (-8) e) -10 + (-12) f) -7 + (-3)
g) -9 + (-6) h) -10 + (-5) i) -8 + (-2) j) -3 + (-2) k) -9 - (+5) l) -6 - (+10)
m) -3 - (+5) n) -4 - (+7) o) -6 - (+3) p) -5 - (+1) q) -1 - (+4) r) 8 - (-7)

5) a) -7 - (-5) b) -3 - (-2) c) -6 - (-4) d) -10 - (-8) e) -5 - (-3) f) -7 - (-1)
g) -6 - (-13) h) -7 - (-3) i) -13 - (-20) j) -6 - (-7) k) -100 + (-30)

6) Work out the missing numbers.

a) 18 - □ = 5   b) 15 - □ = 10   c) 7 - □ = 3   d) 19 - □ = 7

e) 112 - □ = 93   f) 7 - □ = -3   g) 6 - □ = -4   h) □ + 8 = 3

7) Copy and complete the addition square below.

+	-4	-3	-2	-1	0	+1	+2
+4							
-5							
+3							
-2							
0							
+8							
-10							
-7							
-6							

8) Simplify

a) -8 + (-5) - 1

b) -5 - (- 4) - 1 (-2)

c) 9 + (-6) - 7

d) -10 - (- 4) - - (+5)

9) Tunde's bank account reads ₦54 283. He owes Babalola ₦76 542. How much will Tunde be overdrawn if he writes a cheque for the whole amount he owes Babalola?

11.3 MULTIPLYING AND DIVIDING WITH NEGATIVE NUMBERS

RULES:

1) Multiplying two numbers with the **same** sign always gives a **positive (+)** answer.

Example 1: a) $(+4) \times (+5)$ this is the same as $4 \times 5 = 20 = \mathbf{20}$
b) $(-3) \times (-5)$ the answer is $3 \times 5 = \mathbf{15}$
c) $(-6) \times (-2)$ the answer is $6 \times 2 = \mathbf{12}$
d) $(-4)^2 = -4 \times -4 = \mathbf{16}$

2) Multiplying two numbers with **different** signs always gives a **negative (-)** answer.

Example 2: a) $-3 \times 4 = \mathbf{-12}$ b) $-6 \times 10 = \mathbf{-60}$ c) $3 \times -4 = \mathbf{-12}$ d) $6 \times -10 = \mathbf{-60}$

3) The rules for multiplication are the same for the division.

Example 3: a) $(+4) \div (+2) = \mathbf{2}$ b) $(-6) \div (-2) = \mathbf{3}$ } ***Same signs always give positive answers***

c) $(-20) \div (+5) = \mathbf{-4}$ d) $(+6) \div (-3) = \mathbf{-2}$ } ***Different signs always give negative answers***

Generally, when multiplying negative/directed numbers,

(+) × (+) = Positive (+)

(-) × (-) = Positive (+)

(+) × (-) = Negative (-)

(-) × (+) = Negative (-)

Also, the same rules apply exactly when dividing with negative numbers.

(+) ÷ (+) = Positive (+)

(-) ÷ (-) = Positive (+)

(+) ÷ (-) = Negative (-)

(-) ÷ (+) = Negative (-)

EXERCISE 11D

Work out the following:

1) a) $-2 \times (-3)$ b) $-10 \times (-2)$ c) $-7 \times (-1)$ d) $-6 \times (-5)$
e) $-3 \times (-7)$ f) $-5 \times (-5)$ g) $-12 \times (-10)$ h) $-15 \times (-3)$
i) $-7 \times (-2)$ j) $-8 \times (-9)$ k) $-2 \times (-4)$ l) $-8 \times (-7)$

2) a) -2×3 b) -7×10 c) -3×5 d) -6×6
e) -9×4 f) -4×2 g) -10×3 h) 6×-3
i) 9×-7 j) 3×-2 k) 4×-4 l) 11×-3

3) a) $-16 \div (-8)$ b) $-9 \div (-3)$ c) $-30 \div (-10)$ d) $-6 \div (-2)$
e) $-50 \div (-5)$ f) $-8 \div (-2)$ g) $9 \div (-3)$ h) $40 \div (-4)$
i) $8 \div (-2)$ j) $-8 \div 2$ k) $-45 \div 9$ l) $-35 \div 7$

4) Copy and complete the multiplication grid below.

×	8	-2	-5	7	-3
-3					
4			**-20**		
5					
-10					
-2					
-7					

5) Match the following calculations with the answers.

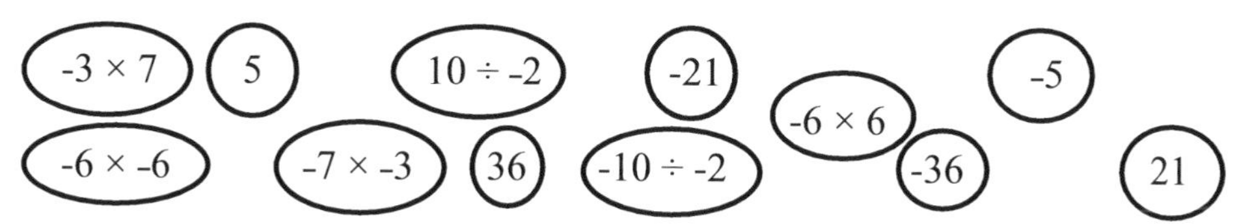

6) Work out without a calculator.

a) $2 \div 2$ b) $-18 \div (-3)$ c) $-10 \div (-5)$ d) $40 \div (-4)$

e) -4×8 f) 9×-3 g) $-7 \times (-8)$ h) $-126 \div 14$

Chapter 11 Review Section

Assessment

1) Which is greater -5 or 2? .... **1 mark**
2) What do you subtract from -4 to make -9? .... **1 mark**
3) Simplify the following
   a) (+ 3) + (- 6) .... **1 mark**
   b) (- 7) - (- 4) .... **1 mark**
   c) - 18 ÷ 6 .... **1 mark**
   d) - 9 - (+5) .... **1 mark**

4) The temperature in a room in winter was -5°C. The temperature fell by a further 4°C. What is the new temperature? .... **1 mark**

5) Look at the number line below.

-6 -5 -4 -3 -2 -1 **0** +1 +2 +3 +4 +5 +6

Find the difference between

a) -5 and 0 b) -2 and -4 c) 3 and -6 d) -6 and 5

.... **4 marks**

6) Write these temperatures (°C) in ascending order.
   a) -2, -3, -5, 7, 10, -1
   b) 5, -5, 3, -4, -7, 8, 0
   c) -2, 1, -17, 4, -5, 5, 2

7) Copy and complete the multiplication table below.

×	-4	+3	-8	+2
-2				
-7				
+8				
+5				

12 Percentages 2

This section covers the following topics:

- Percentage of a quantity
- Proportions using percentages
- Find percentage change

LEARNING OBJECTIVES

By the end of this unit, you should be able to:

a) Find percentage of a quantity
b) Compare proportions using percentages
c) Find a percentage increase
d) Find a percentage decrease
e) Find percentage change

KEYWORDS

- Percentage
- Change
- Increase
- Decrease
- Quantity
- Proportion

12.1 PERCENTAGE OF A QUANTITY

Finding the percentage of a quantity is the same as finding the fraction of the quantity. Remember, 10% means $\frac{10}{100} = \frac{1}{10}$, therefore, finding 10% of a number simply means **dividing the number by 10**.

Example 1: Find 10% of 20.
Since 10% means $\frac{1}{10}$, we have to find $\frac{1}{10}$ of 20 which means $\frac{20}{10} = \mathbf{2}$

Example 2: Find 10% of 70.........................It means 70 ÷ 10 = **7**
Example 3: Find 10% of 635......................It means 635 ÷ 10 = **63.5**

From 10%, a lot of percentages can be obtained.

- To find 20%, find 10% and multiply by 2
- To get 30%, find 10% and multiply by 3
- To get 70%, find 10% and multiply by 7, and so on.
- To get 5%, find 10% and divide by 2.
- To get 50%, find 10% and multiply by 5 **or** simply divide the number by 2.

As you get familiar with percentages, there are other ways you could work out any percentage of an amount. For example: to get 70%, you could work out 50% and then add it on to 20%.

Also remember, 1% is the same as $\frac{1}{100}$. Therefore, finding 1% of a number simply means **dividing the number by 100.**

Example 3: Find 1% of 34.........................It means $\frac{34}{100} = \mathbf{0.34}$
Example 4: Find 1% of 436......................It means $\frac{436}{100} = \mathbf{4.36}$

Mixed examples

Example 5: Find 20% of 70
Find 10% and multiply by 2. 10% of 70 = 7 7 × 2 = **14**

Example 6: Find 40% of ₦120 10% of 120 = 12 12 × 4 = **₦48**

Example 7: Find 5% of 30 10% of 30 = 3 3 ÷ 2 = **1.5** or $\mathbf{1\frac{1}{2}}$
Example 8: Find 15% of 80
10% of 80 = 8 5% of 80 = 10% ÷ 2 = 8 ÷ 2 = 4 **15%** of 80 = 8 + 4 = **12**

EXERCISE 12A

Work out without a calculator. Show all working out.

1) a) 10% of 30 b) 10% of 4 c) 10% of 60 d) 10% of 8
e) 10% of 16 f) 10% of 40 g) 10% of ₦80 h) 10% of ₦300
i) 20% of 30 j) 20% of 4 k) 20% of 60 l) 20% of 8
m) 40% of 40 n) 50% of 36 o) 80% of 20 p) 100% of 50

2) a) 5% of 20 b) 5% of 140 c) 60% of 100 d) 1% of 200
e) 5% of 38 f) 1% of 24 g) 50% of 2440 h) 35% of 80kg

Example 9: Find 13% of 30

Method 1

10% of 30 = 30 ÷ 10 = 3
\+ 1% of 30 = 30 ÷ 100 = 0.3
1% of 30 = 30 ÷ 100 = 0.3
1% of 30 = 30 ÷ 100 = 0.3
13% of 30 = **3.9**

Method 2

13% means $\frac{13}{100}$

Therefore, $\frac{13}{100} \times 30$

$= 0.13 \times 30 =$ **3.9**

Example 10: Find 7% of \$150

1% of 150 $= \frac{1}{100} \times 150 = \frac{150}{100} = 1.5$

$1.5 \times 7 =$ **\$10.50**

Example 11: Find 11% of ₦200

10% of ₦200 = 200 ÷ 10 = 20
\+ 1% of ₦200 = 200 ÷ 100 = 2
11% of ₦200 = **₦22**

Method 2

11% means $\frac{11}{100} = 0.11$
$0.11 \times 200 = 22$
Therefore, 11% of 200 = **₦22**

Example 12: Work out 2.3% of 19.35

2.3% means $\frac{2.3}{100} = 0.023$

$0.023 \times 19.35 =$ **0.44505**

Example 13: Work out $4\frac{1}{2}$% of 90

$4\frac{1}{2}$% means $\frac{4.5}{100} = 0.045$

$0.045 \times 90 =$ **4.05**

EXERCISE 12B

Work out the following without using a calculator.

1) a) 1% of 130 b) 3% of 35 c) 20% of 80 d) 80% of 30
e) 21% of 33 f) 14% of 55 g) 4% of 130 h) 7% of ₦300
i) 55% of £60 j) 90% of 750 k) 15% of 8 l) 35% of 80
m) 61% of 40 kg n) 7% of 890 o) 23% of ₦2000 p) 4% of 9

2) a) 2.7% of £70 b) 3.2% of 56 c) 2.8% of 200 d) 17.5% of 460
e) $2\frac{1}{2}$% of 30 f) $7\frac{1}{2}$% of 120 g) 3.5% of 68kg h) $6\frac{1}{2}$% of ₦5000

12.2 COMPARING PROPORTIONS USING PERCENTAGES

Anthony sat tests in Physics, Chemistry, Biology, and Mathematics. His results were:

Physics: $\frac{20}{30}$, Chemistry: $\frac{77}{90}$, Biology: $\frac{16}{25}$, Mathematics: $\frac{18}{20}$

Which subject did Anthony do best?

Solution: Before making any decision(s), we can write each fraction as a percentage. (Refer to section 6.4).

Physics: $\frac{20}{30} \times 100 = 66.7\%$ Biology: $\frac{16}{25} \times 100 = 64\%$

Chemistry: $\frac{77}{90} \times 100 = 85.6\%$ Mathematics: $\frac{18}{20} \times 100 = 90\%$

From the above calculations, Anthony did best in Mathematics test.

12.3: PERCENTAGE INCREASE AND DECREASE

A quantity can be increased or decreased by a percentage.

Example 1: Increase 100 by 10%

First, find 10% of 100 which is $\frac{100}{10} = 10$.

Therefore, increasing 100 by 10 means

100 + 10 = **110**

Example 2: Decrease ₦510 by 20%

First, find 20% of ₦510

Find 10% and multiply by 2

10% of ₦510 = $\frac{510}{10}$ = ₦51

20% of 510 = 51 × 2 = ₦102

Decreasing ₦510 by 20% means 510 - 102 = **₦408**

12.4 FINDING A PERCENTAGE CHANGE

Percentage change is calculated by using the formula below.

$$\text{Percentage change} = \frac{\text{Change (increase or decrease)}}{\text{Original amount}} \times 100$$

Example 1: The price of a TV set was increased from £80 to £120. Work out the percentage increase.

Solution: First work out the increase.

$120 - 80 = 40$

Percentage increase $= \frac{40}{80} \times 100 = \mathbf{50\%}$

Example 2: The original price of a carpet was ₦2 400. The price is now ₦2 000. Find the percentage decrease.

Solution: Decrease = 2400 – 2000 = ₦400

Percentage decrease $= \frac{400}{2400} \times 100 = \mathbf{16.7\%}$

EXERCISE 12 C

1) Increase the following amounts by 10%.

a) 20
b) 80
c) ₦200
d) ₦340
e) 840
f) 3000

2) Increase the amounts in question 1 by 15%.

3) Decrease the following quantities by 20%.

a) 40 kg
b) 92 litres
c) £123
d) $183.52
e) 4000 g
f) 245 cm

4) Alton earns ₦25 000 for selling CD's a day. How much **extra** will he get if he has a 10% salary increase?

5) The price of a Renault car is £4 200. There is a discount of 25% on all Renault cars. How much will Tochukwu pay for a Renault car?

6) Which of the following statements represent a 40% change?

a) 35 to 40 b) 50 to 70 c) 70 to 80.

7) *Shoprite* is having a 35% sale on all men's shirts. The original prices for three shirts are £15, £32 and £45 respectively. Work out the sale price for each shirt.

8) A dog weighs 2 kg and while running away from a fox, it loses one of its legs. As a result, the weight was reduced by 3%. What is the weight of the dog now?

9) Emma buys a car for ₦540 000 and sold it two weeks later for ₦720 000. Work out the percentage profit.

10) Increase ₦5500 by 2%

11) Decrease ₦6000 by 13%

12) Increase 72 kg by 15%

13) Decrease £780 by 1%

Chapter 12 Review Section

Assessment

1) Work out without a calculator.

a) 10% of 130 b) 1% of 34 c) 30% of 600 d) 21% of ₦6000

.... 6 marks

2) Decrease ₦4500 by 20% .... 2 marks

3) Increase 45kg by 50% .... 2 marks

4) Work out the percentage change for each of the following.

a) Increasing ₦200 by ₦50

b) Decreasing ₦5 000 by ₦200

c) Increasing 72 kg by 3 kg

.... 6 marks

5) A man bought a Ferrari car. The value depreciates from £200 000 when new to £180 000 two years later. Calculate the percentage depreciation. .... 2 marks

6) 5% increases a price of ₦960 and then three weeks later, it increased by a further 6%. Find the final price. .... 3 marks

7) 30% reduces the price of a bicycle in a sale. The original price was ₦35 000.

a) What is the sale price?

b) What would ten bicycles cost before the sale?

c) Christiana bought three bicycles while on sale, what would be her loss if she had bought the three bicycles before the sale?

13 Algebra 3

This section covers the following topics:

- Removing brackets
- Multiplying expressions and terms
- Order of operations
- Substitution
- Identifying equations
- Solving simple equations

LEARNING OBJECTIVES

By the end of this unit, you should be able to:

a) Remove brackets from basic algebraic expressions
b) Expand/multiply algebraic terms
c) Divide algebraic terms
d) Work out calculations using BIDMAS
e) Identify and solve simple equations

KEYWORDS

- Brackets
- Algebraic terms
- Multiply
- BIDMAS
- Algebraic expressions
- Equations

13.1 BASIC ALGEBRA AND BRACKETS

In algebra, terms inside brackets are simplified first before any other term(s). The brackets can be removed or ignored if they will not simplify.

POSITIVE SIGN BEFORE THE BRACKET

Example 1: Simplify the following

a) $4 + (3 + 8) \longrightarrow$ Do bracket first, so $3 + 8 = 11$. Therefore $4 + 11 =$ **15**
b) $6 + (7 - 3) \longrightarrow 6 + 4 =$ **10**
c) $8c + (4x + 5c) \longrightarrow 8c + 4x + 5c$ (collect like terms) = **13c + 4x**
d) $p + (q - r) \longrightarrow$ **p + q - r**
e) $(8j - 4) + (5j + 9) \longrightarrow 8j - 4 + 5j + 9 = 8j + 5j - 4 + 9 =$ **13j + 5**
f) $7 + (3 - 5) \longrightarrow 7 + (-2) = 7 - 2 =$ **5**

EXERCISE 13A

1) Simplify by removing the brackets.

a) $6 + (8 - 2)$
b) $7 + (3 + 8) - 2$
c) $16 + (4 - 5)$
d) $(1 + 9) + (8 - 3)$
e) $5 + (2 - 1)$
f) $(4 + 5) + (12 - 7) - 3$
g) $6 + (4 - 2) - 1$
h) $9 + (4 + 1) - 6$
i) $11 + (11 + 5)$
j) $20 + (6 - 2)$
k) $(4 + 1) + (4 - 2)$
l) $3 + (7 - 3) + 5$

2) Write without brackets and simplify where possible.

a) $p + (c + e)$
b) $x + (2x + 4)$
c) $(m + n) + (o + p)$
d) $3x + (x + y) + 5x$
e) $y + (n - m)$
f) $4k + (k + 3)$
g) $(3c - d) + (e - 7f)$
h) $(p - q) + (r - s)$
i) $n + (n - m)$
j) $6g + (g - 4c)$
k) $7w + 5y + (y + 3w)$
l) $d + (e - f) - g$

NEGATIVE SIGN (-) BEFORE THE BRACKET

If you notice a negative sign before a bracket, the sign inside the bracket is reversed when the bracket is removed. Look at the examples below for clarity.

- $10 - (4 + 5)$ ***Reverse the sign inside the bracket*** $10 - 4 - 5 =$ **1**
- $9 - (3 - 4)$ $= 9 - 3 + 4 =$ **10**
- $n - (c - d)$ = **n – c + d**
- $8x - (4y + 3x) = 8x - 4y - 3x = 8x - 3x - 4y =$ **5x - 4y**

EXERCISE 13 B

1) Simplify the following.

a) 6 - (4 + 1)
b) 8 - (7 - 2)
c) 3 - (1 + 2)
d) 13 - (4 + 3)
e) 20 - (7 - 5)
f) 9 - (3 + 3)
g) 5 - (5 - 5)
h) 80 - (60 + 7)
i) 4 - (6 - 2)
j) 11 - (3 - 1)
k) 19 - (11 - 4)
l) 16 - (12 - 9)

2) Remove the brackets and simplify where possible.

a) d - (e - f)
b) t - (h + g)
c) p - (e -w)
d) 3x - (2x - 3)
e) 8b - (5 - 3b)
f) f - (f -5)
g) (5f - b) - (4f - 5b)
h) 10 - (7y -3)
i) 8v - (2 - v)
j) (2c + 3d) - (c + 5d)
k) - (3m + n) + 8n
l) 7a - (3d - 2e)

13.2 MULTIPLYING ALGEBRAIC TERMS

Multiplying algebraic terms is similar to multiplying whole and decimal numbers.

4 × 5 = 20 and 5 × 4 = 20.
In algebra, 10c means 10 × c
3d means 3 × d and 5y means 5 × y.

Likewise, xy means x × y, bc means b × c and xyz means x × y × z.

As a standard, we usually write numbers before letters….. **5y** and not y5, though it has the same end value.

Also as a standard, we do not insert a multiplication sign (×) between a number and a letter. **5y** is preferred instead of 5 × y.

As a summary, these are the basic rules to follow when multiplying algebraic expressions.

Expressions ✓	**DO NOT WRITE** X
ab	a × b
2ac	2 × a × c
4h	4 × h or h × 4
a^2	a × a

Example 1: Simplify

a) 6 × d = 6d
b) 3n × 7 = 21n
c) 2m × 5n = 10mn
d) 5x × 3x = $15x^2$
e) m^2 × 4dm = $4dm^3$
f) 6cd × d = $6cd^2$

EXERCISE 13C

1) Simplify the following.

a) 6 × n
b) 12 × m
c) n × 7
d) m × n
e) 3 × 2w
f) 5x × 3c
g) 5rs × r
h) 9 × y^2
i) 13a × 10
j) 4ac × 2a
k) 15 × 3ab
l) 7v × 3v
m) 4c × 10c
n) 6 × $3c^2$
o) 2d × 2d
p) $(3m)^2$
q) s × s
r) 3y × 8py
s) 15 × 2d
t) 6q × 7q
u) 7cd × 8
v) 11ef × 2eg
w) a × 3w
x) 4bc × b
y) abc × ab
z) 20gh × 10

13.3 DIVIDING ALGEBRAIC TERMS

A letter can be divided by the same letter in algebra. 8 ÷ 8 = 1 and so is n ÷ n = 1. Also, 2x ÷ 2x = 1

Examples:

1) $10y \div 2 = \frac{10y}{2} = \frac{\overset{5}{\cancel{10}} \times y}{\underset{1}{\cancel{2}}} = \mathbf{5y}$

2) $12w^2 \div 3$

$= \frac{12 \times w \times w}{3} = 4 \times w \times w = \mathbf{4w^2}$

3) $\frac{1}{5}$ of $15c = \frac{1 \times 15 \times c}{5} = \mathbf{3c}$

4) $50a^3b^2 \div 10ab$

$= \frac{\overset{5}{\cancel{50}} \times \overset{1}{\cancel{a}} \times a \times a \times \overset{1}{\cancel{b}} \times b}{\underset{1}{\cancel{10}} \times \underset{1}{\cancel{a}} \times \underset{1}{\cancel{b}}}$

$= \frac{5 \times 1 \times a \times a \times 1 \times b}{1 \times 1 \times 1} = \mathbf{5a^2b}$

EXERCISE 13D

Simplify

1) a) $3c \div 3$
b) $8s \div 2$
c) $7x \div 7$
d) $24ef \div 8$
e) $\frac{1}{4}$ of $36w$
f) $\frac{16ab}{8}$
g) $\frac{14r}{7r}$
h) $72y^2 \div 9y$
i) $4d \div 4$
j) $\frac{1}{2}$ of $6m$
k) $20wx \div 4$
l) $6d \div 3d$
m) $\frac{1}{8}$ of k
n) $\frac{9b}{3}$
o) $\frac{16pqr}{8r}$
p) $40jk^2 \div 8jk$

13.4 ORDER OF OPERATIONS

When there are several operations in one calculation, the correct order must be followed to get the calculation right.

$4 + 2 \times 3$ will have two answers; 18 and 10 if the correct order is not followed. To make sure that everybody is doing the same thing, we use BIDMAS or BODMAS to help us.

↓
Brackets
Other things **or Indices** like **(3^2), (2^3), √, ……**
Division and Multiplication
Addition and Subtraction

In a calculation, do the brackets first, followed by other things like powers (3^2) and roots (√), then division and multiplication comes next – ***do them in the order they appear***, and finally, addition and subtraction - **do them in the order they appear.**

Also, you may use a **bracket** to indicate the calculation you want to perform first.

Example 1: Work out the values of

a) $4 + 2 \times 3$
Multiplication (×) comes first before addition (+). $4 + (2 \times 3) = 4 + 6 =$ **10**
Notice that 4 retained the position on the left even though we had to multiply first.

b) $4 - 2 \times 3 = 4 - (2 \times 3) = 4 - 6 =$ **-2**

c) $6 + 3 \times 5 - 4 = 6 + (3 \times 5) - 4 = 6 + 15 - 4 =$ **17**

d) $6 - 4 + 3$ …Since the operations are addition and subtraction, perform the calculation in the order they appear in the question. $6 - 4 = 2$ and $2 + 3 =$ **5**

e) $20 - 6 + 3 \times 4 = 20 - 6 + (3 \times 4) = 20 - 6 + 12 =$ **26**

f) $7 \times 10 \div 2$ …Since the operations are multiplication and division, perform the calculation in the order they appear in the question. $7 \times 10 = 70$ and $70 \div 2 =$ **35**

g) $3x + 8x \div 4 = 3x + (8x \div 4) = 3x + 2x =$ **5x**

h) $24 \div 2 + 5 - 3 \times 4 = (24 \div 2) + 5 - 3 \times 4$
$= 12 + 5 - 3 \times 4 = 12 + 5 - (3 \times 4) = 12 + 5 - 12 =$ **5**

EXERCISE 13 E

1) Calculate

a) $4 \times 3 + 10$
b) $3 + 4 \times 2$
c) $5 \times 10 - 3$
d) $5 + 18 \div 6$
e) $3 + 2 - 1$
f) $3 - 2 + 4$
g) $10 \div 5 \times 9$
h) $3 \times 4 + 5 \times 2$
i) $5 \times 8 - 3 + 7$
j) $4 + 4^2 \div 2 - 3$
k) $9 + 3 \times (17 - 7)$
l) $54 \div (20 - 5 + 3)$
m) $27 - (3^2 \div 1)$
n) $7 - 4 \div 2 + 17$
o) $35 - (7 \times 3)$
p) $4 \times 5 - 5 - 10 \div 2$
q) $5^2 + (8 - 2 + 4) - 20$
r) $2 + 3 \times (8 - 5)$
s) $6 - 4 \times 2$
t) $100 + (34 - 14) \times 10$

2) Copy the calculations below and put in **brackets** where necessary to make the answer correct.

a) $5 + 6 \div 2 = 8$
b) $21 \div 3 + 4 = 3$
c) $2 \times 7 - 4 = 6$
d) $8 - 4 \div 4 = 7$
e) $9 + 3 + 3 \times 2 = 21$
f) $40 - 8 \times 7 = 224$

3) Simplify

a) $4w \times 3 + 10$
b) $3 + 4 \times 2y$
c) $5 \times 10c - 3c$
d) $5 + 18x \div 6$
e) $3n + 2n - 1$
f) $3w - 2 + 4w$
g) $10s \div 5s \times 9$
h) $3 \times 4m + 5 \times 2$
i) $5p \times 8 - 3 + 7p$
j) $4w + 4^2 \div 2 - 3$
k) $9 + 3 \times 17x - 7$
l) $200x \div 20 - 5 + 3$
m) $27p - 3^2 \div 1$
n) $7 - 4n \div 2 + 17$
o) $35k - 7 \times 3$
p) $4 \times 5 - 5 - 10k \div 2k$
q) $8b \times 8 - 5 \times 4b$
r) $2 + 3g \times 8g - 5$
s) $6x - 4x \times 2$
t) $4v \times 3 + 4 \times 6v - 4v \times 2$

13.5 SUBSTITUTION AND FORMULAE

Substitution simply means putting numbers in place of letters in expressions and formulae and performing the calculation(s).

For example, 6x + 3y is an algebraic expression. 6x and 3y cannot be added together because they are not like terms. However, if the values of the letters x and y are known, we can then substitute (replace) them for the letters in the expression and find its value.

Remember: $p \times p \times p = p^3$, $p + p + p = 3p$, $2y = 2 \times y$, $abc = a \times b \times c$, $2x^2 = 2 \times x \times x$

Example 1: If x = 10 and y = 2, the value of the expression 6x + 3y will be
$6 \times \mathbf{10} + 3 \times \mathbf{2} = 60 + 6 = \mathbf{66}$

Example 2: If n = 3, y = ½ and t = 9, find the values of

a) t - n ⟶ $9 - 3 = \mathbf{6}$
b) n^2 ⟶ $3 \times 3 = \mathbf{9}$
c) 4t ⟶ $4 \times 9 = \mathbf{36}$
d) 4(n + 9) ⟶ $4 \times (3 + 9) = 4 \times 12 = \mathbf{48}$
e) 8y ⟶ $8 \times \frac{1}{2} = \mathbf{4}$

EXERCISE 13F

1) If m = 3, n = 10 and y = ½, work out the value of

a) $2n$
b) $10m$
c) $m + n$
d) $n - m$
e) $20y$
f) $4y + m$
g) n^2
h) $6n + \frac{1}{2}y$
i) y^2
j) $2m^2$
k) $13n$
l) $15 - n$

2) If q = 7 and c = 0, work out the value of

a) $3q$
b) $40 – q$
c) $145c + 37$
d) q^2
e) $q - q$
f) $\frac{1}{3}c + 10q$
g) $5q + c + 7$
h) $8 – 2q$
i) $3(q + c)$
j) $16 + 10c - q$
k) $91 – 6q$
l) $20 - \frac{3}{4}c$

3) Work out the perimeter of the shapes below if
i) $x = 3$ cm ii) $x = 1.5$ m

a)

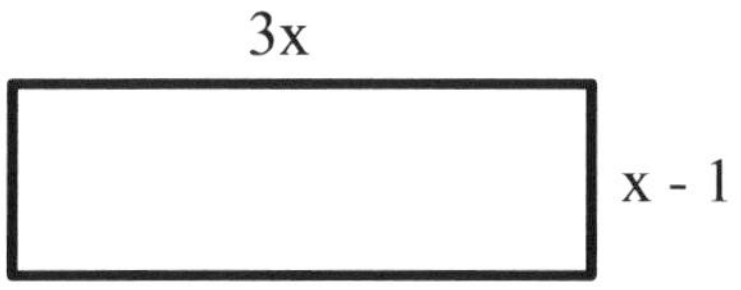

b)

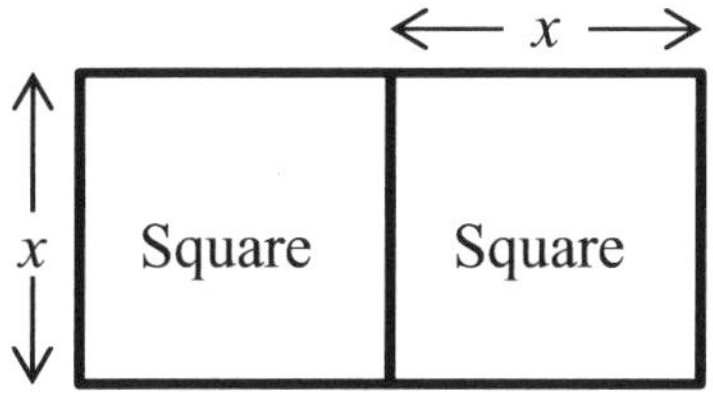

c)

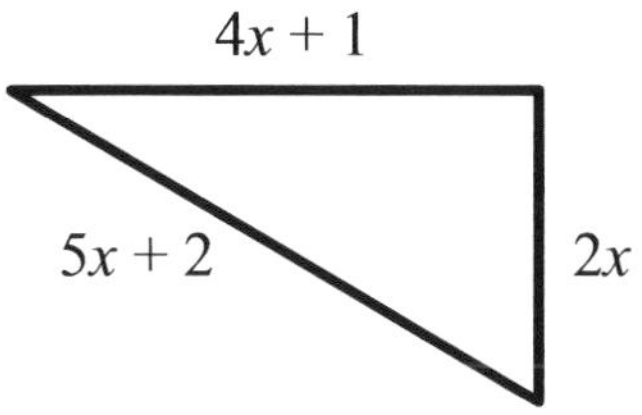

d)

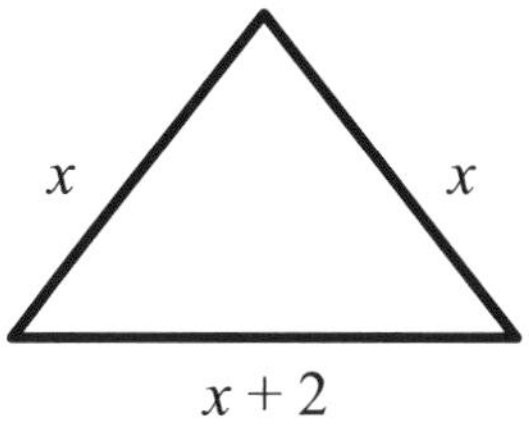

4) If c = 4 and d = 10, find the value of

a) $3(d - c)$
b) $4c + 10d$
c) $40 \div c$
d) $10(d + c)$
e) $\frac{42}{c + d}$
f) $c(c - d)$
g) $\frac{cd}{20}$
h) $\frac{100}{d}$

13.6 SUBSTITUTION INVOLVING NEGATIVE NUMBERS

In algebra, it is important to understand the operations with negative numbers. A negative number multiplied by a positive number always gives a **negative result**. A negative number multiplied by another negative number always gives a **positive result**.

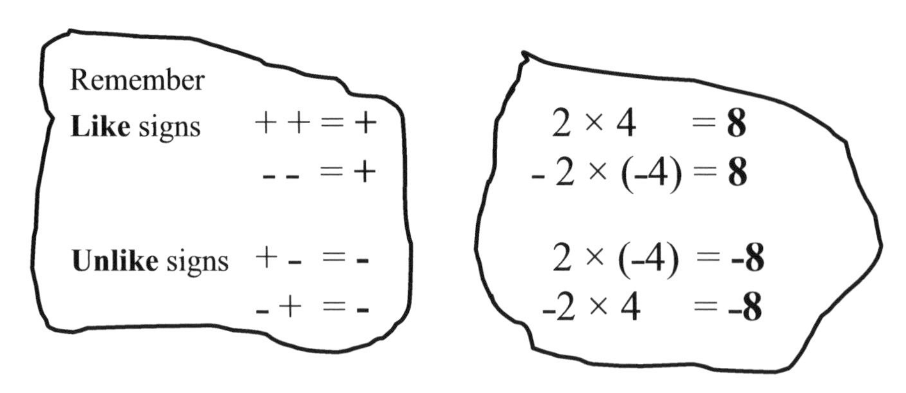

Example 1: If $x = 10$ and y = -5, find the value of

a) $x + y$ — $10 + (-5) = 10 - 5 = \mathbf{5}$
b) $2x - y$ — $2 \times \mathbf{10} - (-5) = 20 + 5 = \mathbf{25}$
c) y^2 — $(-5) \times (-5) = \mathbf{25}$
d) x^2 — $10 \times 10 = \mathbf{100}$
e) $4x^2$ — $4 \times x^2 = 4 \times x \times x = 4 \times 10 \times 10 = \mathbf{400}$

EXERCISE 13G

1) If p = -3, find the value of

a) 3p
b) 4p + 6
c) p + 1
d) 10 - p
e) 2p - 4
f) 18 - 2p
g) p - 100
h) p + p + p
i) $p^2 + 12p$
j) p - p
k) p ÷ p
l) 16p - 10

2) If $x = 3$ and y = -5 find the value of

a) $x + y$
b) $3x + 4y$
c) $6x - 2y$
d) $2x + y$
e) $x + 1 + y$
f) x - y
g) 70 - 5y
h) $3x + 10y + 20$
i) 9y - 3
j) y + 3x
k) $x^2 + y^2$
l) 19 + y - y

3) If a = -12 and b = -6, work out the value of

a) a + b
b) b + a
c) a + a + b
d) b – a
e) a – b
f) 2a
g) 3b + 5
h) 3a – 2b
i) 3 – b
j) a ÷ b
k) a ÷ 2b
l) 40 + b

13.7 LINEAR EQUATIONS

In expressions, there are **no equal** signs. Examples $5x$, $2x + 5$, $n + 6y$….
In equations, there are equal signs and letters stand for particular numbers (the solution of the equation).

Linear equations have only the first power of the unknown quantity.
Examples of linear equations are: $2x = 10$, $x + 5 = 20$…

RULES FOR SOLVING EQUATIONS

Do the same thing to both sides. You may

- Add the same number to both sides
- Subtract the same number from both sides
- Multiply both sides by the same number
- Divide both sides by the same number

Example 1: Solve $x + 5 = 9$

To solve means to find the value of the unknown. The left side of the equation MUST equal the right side. $x + 5$ must equal 9. To make the left side equal to 9, find the opposite of **+ 5**, which is -5. Subtract 5 from both sides as explained below.

So in $x + 5 = 9$
(-5) (-5) ………………….. Subtract 5 from both sides
$x + 5 - \mathbf{5} = 9 - \mathbf{5}$
$x = 4$ ✓

Check that $x = 4$ is the right answer by replacing x with 4 in the original equation. If it equals 9, then 4 is the solution. From the original equation
$x + 5 = 9$
$\mathbf{4} + 5 = 9$
$9 = 9$…. this is the confirmation needed. The left side is now equal to the right side.

Example 2: Solve $2x = 10$
This means 2 multiply by a number gives 10. We must, therefore, find that number.
Find the opposite of times 2 (×2). This is (÷ 2). We, therefore, divide both sides by 2.

$$\frac{\overset{1}{\cancel{2}}x}{\underset{1}{\cancel{2}}} = \frac{\overset{5}{\cancel{10}}}{\underset{1}{\cancel{2}}}$$

$x = 5$ ✓

Or	$\mathbf{2x}$	=	**10**
	(÷2)		**(÷2)**
	$\mathbf{x}$	=	**5**

Check your answer by multiplying 2 by 5 in the original equation. $2 \times \mathbf{5} = 10$

Example 3: Solve $\frac{x}{5} = 7$

Multiply both sides by 5 to make a whole number.

$\frac{x}{\cancel{5}} \times \cancel{5} = 7 \times \mathbf{5}$ Therefore, $x = 35$ ✓

Check: From the original equation, **35** ÷ 5 = 7

Example 4: Solve x - 3 = -7
Add 3 to both sides. $x - 3 + \mathbf{3} = -7 + \mathbf{3}$ $x = -4$ ✓
Check: - **4** - 3 = -7

Example 5: solve 40 = 8y
Divide both sides by 8 because 8y means 8 × y. 5 = y which implies that y = 5 ✓
Check: 8 × **5** = 40. Left side equals right side.

Example 6: Solve $7 = \frac{n}{3}$

A number divides 3, and the answer is 7. To solve this, find the opposite of divide by 3. It is multiply (×) by 3. So we multiply both sides by 3.

$7 \times \mathbf{3} = \frac{n}{\cancel{3}} \times \cancel{3}$ 21 = n............ Therefore, n = 21 ✓

Check: **21** ÷ 3 = 7

Example 7: Solve 2x = -8

Divide both sides by **2** $\frac{\cancel{2}x}{\cancel{2}} = \frac{\overset{-4}{\cancel{-8}}}{\cancel{2}}$ $x = -4$ ✓

Check: 2 × - **4** = - 8

Equations could be thought of as a set of balanced scales. The middle section represents the equal sign.

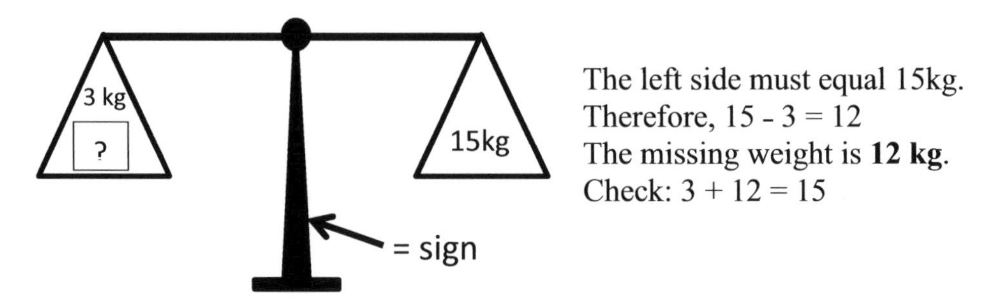

The left side must equal 15kg.
Therefore, 15 - 3 = 12
The missing weight is **12 kg**.
Check: 3 + 12 = 15

EXERCISE 13 H

1) Solve the equation below.

a) $c + 5 = 7$
b) $c - 4 = 9$
c) $c + 5 = 20$
d) $9 - x = 8$
e) $x + 11 = 22$
f) $x - 90 = 1$
g) $x + 19 = 37$
h) $w + 6 = 16$
i) $w - 6 = 16$
j) $w + 7 = -2$
k) $u + 5 = -8$
l) $6 + x = -10$
m) $u + 0 = -7$
n) $r - 49 = 4$
o) $g - 3 = -8$
p) $w - 7 = 16$
q) $13 + y = -3$
r) $56 = r - 8$
s) $4 + y = 34$
t) $n - 7 = 15$
u) $9 + e = 2$
v) $20 = x - 4$
w) $c - 20 = -10$
x) $x + 6.8 = 13.6$
y) $x - 5.5 = 2.3$
z) $u + 256 = 600$

EXERCISE 13I

1) Solve

a) $8x = 16$
b) $7x = 21$
c) $42 = 7x$
d) $63 = 9x$
e) $5x = 55$
f) $13w = 39$
g) $3w = 27$
h) $-2c = 4$
i) $-7c = 14$
j) $20 = -5w$
k) $8x = 64$
l) $-7x = 70$
m) $-x = -7$
n) $-3n = -9$
o) $\frac{n}{7} = 6$
p) $\frac{x}{3} = 10$
q) $\frac{w}{9} = -2$
r) $\frac{3}{n} = 1$
s) $-9x = -63$
t) $-n = 9$
u) $18 = -9n$
v) $-20m = -60$
w) $\frac{w}{1.5} = 3$
x) $9z = 3$
y) $67n = 670$
z) $\frac{y}{2} = -47$

13.8 SOLVING EQUATIONS IN TWO-STEPS

Example 1: Solve $2x + 5 = 11$
Solution: Get rid of (+5) by doing the exact opposite. Therefore, subtract 5 from both sides.

$2x + 5 - \mathbf{5} = 11 - \mathbf{5}$
$2x + 0 = 6$ Therefore, $2x = 6$
Divide both sides by 2 to give $\boldsymbol{x = 3}$ ✓

Check: Put $x = 3$ in the original equation
$2 \times \mathbf{3} + 5 = 11$ ……………..this gives $6 + 5 = 11$
11 = 11 (*Left hand side is equal to right hand side)*

So, the solution to the above equation is x = 3

Example 2: Solve $7x - 1 = -8$
Add 1 to both sides
$7x - 1 + \mathbf{1} = -8 + \mathbf{1}$
$7x = -7$
Divide both sides by 7
x = -1 ✓

Example 3: Solve $\frac{2x}{5} = 10$
Multiply both sides by 5 to get
$2x = 50$
Divide both sides by 2
***x* = 25** ✓
Check: $2 \times 25 = 50$, $50 \div 5 = 10$

EXERCISE 13J

1) Solve the equations

a) $2c + 5 = 13$
b) $5c + 10 = 20$
c) $5x - 10 = 10$
d) $4w + 1 = 9$
e) $5 + 3x = 8$
f) $8 + 2x = 6$
g) $7x + 5 = 26$
h) $20w - 7 = 13$
i) $8y + 3 = 51$
j) $6x - 16 = -4$
k) $3u - 7 = -10$
l) $10x - 10 = -20$
m) $4y + 6 = 14$
n) $16y - 5 = 155$
o) $\frac{4n}{3} = 8$
p) $\frac{6v}{9} = 2$
q) $7 = 6x - 5$
r) $19 = 5x + 4$
s) $8c + 9 = 49$
t) $40w - 4 = 76$
u) $3x + 20 = 44$
v) $2n - 5 = 9$
w) $8w + 2 = 66$
x) $8 + 4x = 20$
y) $13c - 6 = 46$
z) $4 + 17x = -13$

13.9 FRACTIONAL EQUATIONS

Example 1: Solve $\frac{x}{3} + 7 = 5$

Subtract 7 from both sides

$\frac{x}{3} + 7 - \mathbf{7} = 5 - \mathbf{7}$

$\frac{x}{3} = -2$

Multiply both sides by 3

$\boldsymbol{x = -6}$ ✓

It could also be viewed as a function machine

x → [÷3] → $\frac{x}{3}$ → [+7] → $\frac{x}{3} + 7$

Like all function machines, the inverse of the above operation is subtract 7, then multiply by 3.

$\frac{x}{3} + 7 = 5$

(-7) (-7)............Inverse of +7 is -7

$\frac{x}{3} = -2$

$\frac{x}{3} \times \mathbf{3} = -2 \times \mathbf{3}$ ....Inverse of ÷3 is ×3

$x = -6$

Example 2: Solve $\frac{2x}{5} = 10$

Multiply both sides by 5 to get

$2x = 50$

Divide both sides by 2

$\boldsymbol{x = 25}$ ✓

Check: $2 \times 25 = 50$, $50 \div 5 = 10$

Example 3: Solve $\frac{c}{2} = \frac{1}{4}$

Multiply both sides by 2 and 4 to make the numbers horizontal.

$\frac{c}{2} \times \mathbf{2} \times \mathbf{4} = \frac{1}{4} \times \mathbf{2} \times \mathbf{4}$

$\frac{8c}{2} = \frac{8}{4}$ ....cancelling down gives

$4c = 2$....divide both sides by 4

$c = \frac{2}{4}$

$c = \frac{1}{2}$ ✓

EXERCISE 13K

1) Solve the equations below

a) $\frac{n}{5} + 1 = 4$

b) $\frac{n}{4} - 3 = 7$

c) $\frac{n}{2} + 5 = 10$

d) $8 = \frac{40}{n}$

e) $\frac{n}{-3} = 3$

f) $\frac{n}{10} - 8 = 22$

g) $\frac{n}{-7} = 9$

h) $\frac{3n}{5} = 3$

i) $6 + \frac{n}{7} = 8$

j) g) $\frac{n}{8} + 8 = 23$

k) $\frac{4n}{3} = 4$

l) $\frac{n}{6} - 2 = 11$

m) $8 = \frac{1}{4}n$

n) $\frac{n}{6} = \frac{2}{3}$

o) $\frac{n}{5} = \frac{8}{20}$

p) $\frac{3n}{6} = \frac{2}{5}$

q) $\frac{n}{-10} = \frac{2}{5}$

r) $\frac{n}{6} + 5 = 2$

s) $15 + \frac{n}{5} = 16$

t) $\frac{20}{n} = 5$

u) $\frac{35}{n} = 7$

v) $3 - \frac{3}{n} = 0$

w) $\frac{1}{2} = \frac{4n}{8}$

x) $\frac{n}{6} = \frac{2}{3}$

y) $20 = 6 + \frac{n}{2}$

z) $\frac{n}{4} - 1 = -3$

EXTENSION QUESTIONS

1) The rectangle below has a perimeter of 26 cm. Work out the length of the longest side of the rectangle.

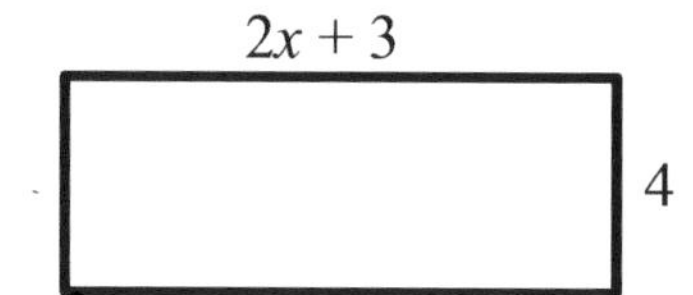

2) Work out the area of the rectangle in question 1 above.

3) Solve the equations below.

a) $\frac{d}{5} - 3 = -6$

b) $\frac{4}{5}w = -8$

c) $\frac{7-n}{3} = 2$

Chapter 13 Review Section

Assessment

1) Simplify by removing brackets

a) $8 + (13 - 5)$ b) $10 - (6 + 7)$ c) $(4 + 2) + (13 - 4) - 8$

......................3 marks

2) Write without brackets and simplify where possible.

a) $(c- d) + (e - f)$ b) $9w + 5n + (n + 4w)$ .... 2 marks

3) Simplify

a) $(5f - b) - (4f - 5b)$ b) $7a - (3d - 2e)$

c) $8 \times w^2$ d) $30a^3b^2 \div 10ab$ ....4 marks

4) Work out the values of

a) $6 + 2 \times 3$ b) $5 \times 8 - 3 + 7$ c) $4^2 + (5 - 2 + 4)$ .... 3 marks

5) If $x = 5$ and $y = -2$, work out the value of

a) $x + y$ b) $x - y$ c) y^2 d) $7 - y$ .... 4 marks

6) Solve the equations

a) $n + 8 = 10$ e) $x + 4 = -5$

b) $2 + n = 13$ f) $3x = -3$

c) $n - 7 = 20$ g) $17 - y = 3$

d) $6x = 24$ h) $x - 14 = -4$

.... 8 marks

7) Solve the equations

a) $\frac{n}{5} = 3$ e) $\frac{x}{20} = -3$

b) $\frac{x}{7} = 10$ f) $13 + \frac{x}{3} = 19$

c) $\frac{2x}{7} = 2$ g) $11 - \frac{y}{6} = 3$

d) $\frac{x}{9} = -5$ h) $5x - 4 = 11$ .... 16 marks

14 Volume of Prisms

This section covers the following topics:

- Volume of Prisms
- Appropriate units for volume
- Understanding Capacity

LEARNING OBJECTIVES

By the end of this unit, you should be able to:

a) Calculate the volume of Cuboids and Cubes
b) Find volume by counting cubes
c) Find the volume of triangular prisms

KEYWORDS

- Volume
- Prism
- Capacity
- Cubic metre
- Cubic centimetre

14.1 UNDERSTANDING VOLUME

Area and volume are sometimes misused. It is vital that students understand the correct meaning of volume. The volume of a solid (3D shape) is the amount of space it fills.

Unfortunately, there is no equipment to measure volume directly. Hence, we have to work out the volume of solids given their dimensions.

Volume is measured in cubes, and the units used depend on the initial measurements for the dimensions. The units could be in a cubic centimetre (**cm^3**), cubic metre (**m^3**) or other appropriate units.

14.2 VOLUME BY COUNTING CUBES

If a solid is made up of cubes, the volume can be found by counting the cubes.

Example 1: The shape below is made up of 7 cubes.

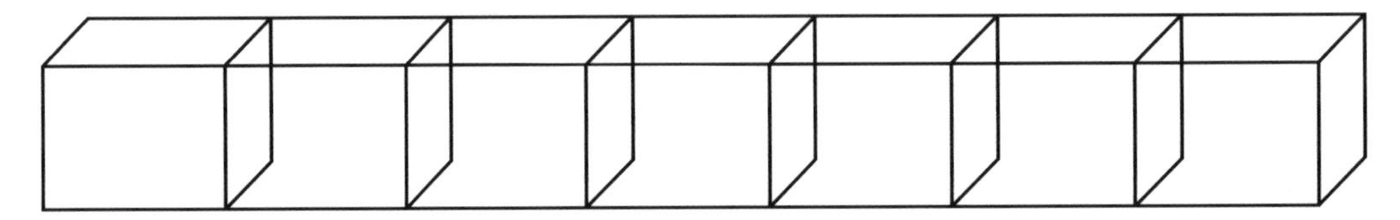

If each cube has a volume of 1 cm^3, the total volume will be **7 cm^3** by counting the number of cubes.

Example 2: The shape below was built with 1 cm3 cubes. Work out the volume.

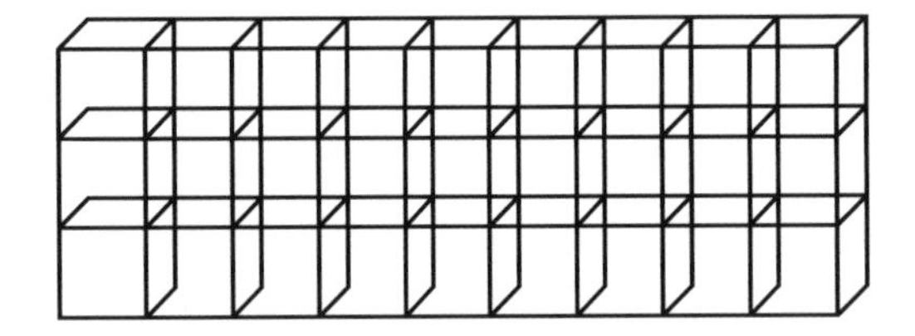

By counting, there are 27 cubes. Therefore, the volume of the solid is **27 cm^3**

CONVERTING UNITS OF VOLUME

The basic unit of volume is the cubic metre, **m^3**.

1 cm = 10 mm
1 cm^3 = (10 × 10 × 10) mm^3
1 cm^3 = 1 000 mm^3

1 m = 100 cm
1 m^3 = 100 × 100 × 100
1 m^3 = 1 000 000 cm^3

EXERCISE 14 A

1) Each cube has a volume of 1 cm^3. Find the volume of each shape/solid.

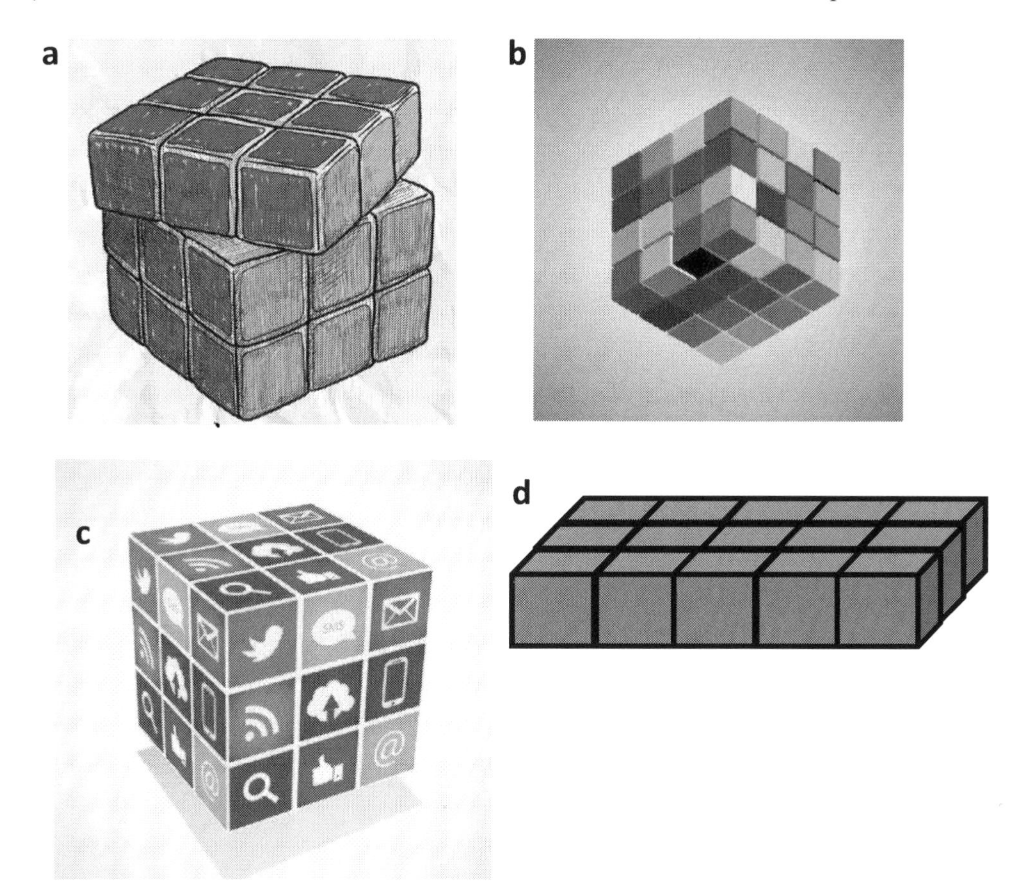

14.3 VOLUME OF A CUBOID BY USING A FORMULA

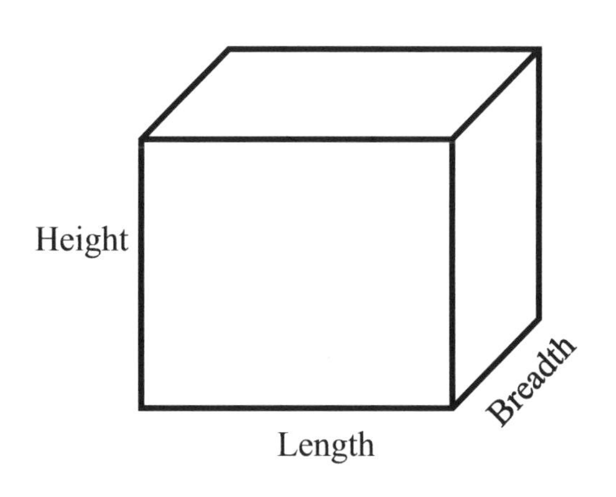

Volume = Base area × height
= **Length × breadth × height**

Example 1: Find the volume of the cuboid

Volume = $5 \times 3 \times 6$
= $\mathbf{90\ cm^3}$

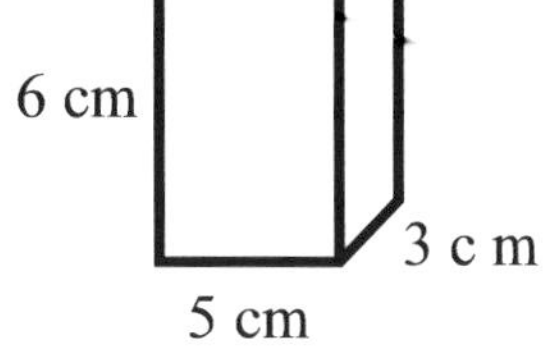

Example 2: Calculate the volume of a rectangular solid with the following dimensions: 2 cm by 4 cm by 10 cm.
Volume: $2 \times 4 \times 10 = \mathbf{80\ cm^3}$

Example 3: The diagram represents a cornflakes pack.

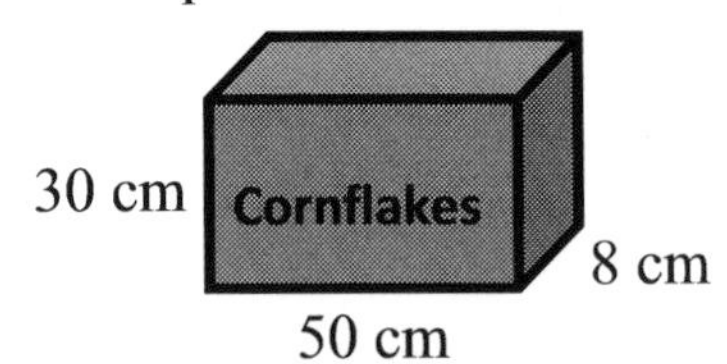

a) Calculate the volume of the pack.
b) 10 of these corn flakes packs are stacked in a carton with base 12 cm by 8 cm, and they fit exactly. Calculate
i) the volume of the carton
ii) the height of the carton

Solution:
a) Volume = $50 \times 8 \times 30 = \mathbf{12\ 000\ cm^3}$
b) i) $10 \times$ volume of one cornflakes pack = volume of the carton.
$= 10 \times 12\ 000 = \mathbf{120\ 000\ cm^3}$
ii) Volume of carton = $12 \times 8 \times h$
$120\ 000 = 12 \times 8 \times h$
$\frac{120\ 000}{12 \times 8} = h \longrightarrow h = 1\ 250$
Therefore, height of carton = **1 250 cm**

Example 3: The diagram is the net of a cube. Work out its volume.

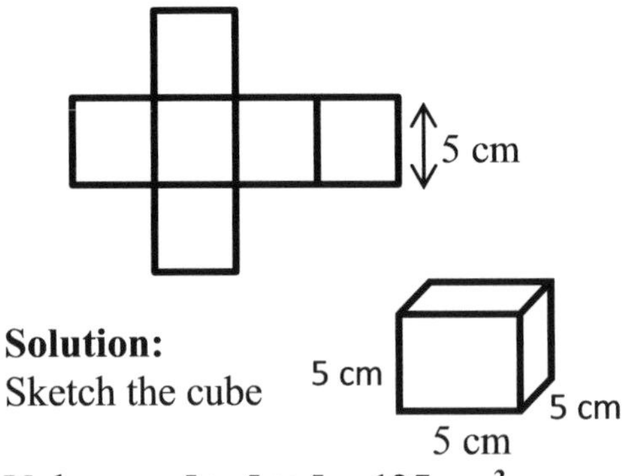

Solution:
Sketch the cube

Volume = $5 \times 5 \times 5 = \mathbf{125\ cm^3}$

CAPACITY AND VOLUME

The measure of space inside a container is its **capacity**.

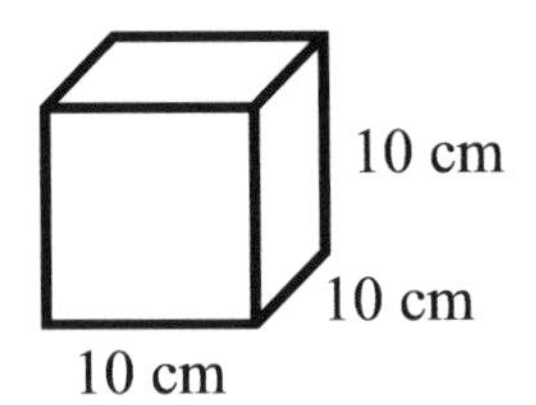

The volume of the cube above is
$10\ cm \times 10\ cm \times 10\ cm = 1\ 000\ cm^3$

The unit of capacity is the **Litre.**

1 litre = $1\ 000\ cm^3$ and
$1\ m^3 = 1000$ litres

1 millilitre (ml) = 1 cubic centimetre ($1\ cm^3$)

1 kilolitre (kl) = $1\ 000,\ 000\ cm^3$

Example 1:

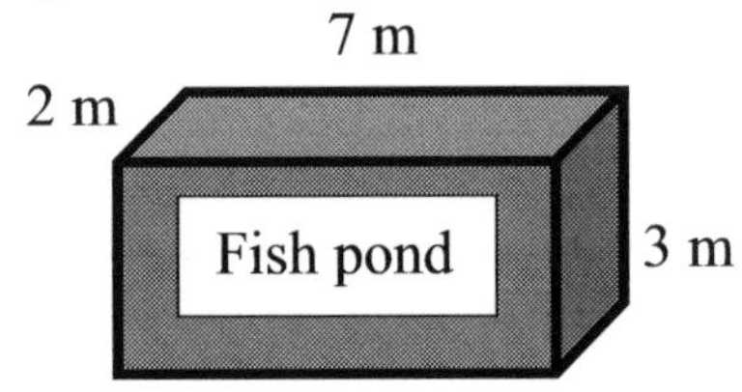

How many litres of water does the fish pond hold?

Solution:

Volume of pond = $7 \times 2 \times 3 = 42\ m^3$

From $1\ m^3 = 1\ 000$ litres, the capacity would be $42 \times 1\ 000 =$ **42 000 litres**

EXERCISE 14 B

In questions 1 to 4, work out the volume of each cuboid. All lengths in cm.

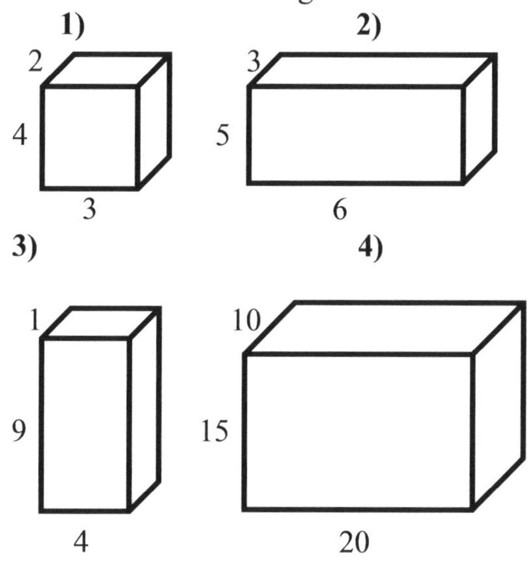

5) A water tank is in the shape of a cuboid with the dimensions 3 m by 4m by 5m.

a) Work out the volume of the tank.

b) One litre of water = 1 000 cm^3

i) How many litres of water will the tank hold?

ii) If 2000 litres of water is consumed every day, how many days will a full tank of water last?

6) How many cubic centimetres (cm^3) are in a cube of edge 4 cm?

7) A fish pond is in the form of a cuboid. On a Friday, the pond was filled with water to a depth of 6 m. Work out the volume of water in the fish pond if the other dimensions are 4m and 11 m.

8) A container is 30 cm wide, 2 m long and 15 cm deep. How many 10 cm by 5 cm by 2 cm cubes can be fitted in the container?

9) Complete the table below.

Volume	Length	Breadth	Height
	2 cm	4 cm	5 cm
90 cm^3		6 cm	3 cm
	7 m	3 m	10 m
192 cm^3	4 cm	8 cm	

10) The end face of a concrete block is rectangular 80 cm by 90 cm. The length of the block is 35 m long.

a) Work out the volume of the block

b) 1 m^3 of concrete has a mass of 3 tonnes. Work out the mass of the block

11) Work out the value of the missing lengths.

a)

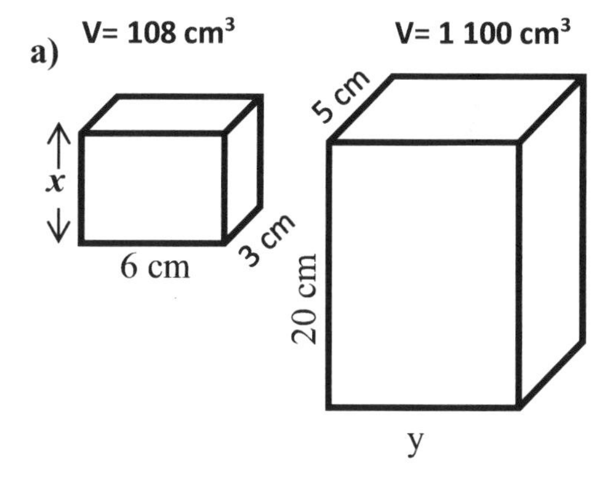

12) A hotel room is 3.5 m high with a volume of 21 m^3. Work out the area of the floor of the hotel room.

14.4 VOLUME OF PRISMS

Refer to section 8.3 for definition and different types of prisms. Volume of any prism is the ***area of the cross section (end face) multiplied by the length (distance between the end faces)***

Example 1: Work out the volume of the right-angled triangular prism below.

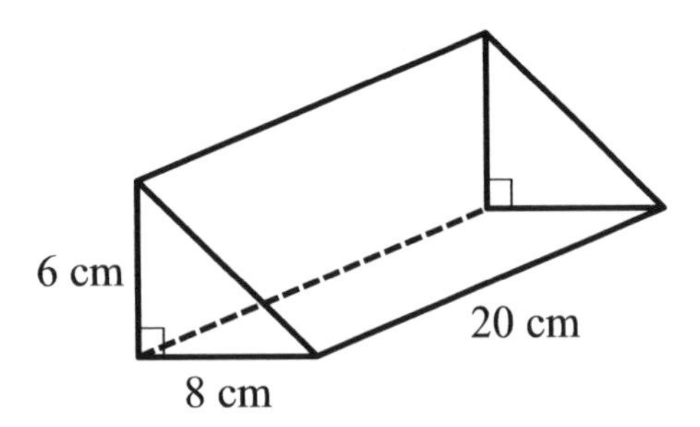

The cross section of the triangular prism is a triangle. Therefore, we work out the area of the triangle.

Area of triangle $= \frac{1}{2} \times 8 \times 6 = 24 \text{ cm}^2$
Volume of the prism $= 24 \times 20$
$= \mathbf{480 \text{ cm}^3}$ ✓

Example 2: The volume of the triangular prism is 72 m^3. Calculate the missing length, x.

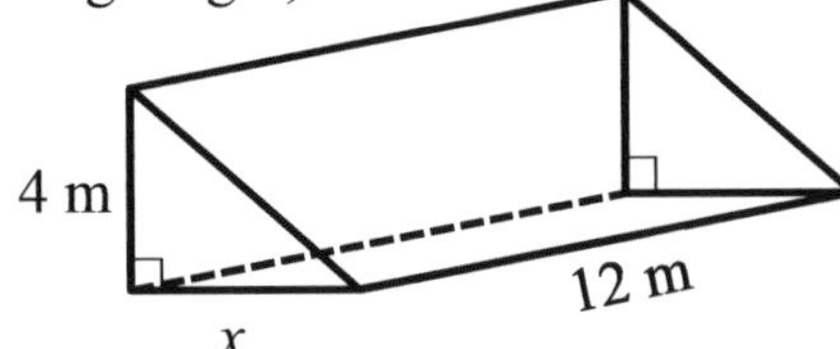

Volume = area of end face × 12
$72 = \frac{1}{2} \times x \times 4 \times 12$
$72 = 24x$
$x = \frac{72}{24} = 3$
The missing length is **3 m** ✓

EXERCISE 14C

Work out the volume of each prism. All lengths are in cm.

1)

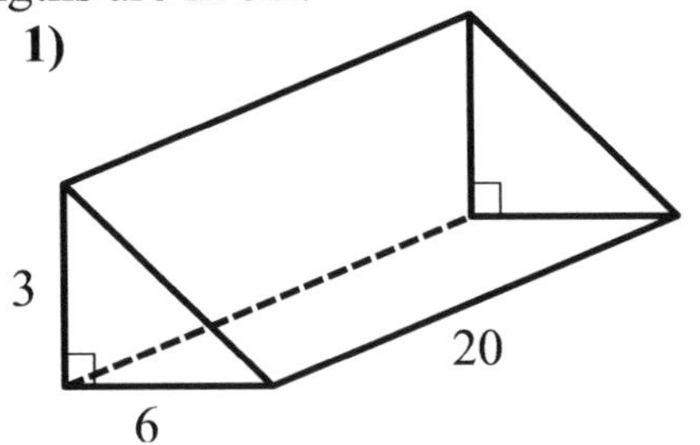

2)

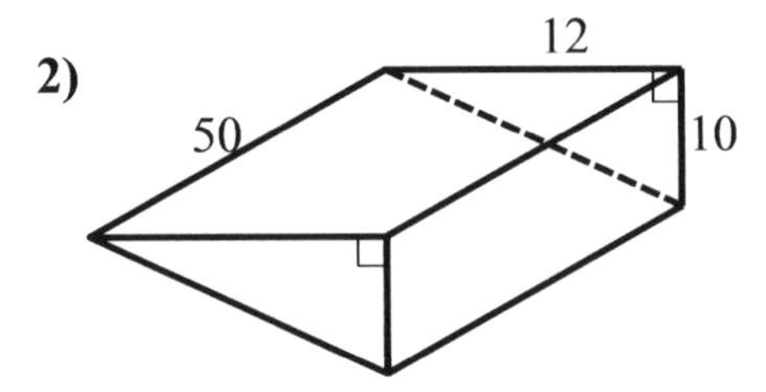

3)

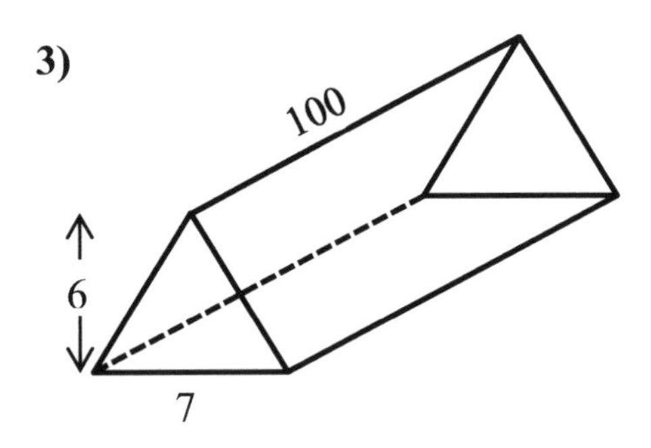

4) The volume of the triangular prism is 720 m^3. Calculate the missing length, y.

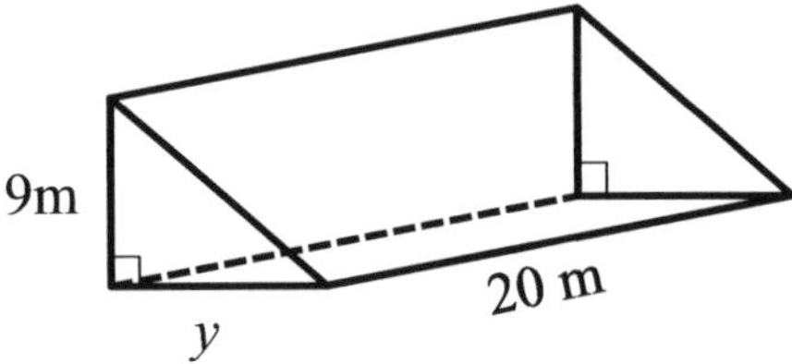

5) Work out the volume of the square-based prism. All lengths are in cm.

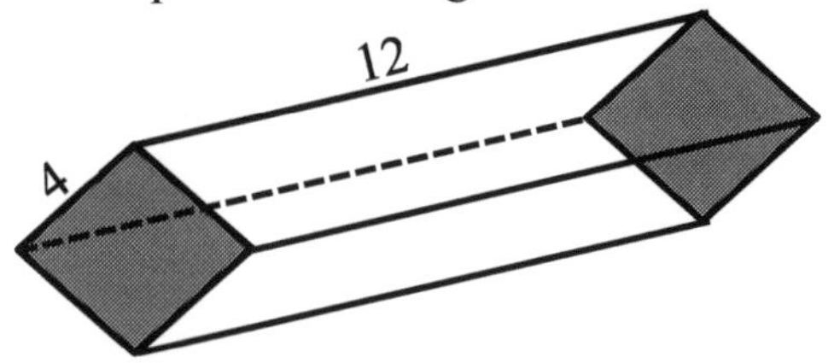

Chapter 14 Review Section

Assessment

1) Each cube has a volume of 1 cm^3. Find the volume of the cuboid.

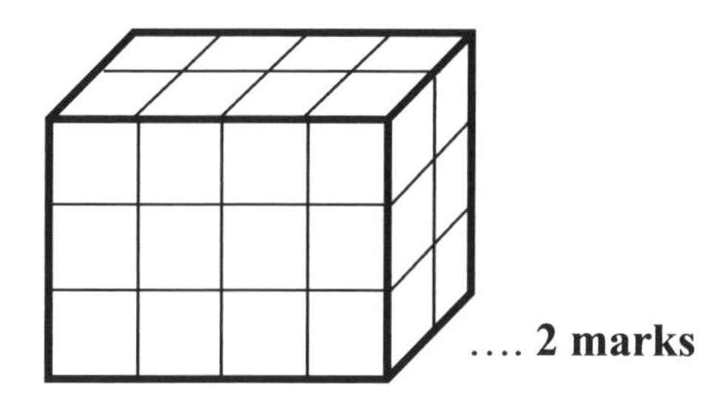

.... **2 marks**

2) Calculate the volume of a rectangular box with the dimensions 3 cm by 7 cm by 7 cm. .... **2 marks**

3) The diagram shows the net of a cube. Work out its volume.

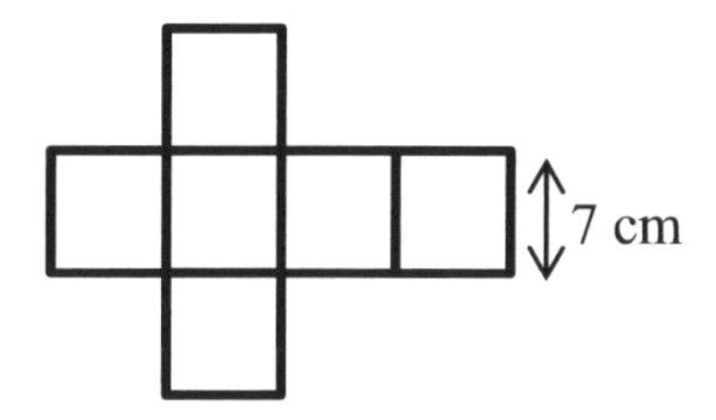

.... **2 marks**

4) Calculate the volume of the cuboid.

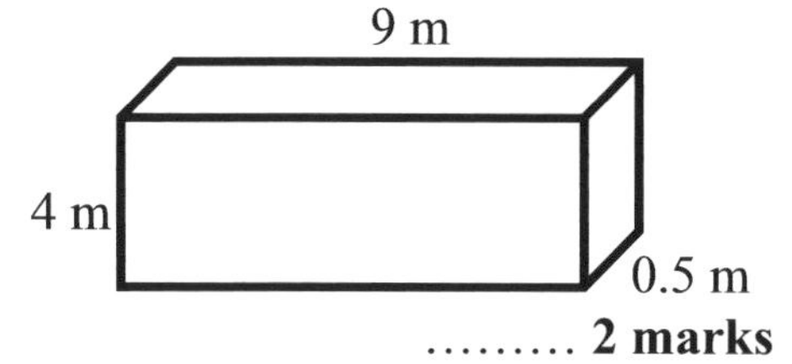

.... **2 marks**

5) Work out the area of each prism. All lengths are in cm.

a)

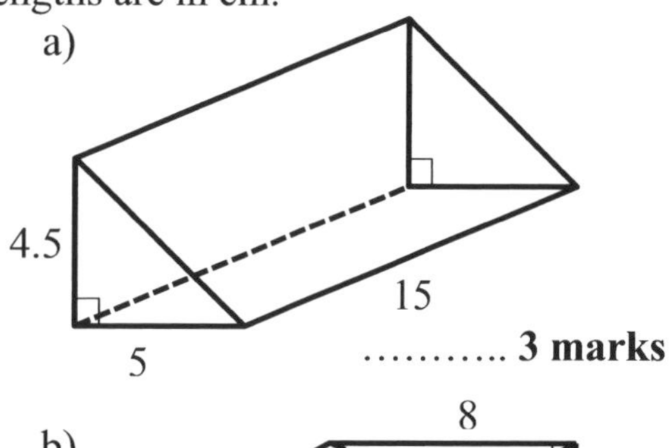

.... **3 marks**

b)

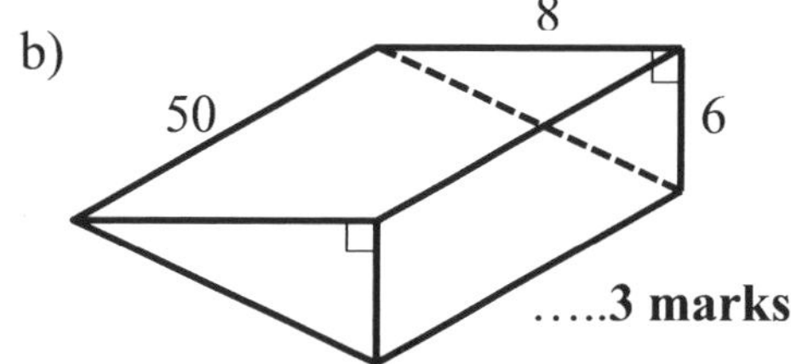

.....**3 marks**

c)

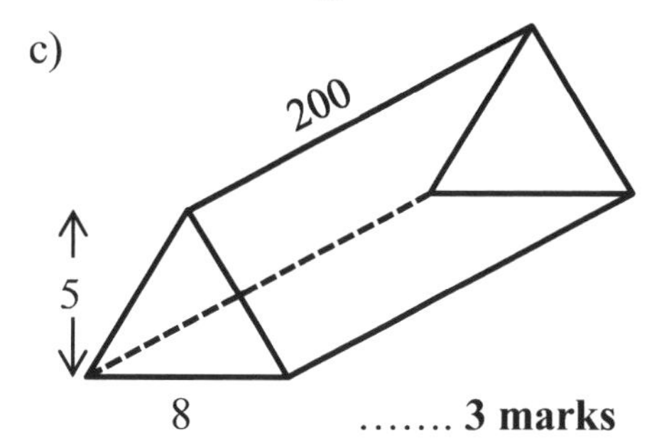

.... **3 marks**

6) The volume of the triangular prism is 270 m^3. Calculate the missing length, w.

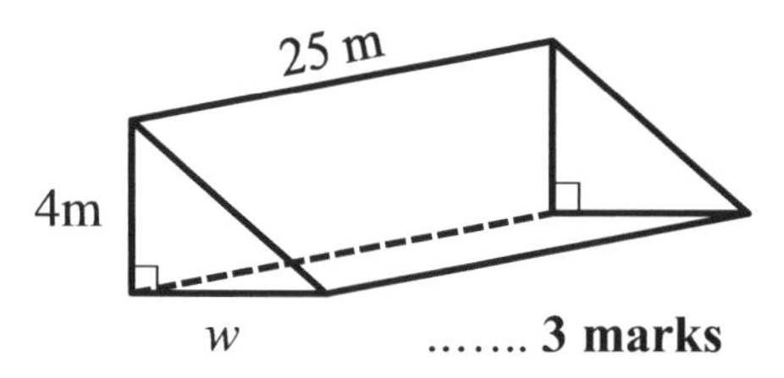

.... **3 marks**

15 Angles 2

This section covers the following topics:

- Geometrical properties of 2D shapes
- Angles in 2D shapes
- Polygons
- Circles

LEARNING OBJECTIVES

By the end of this unit, you should be able to:

a) Understand and use properties of lines and angles
b) Calculate angles on a straight line
c) Calculate angles at a point
d) Work out vertically opposite angles
e) Understand geometrical properties of triangles, quadrilaterals and polygons
f) Calculate angles in parallel lines
g) Calculate angles in triangles and quadrilaterals
h) Understand corresponding and alternate angles
i) Understand the properties of circles

KEYWORDS

- Angles
- Parallel line
- Corresponding and Alternate angles
- Polygons
- Circles
- Quadrilaterals

15.1 LINES AND ANGLES

Angles on one side of a straight line always add to 180 degrees.

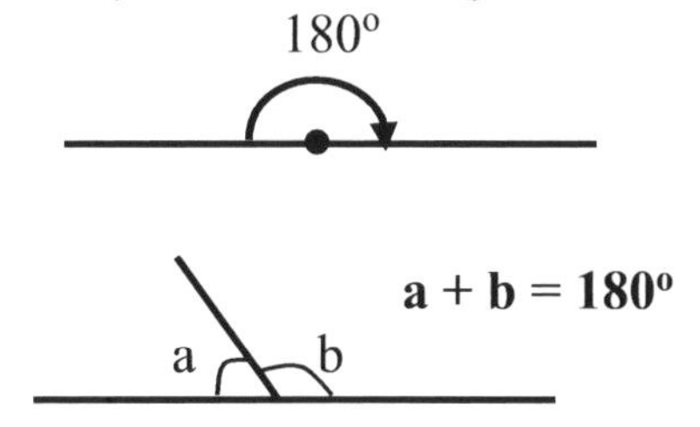

a + b = 180°

Example 1: Calculate the value of the missing angle.

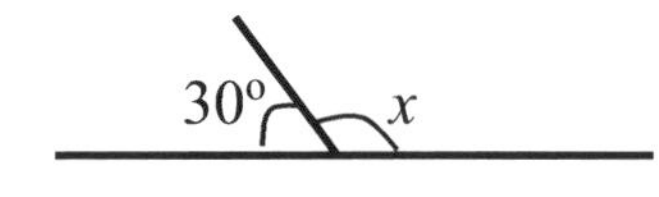

$30 + x = 180$......... $x = 180 - 30 =$ **150°**

Check: 30° + 150° = 180 ...angles on a straight line

Example 2: work out the missing angle.

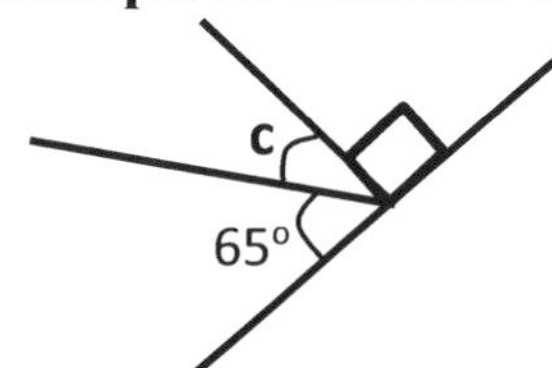

Since angles on a straight line add to 180°,

$65° + c + 90° = 180°$

$155° + c = 180°$

$c = 180 - 155 =$ **25°**

Check: $65° + 25° + 90° = 180°$

Example 3: Work out the value of x.

$x + 3x + 2x = 180°$

$6x = 180°$

$x = 180 \div 6$

$x = 30°$

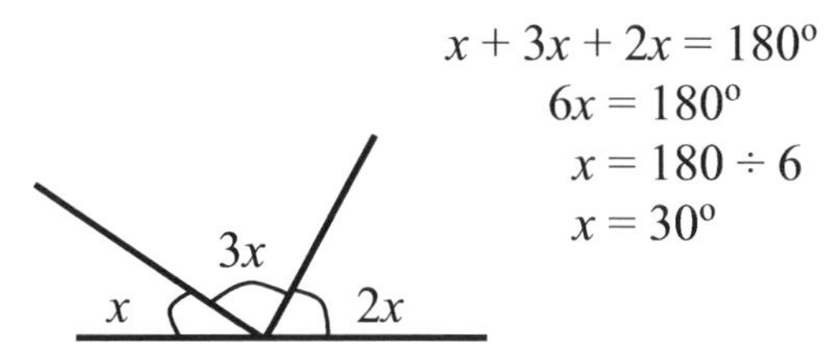

15.2 ANGLES AT A POINT

The sum of angles at a point is 360°.
Note: There are 360° in a ***full*** turn.

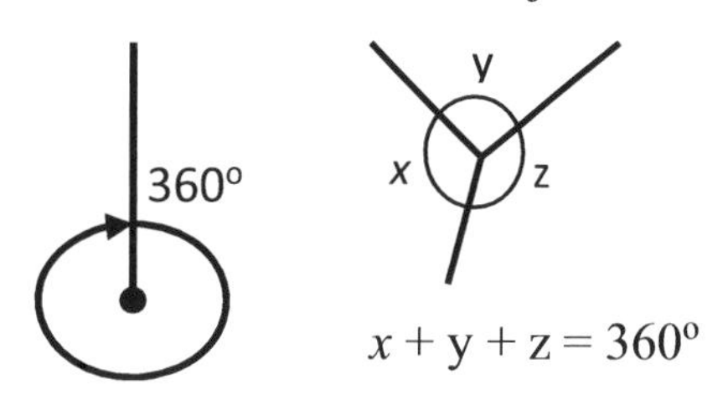

$x + y + z = 360°$

Example 1: Work out the missing angle

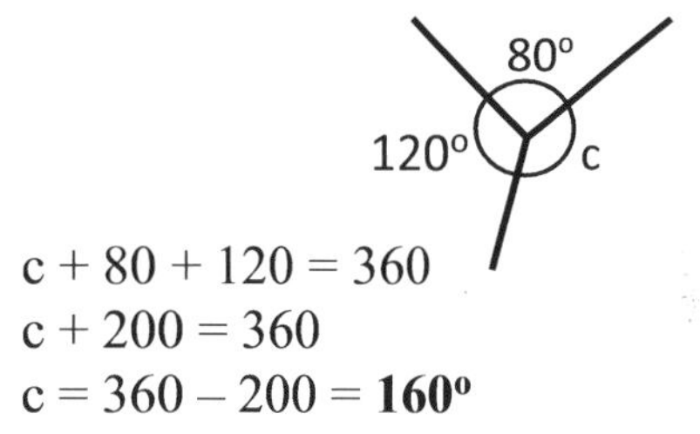

$c + 80 + 120 = 360$

$c + 200 = 360$

$c = 360 - 200 =$ **160°**

Example 2:

$w + w + w = 360°$

$3w = 360$

$w = 360 \div 3$

$w = 120°$

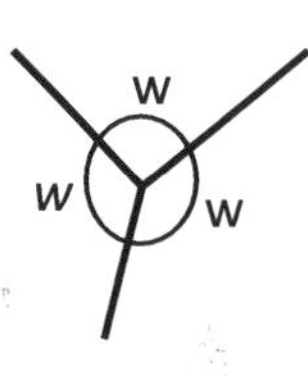

15.3 VERTICALLY OPPOSITE ANGLES

Vertically opposite angles are equal.

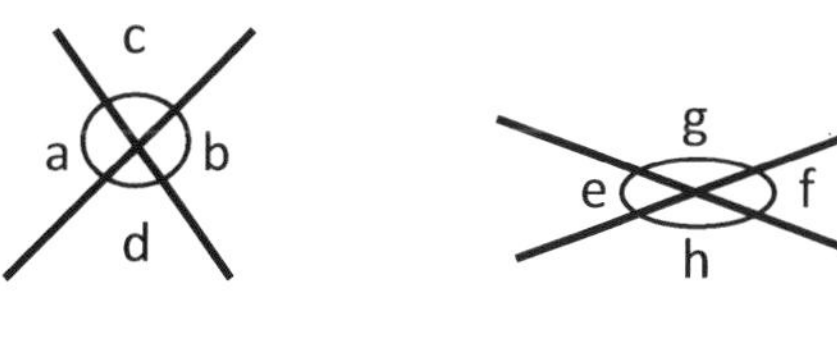

a = b, c = d

e = f and g = h

Example 1: Work out the missing angle

x = **40°**....vertically opposite angles

Example 2 Work out the missing angles

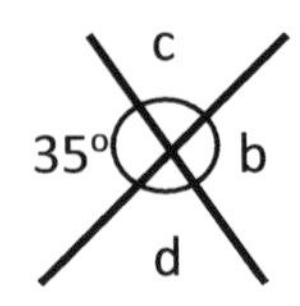

b = **35°** (vertically opposite angles)
c = 180 – 35 = **145°**
(Angle on a straight line)
d must be **145°** (Vertically opposite to angle c)

Example 3 Work out the missing angles

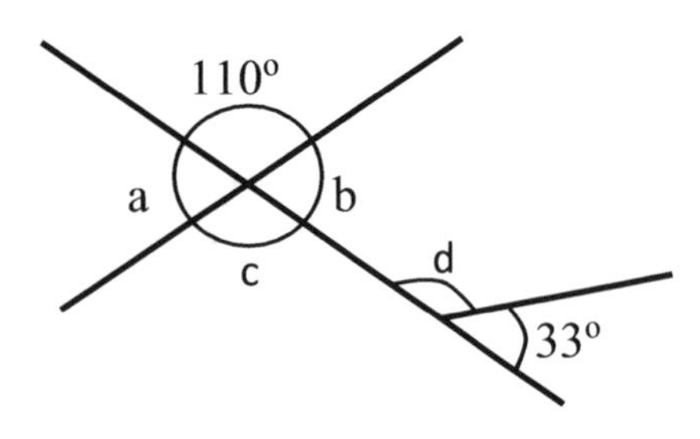

c = 110°......Vertically opposite angles
b = 180 – 110
= 70°........angle on a straight line
a = b = 70°....Vertically opposite angles
d = 180 – 33 = 147°
........angle on a straight line

EXERCISE 15A

1) Calculate the value of the angles marked with letters in each diagram.

a) 150° a

b) b 170°

c)

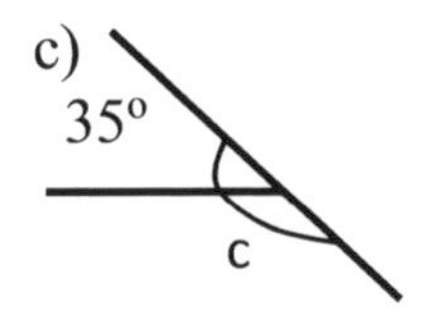

d) d 46°

e)

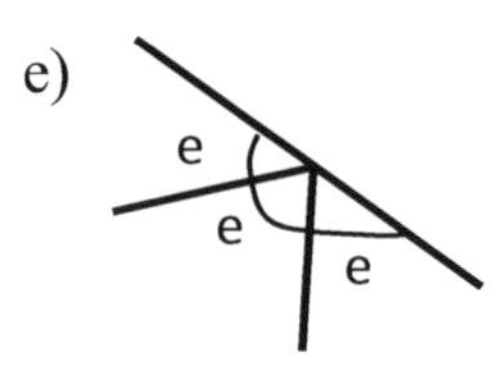

f)

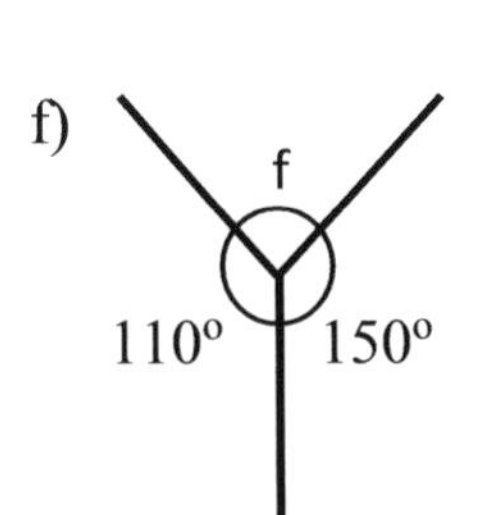

g)

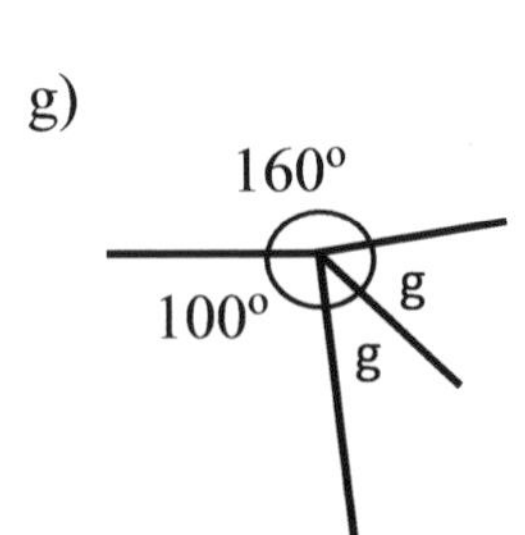

h)

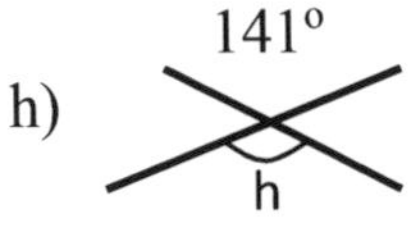

i)

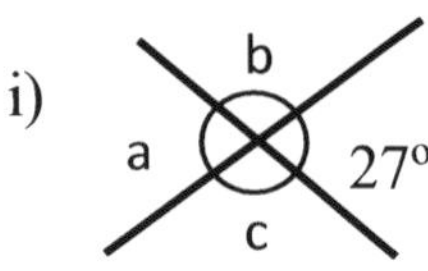

j)

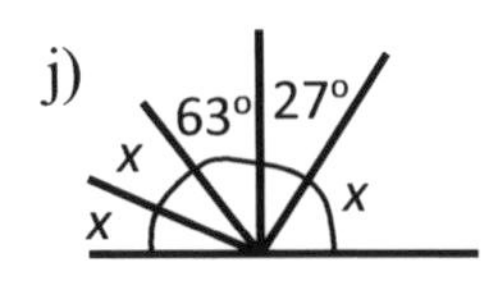

k)

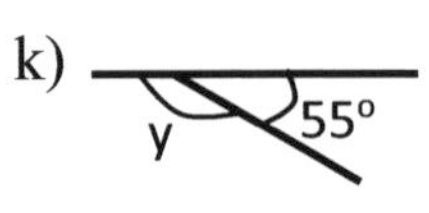

l)

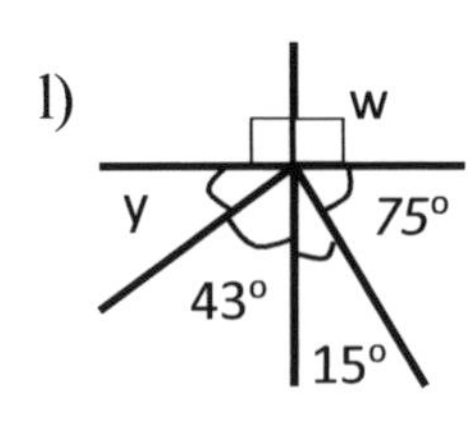

m)

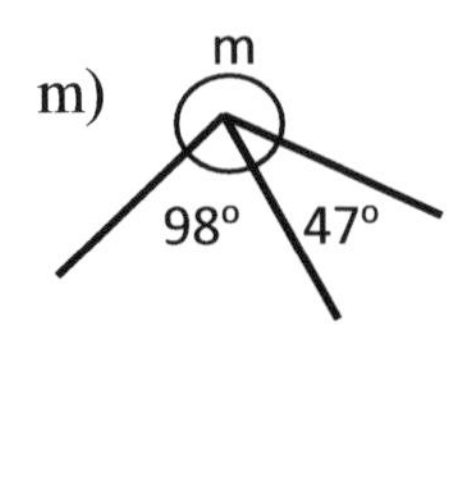

n)

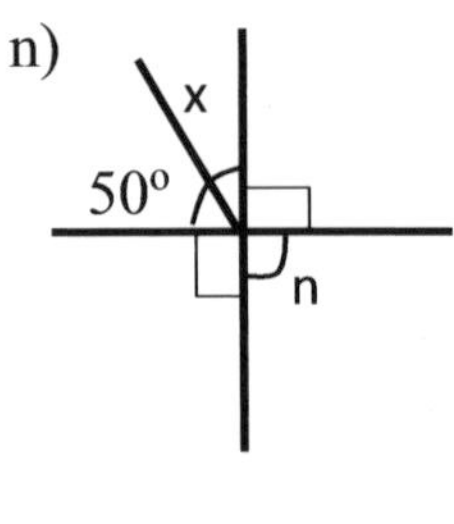

15.4 TRIANGLES AND ITS PROPERTIES

A triangle is a two-dimensional shape made up of three sides and three angles. All the interior (inside) angles in a triangle add to **180°**.

There are four types of triangles.

1) Right-angled triangle
2) Isosceles triangle
3) Equilateral triangle
4) Scalene triangle

RIGHT-ANGLED TRIANGLE

A right-angled triangle has one angle that is 90°. It is usually marked as a square.

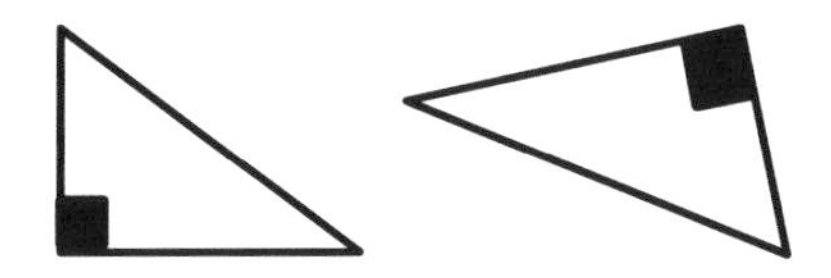

A right-angled triangle has **no line of symmetry** but has rotational symmetry of **order 1**.

However, if the right-angled triangle is also isosceles, there is only **one** line of symmetry.

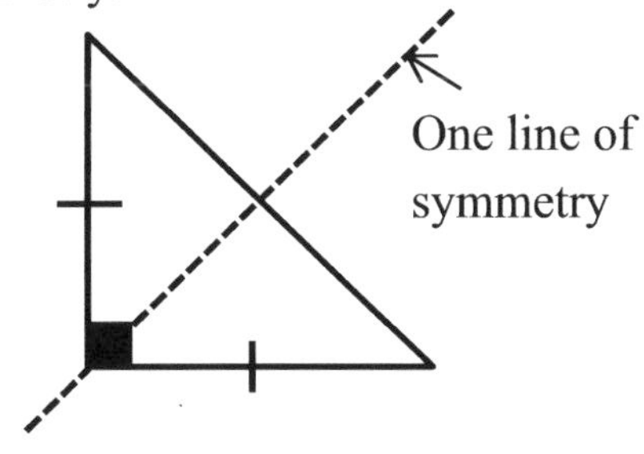

Right-angled isosceles triangle

ISOSCELES TRIANGLE

An isosceles triangle has two equal sides and two equal *base* angles. It has **one line of symmetry** and a rotational symmetry of **order 1**.

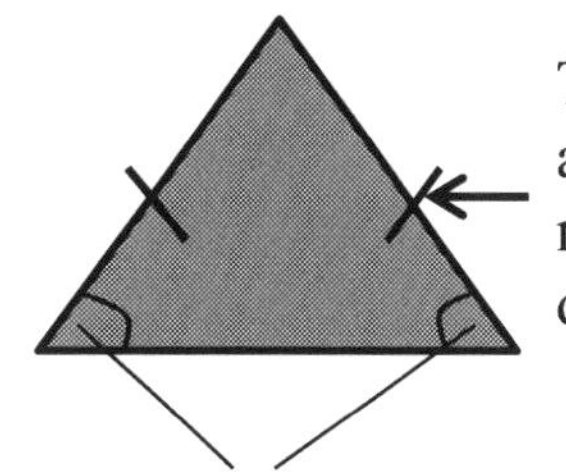

EQUILATERAL TRIANGLE

An equilateral triangle has three equal sides and three equal angles. All the three angles are 60° each.

It has **three lines of symmetry** and rotational symmetry of **order 3**.

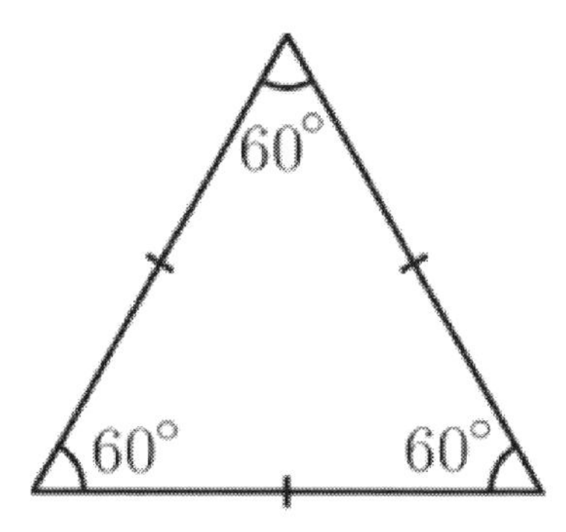

Most times, three dashes are used on the three sides indicating three equal sides.

SCALENE TRIANGLE

A scalene triangle has three unequal sides and angles

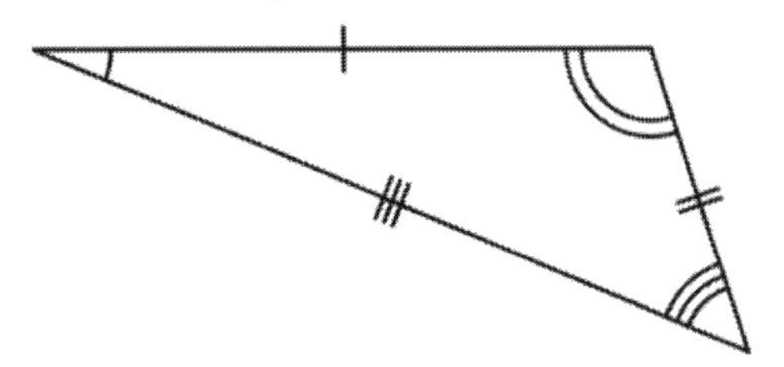

It has **no line of symmetry** and rotational symmetry of **order 1**.

Remember: Rotational symmetry is the number of times a shape fits exactly onto itself in a complete turn (360°).

CALCULATING ANGLES IN TRIANGLES

The interior (inside) angles in any triangle add up to 180°.

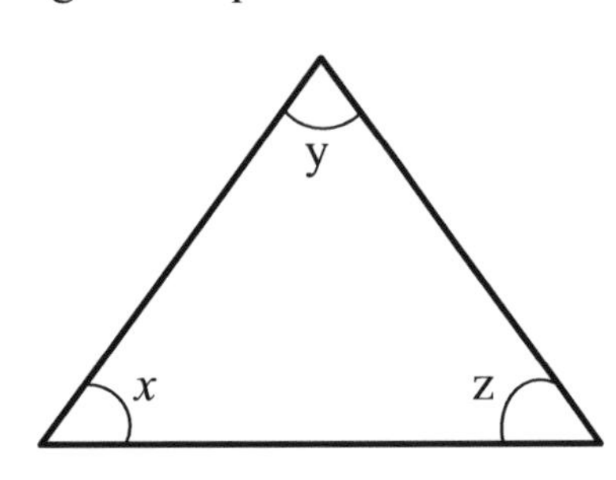

$x + y + z = 180°$

Once two angles are known, the third angle can be calculated.

Example 1: Calculate the missing angle

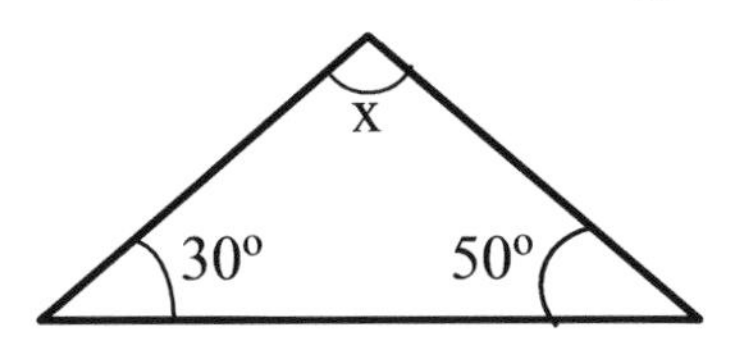

$x + 30 + 50 = 180°$
$x + 80 = 180°$
$x = 180 - 80 = \mathbf{100°}$

Check: 100 + 30 + 50 = 180

Example 2: Calculate the missing

angles $A\hat{C}B$ and $C\hat{A}B$

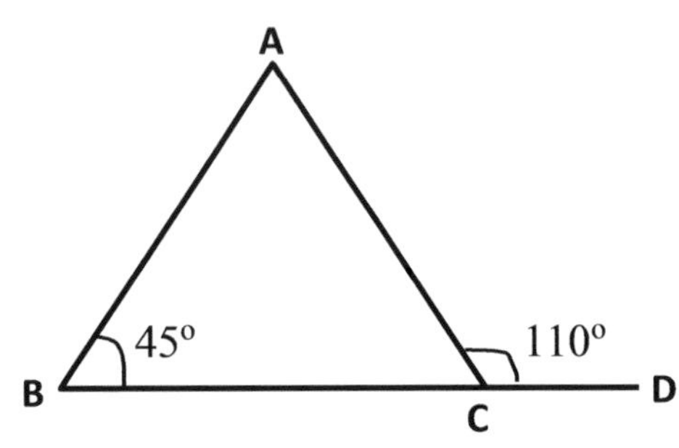

$A\hat{C}B = 180 - 110 = \mathbf{70°}$
(Angles on a straight line)

$C\hat{A}B = 180 - (45 + 70) = \mathbf{65°}$
(Angles in triangles add to 180°)

Example 3: Calculate the missing angles.

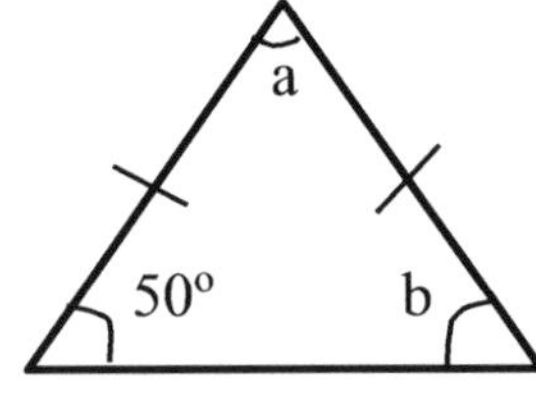

Triangle is isosceles; therefore the base angles must be the same. b = **50°**

Angles in a triangle add to 180°, so angle a = 180 – (50 + 50) = **80°**

EXERCISE 15 B

All diagrams are not accurately drawn.
Calculate the size of angles marked by letters in the diagrams below.

1)

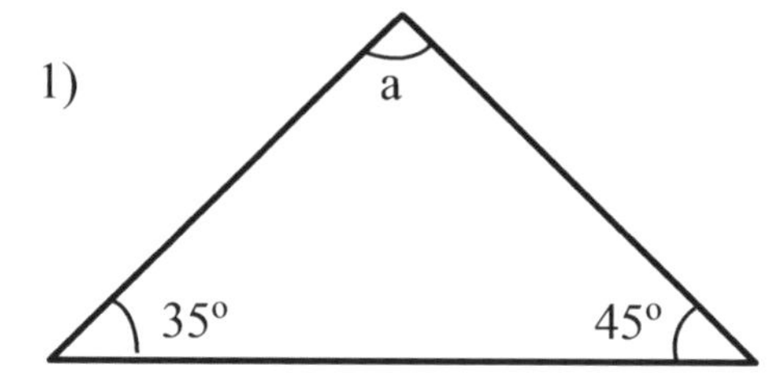

2)

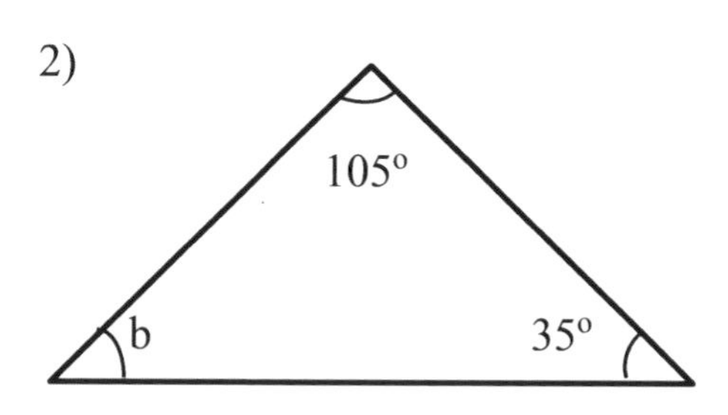

3)

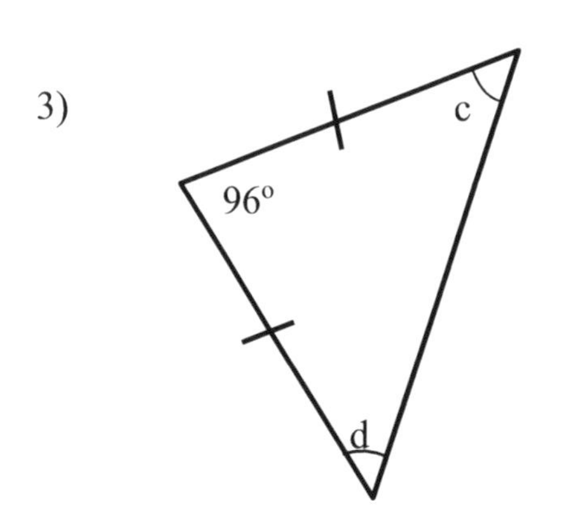

4)

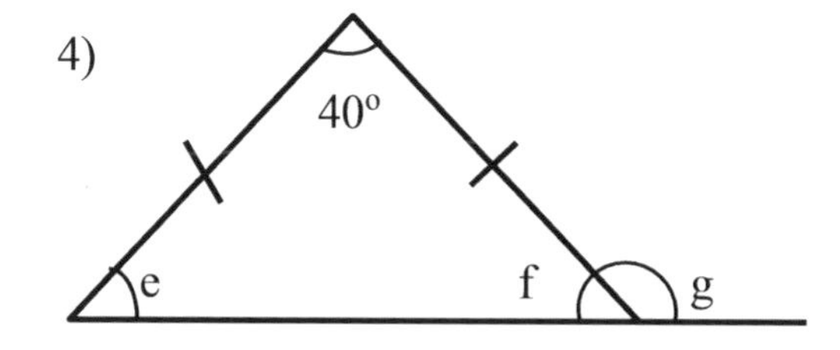

5)

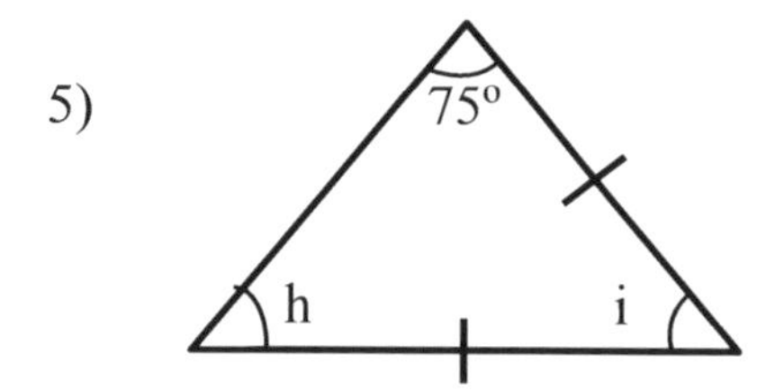

6)

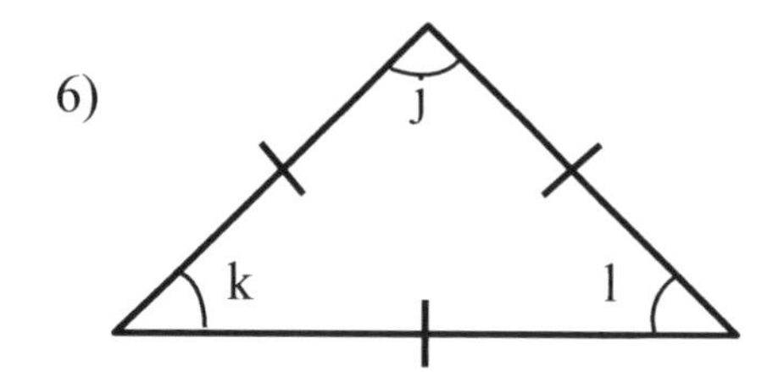

7)

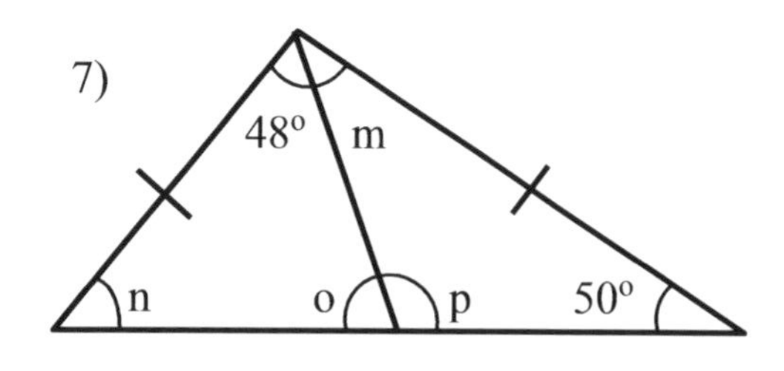

8)

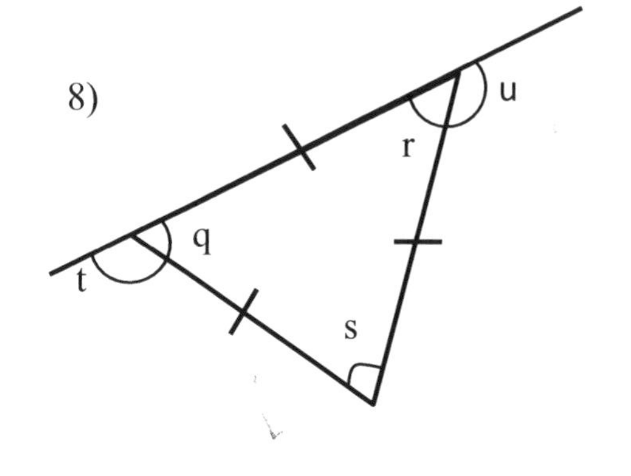

9)

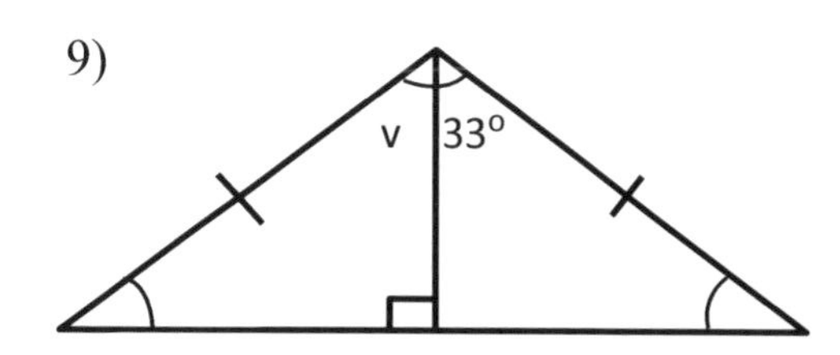

10) Indicate whether the triangles below are isosceles, right-angled, equilateral or scalene. *Give a reason for your answer.*

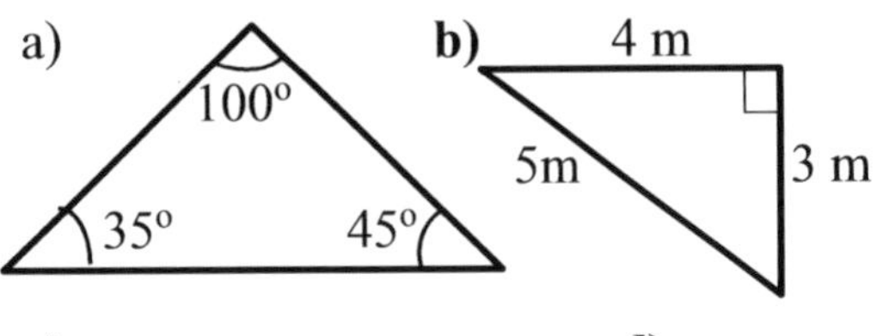

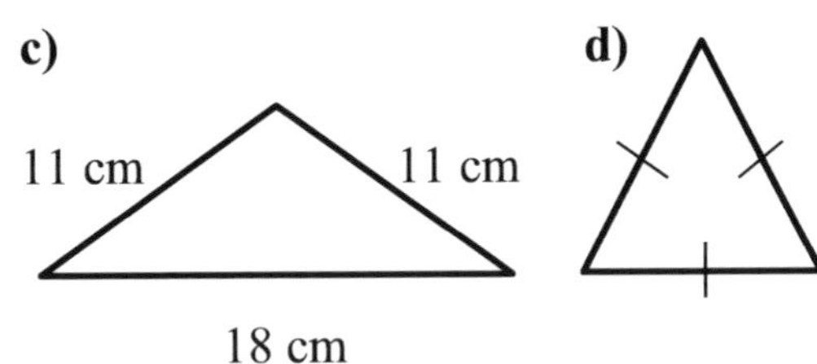

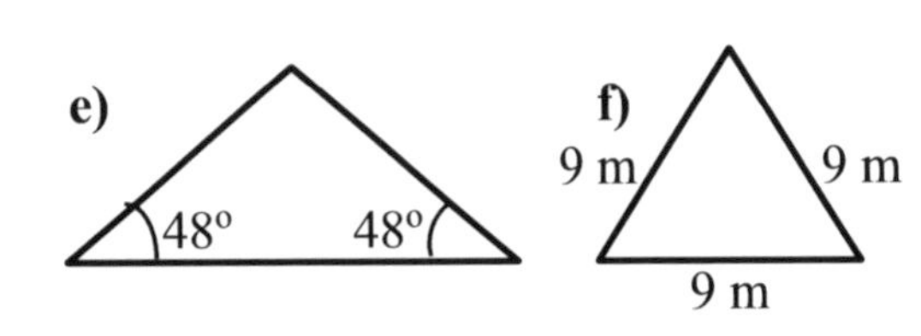

11)

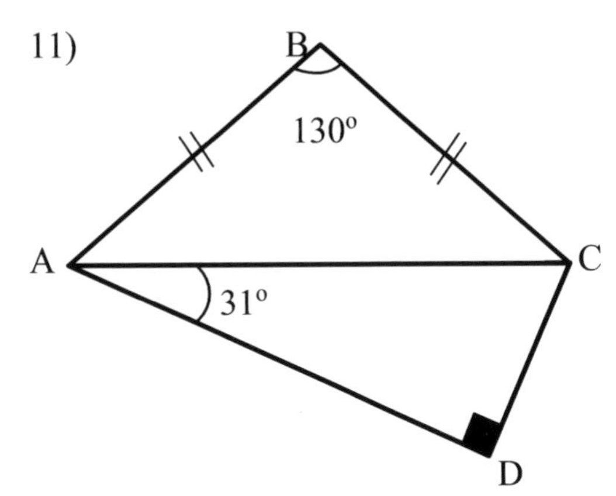

Work out the following angles.

a) $B\hat{A}C$

b) $A\hat{C}B$

c) $A\hat{C}D$

d) $B\hat{A}D$

e) $B\hat{C}D$

12) Obiora says "Angles 54°, 85° and 42° will form a triangle."
Is Obiora correct? Explain fully.

13) If two angles of a triangle are given below, calculate the third angle.

a) 30° and 56°
b) 130° and 20°
c) 23° and 76°
d) 114° and 56°
e) 100° and 40°
f) 60° and 60°
g) 45° and 45°
h) 20° and 79°
i) 56° and 85°
j) 55° and 70°

14) Look at the triangle below.

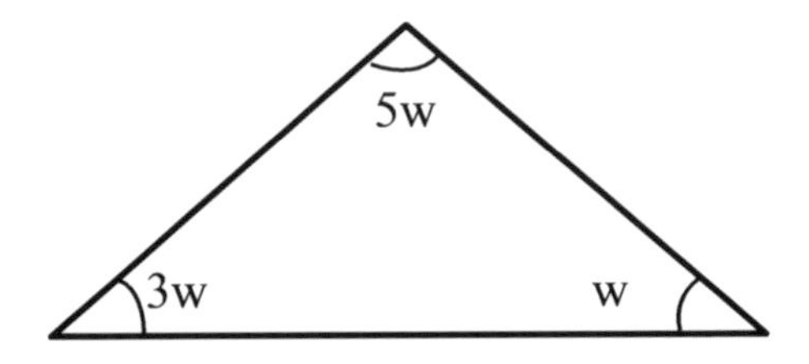

a) Form an equation in w
b) Sole the equation to find w
c) Work out the biggest angle.
d) Work out the range of the angles

15)

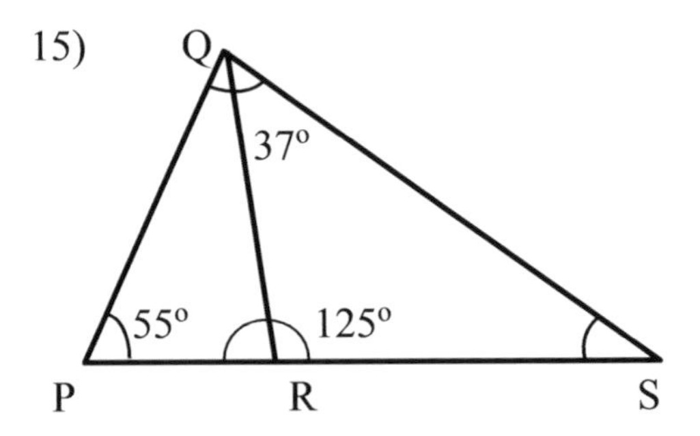

a) Calculate angles

i) $P\hat{R}Q$ ii) $P\hat{Q}R$ iii) $Q\hat{S}R$

b) What type of triangle is

i) $P\hat{Q}R$ ii) $Q\hat{R}S$?
Give a reason for each answer.

15.5 QUADRILATERALS

A quadrilateral is a two-dimensional shape with four straight sides and four angles.

A diagonal is a line joining two opposite corners. By using a diagonal, a quadrilateral may be divided into two triangles. The interior angles of a quadrilateral will always **add to 360°**.

Examples of quadrilaterals are square, rectangle, parallelogram, rhombus, trapezium and kite.

SQUARE

A square has four right angles. All the lengths are equal, and the angles add up to 360 degrees.

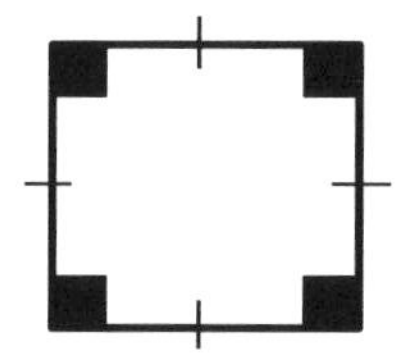

Also, opposite sides are parallel as shown with the arrows.

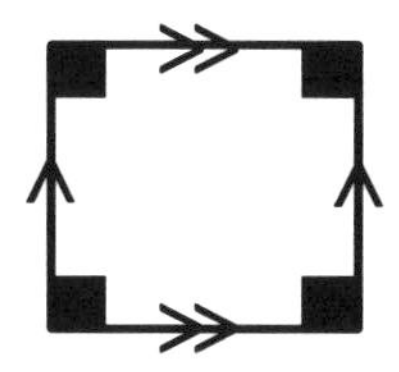

The diagonals are equal in length and bisect each other at 90 degrees.

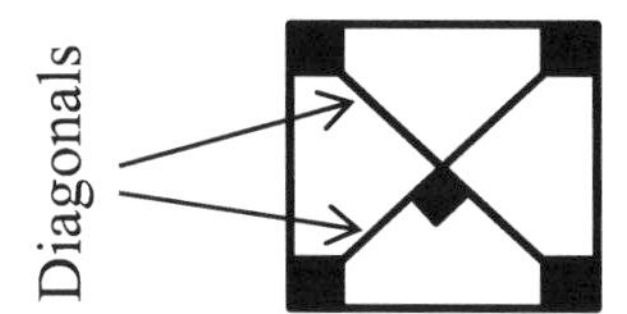

A square has four lines of symmetry and rotational symmetry of order 4.

RHOMBUS

A rhombus is a flat shape with four equal straight sides. None of the angles is a right angle. They are like diamonds.

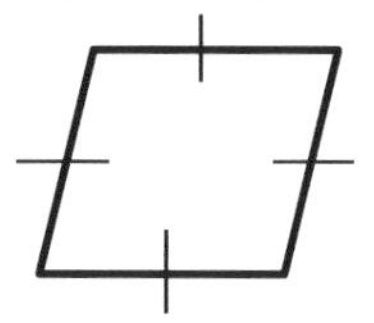

PROPERTIES

Four equal sides
Opposite angles are equal but not 90°
Opposite sides are parallel
Diagonals bisect each other at 90°
All angles add up to 360°
Two lines of symmetry
Rotational symmetry of order 2

RECTANGLE

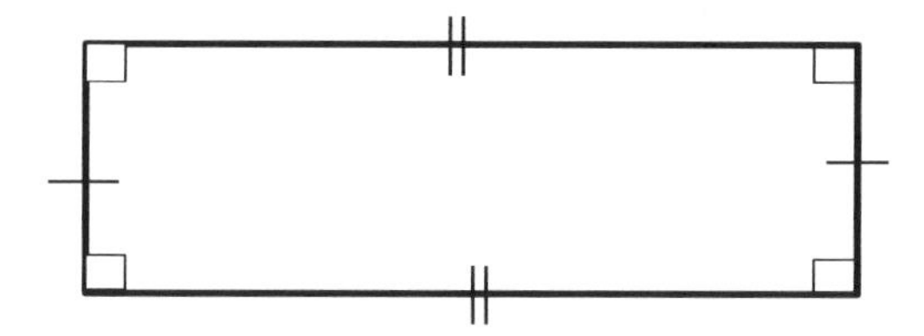

PROPERTIES

Opposite sides are parallel
Opposite sides are equal in length
All angles are 90° each
Diagonals bisect each other and are equal in length
All angles add up to 360°
Two lines of symmetry
Rotational symmetry of order 2

Remember: A square is a special type of rectangle with all the sides equal in length.

PARALLELOGRAM

A parallelogram is like a rectangle pushed out of shape. The angles **are not** 90° each.

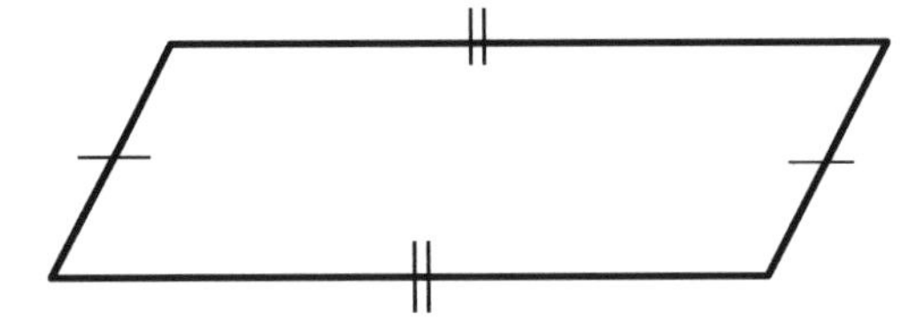

PROPERTIES

Opposite angles are equal
Opposite sides are equal in length
Opposite sides are parallel
Diagonals bisect each other
All interior angles add to 360°
Rotational symmetry of order 2
A general parallelogram like the shape above has **no line of symmetry**.

However, special parallelograms like rhombus, square and rectangle have lines of symmetry.

TRAPEZIUM

A trapezium has **one pair** of parallel sides.

PROPERTIES

One pair of parallel sides
No line of symmetry (unless it is an isosceles trapezium)
Rotational symmetry of order 1

A quadrilateral may also be called **isosceles trapezium** if the non-parallel sides are equal in length.

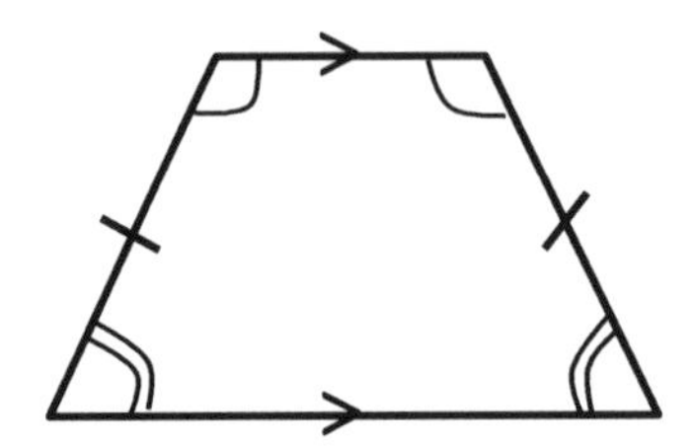

PROPERTIES

Two sets of equal angles
A set of equal sides
One line of symmetry
Rotational symmetry of order 1

KITE

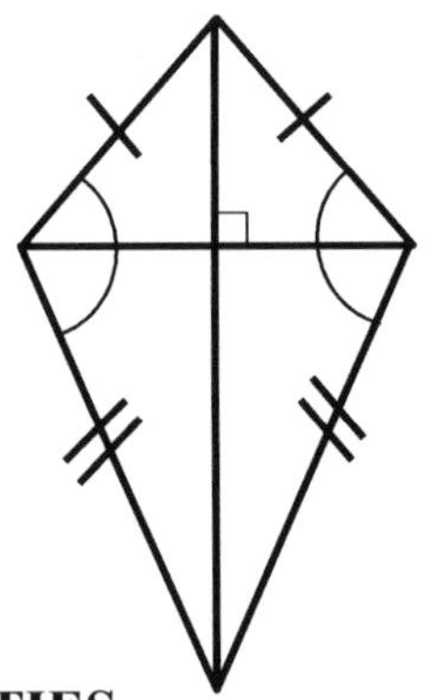

PROPERTIES

A pair of opposite angles is equal
Two pairs of adjacent sides are equal
Diagonal intersect at 90°
Note: A kite is made up of two isosceles triangles with a common base.

EXERCISE 15 C

1) Write down the mathematical name for each shape below.

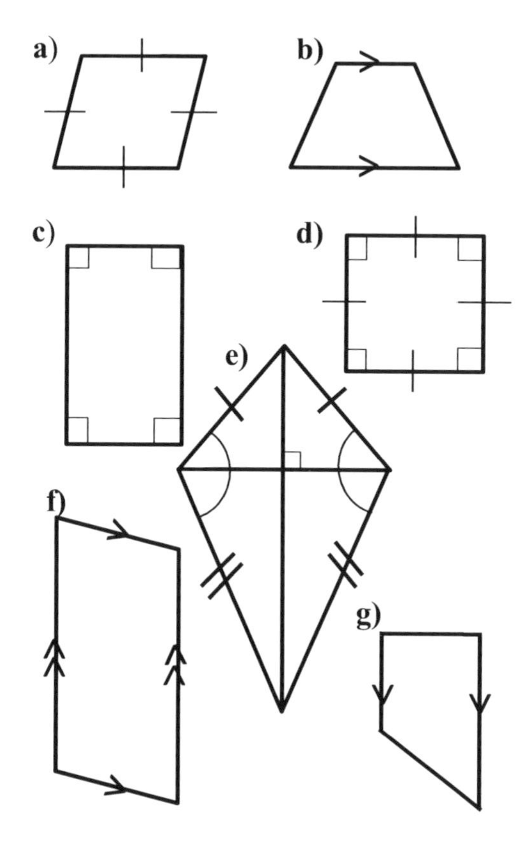

2) Show (by drawing) how two isosceles triangles can be joined to form a kite.

3) Show by drawing how two of these shapes can be joined to make a rectangle.

4) Mention one difference between a square and a rhombus.

5) Is shape B a quadrilateral? Explain.

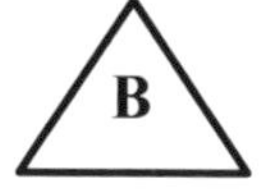

6) From the list of quadrilaterals below, copy and complete each statement.

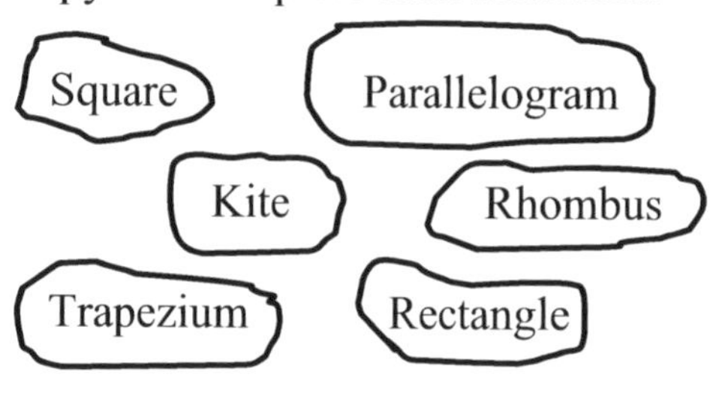

a) I have one pair of parallel sides. Therefore, I am a ……………………..

b) My diagonals bisect each other at right angles. All my sides are equal. All my angles are **not** 90°.
I am a …………………………………

c) I have one pair of opposite angles equal. I am made up of two isosceles triangles. My name is a ………………

d) I am a ……………….......... because all my sides are equal and all my angles are 90°.

e) My opposite sides are parallel and equal in length. All my angles are 90° each. I am a …………………. because I also have two lines of symmetry?

f) My opposite angles are equal. Also, my opposite sides are parallel and equal in length. However, my angles are not 90° and I have **no line** of symmetry. My name is a ……………………

7) Kunle says "The shape below is a trapezium because it has two sets of parallel sides." Is Kunle correct? Explain fully.

15.6 ANGLES IN QUADRILATERALS

The sum of the interior angles of any quadrilateral is 360 degrees.

Example 1: Calculate the size of the missing angle.

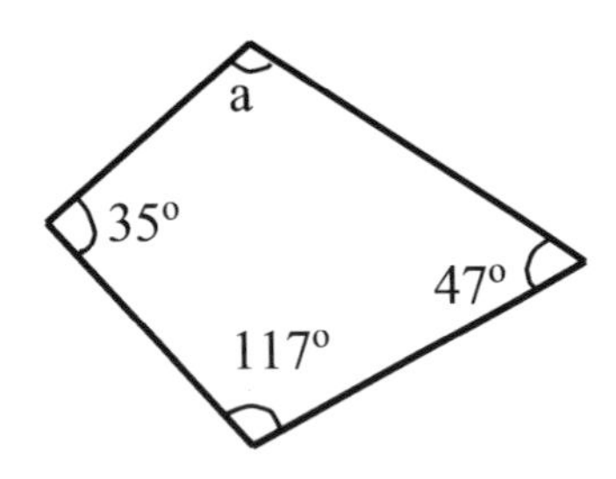

a + 35° + 117° + 47° = 360°
a + 199° = 360°
a = 360° – 199° = **161°** ✓

Check: 161 + 35 + 117 + 47 = 360

Example 2: Calculate the missing angles.

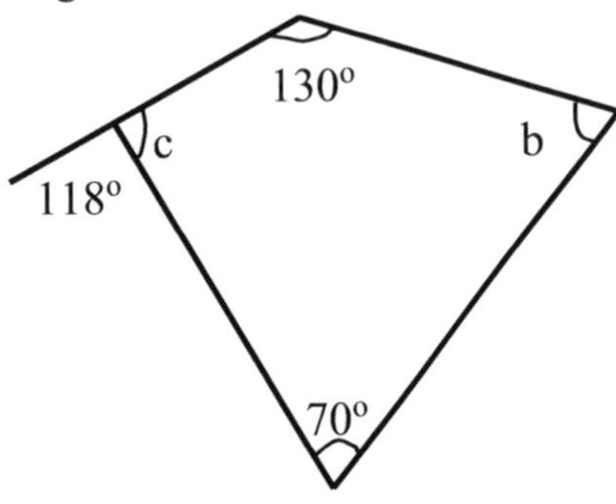

c = 180 – 118 = **62°** ✓
…..angle on a straight line add to 180°

62 + 130 + 70 + b = 360°
…angles in a quadrilateral add to 360°

262 + b = 360
b = 360 – 262 = **98°** ✓

Example 3
a) Form an equation in x.
b) Calculate the value of x.
c) Work out the value of $B\hat{C}D$

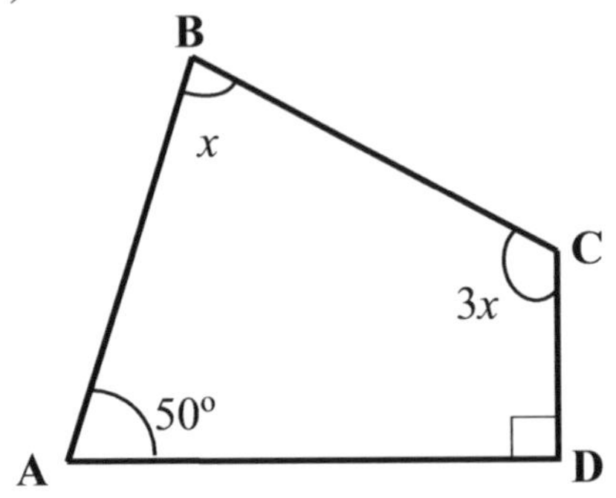

a) Form an equation in x.
$x + 3x + 90 + 50 = 360$
$4x + 140 = 360$
$4x = 360 - 140$
$\mathbf{4x = 220}$ ✓

b) $4x = 220$
$x = \frac{220}{4} = \mathbf{55^o}$ ✓

c) $B\hat{C}D = 3x = 3 \times 55 = \mathbf{165^o}$ ✓

Example 4 Calculate angles w, c and y.

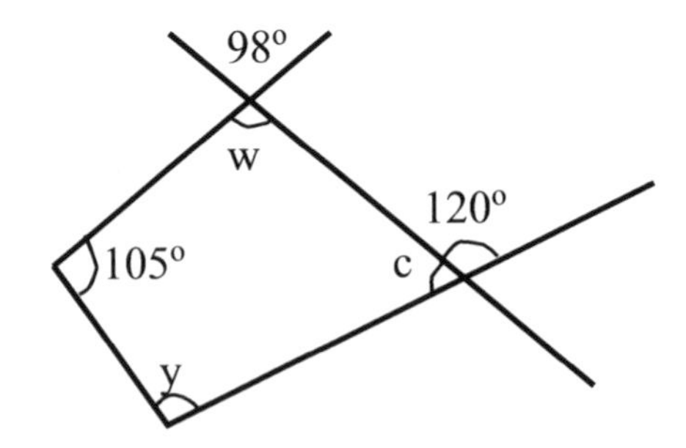

w = **98°**…..vertically opposite angles
c = 180 – 120 = **60°**…….straight line
y = 360 – (105 + 98 + 60)
= 360 – 263 = **97°**

EXERCISE 15 D

1) Work out the size of the unknown angles in each diagram.

a)

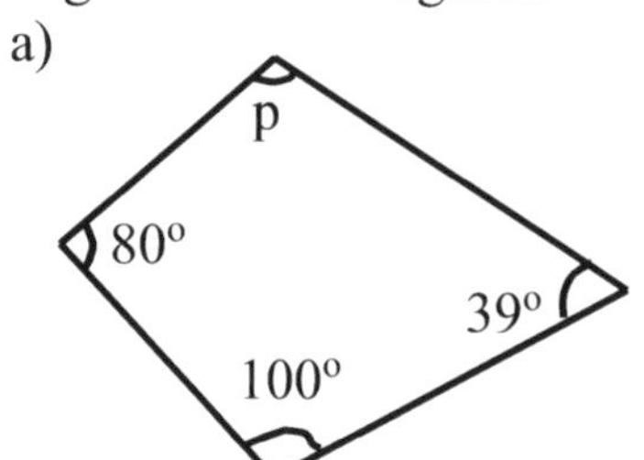

b)

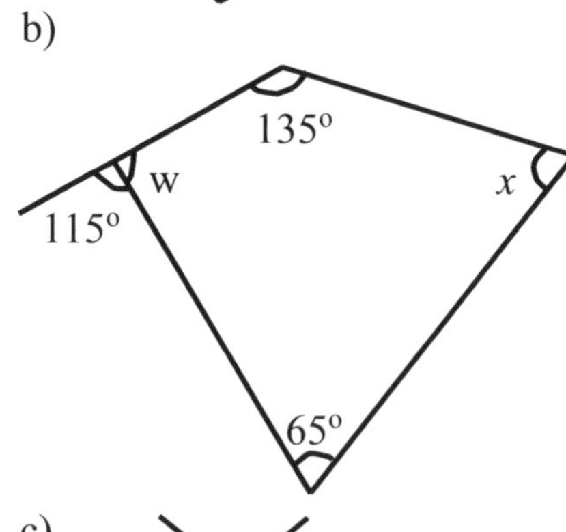

c)

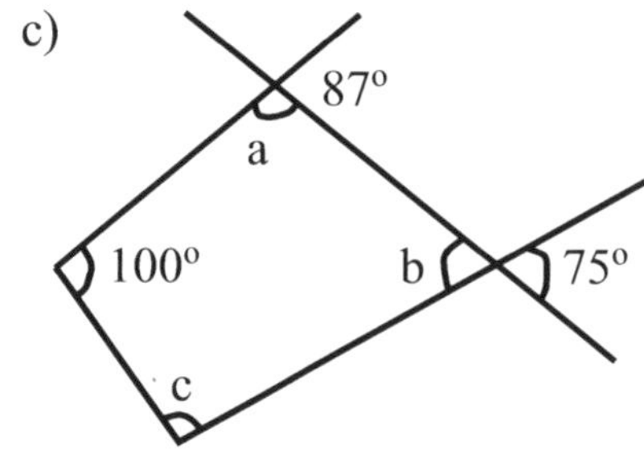

d)

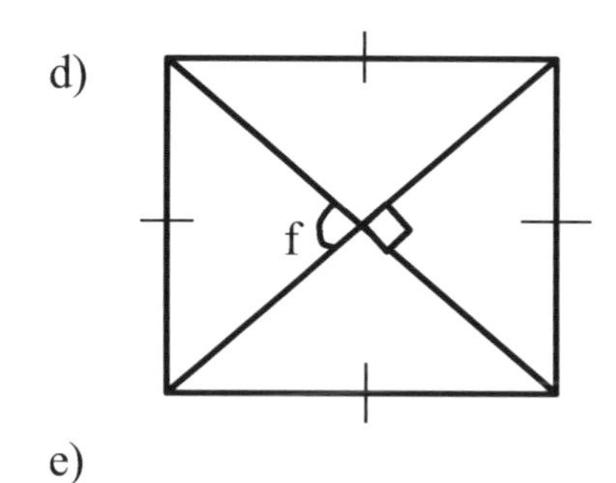

e)

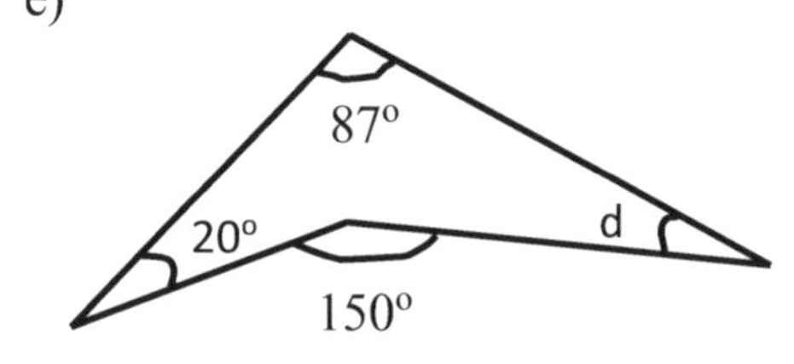

2) The smallest angle is 40°. The opposite angle to the smallest angle is 60° more. The third angle is twice the smallest angle. Calculate the value of the remaining angle.

3) For the quadrilaterals drawn below, work out the missing angles by forming an equation first.

a)

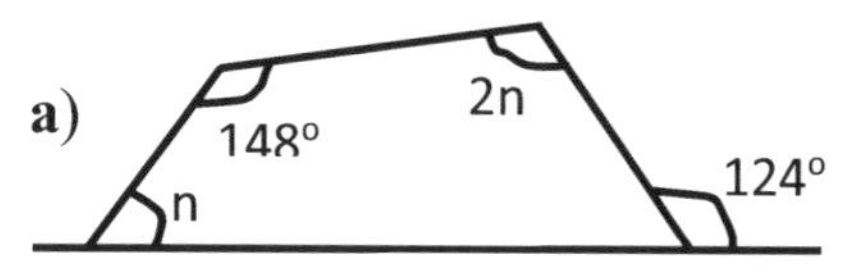

b)

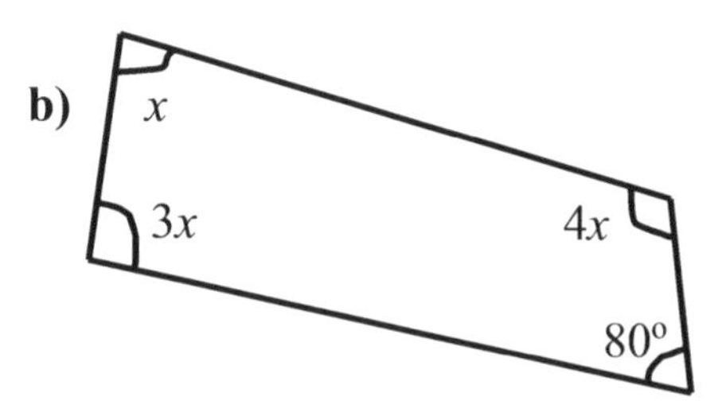

c)

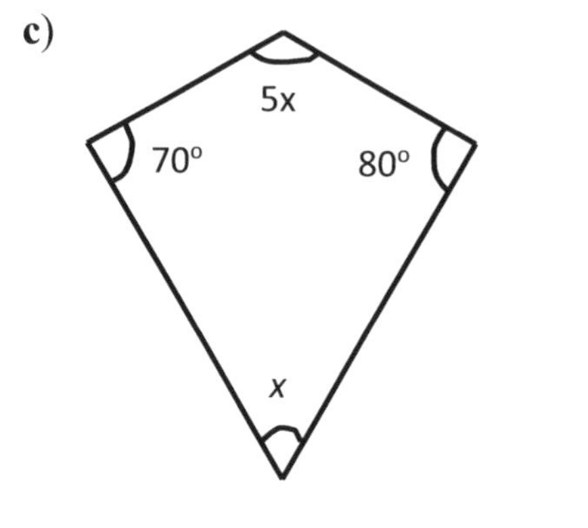

d)

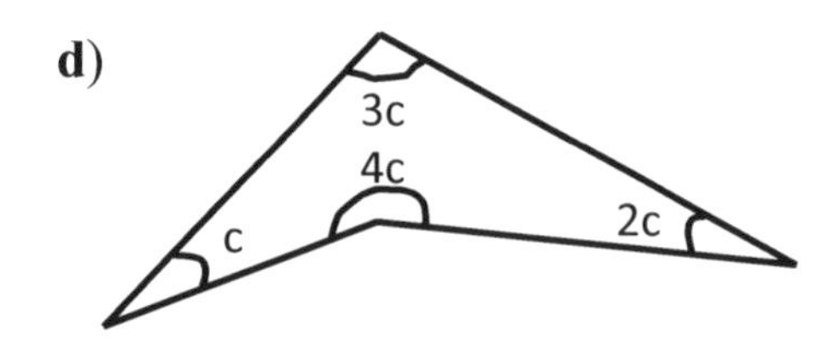

15.7 POLYGONS

Any close two-dimensional shapes with three or more straight sides are called **polygons**.

POLYGON	NUMBER OF SIDES	SUM OF INTERIOR ANGLES
Triangle	3	180°
Quadrilateral	4	360°
Pentagon	5	540°
Hexagon	6	720°
Heptagon	7	900°
Octagon	8	1080°
Nonagon	9	1260°
Decagon	10	1440°

Note: The number of interior angles goes up by 180° each time.

REGULAR POLYGON

A regular polygon has all its angles the same and all the lengths equal. Therefore, squares, equilateral triangles, regular pentagons, etc. are all examples of regular polygons.

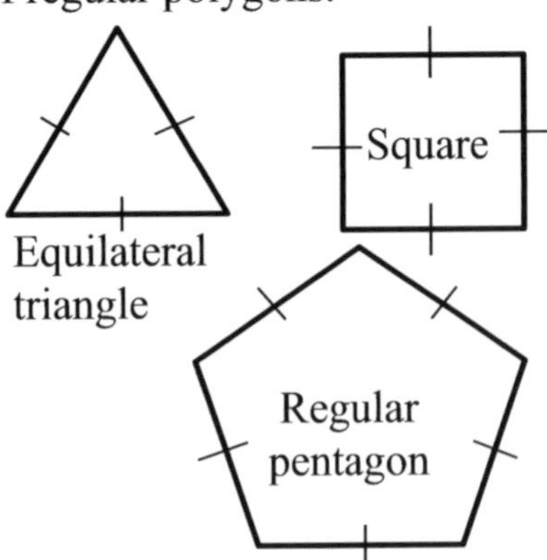

15.8 CIRCLES

We need a pair of compasses to draw a circle accurately.

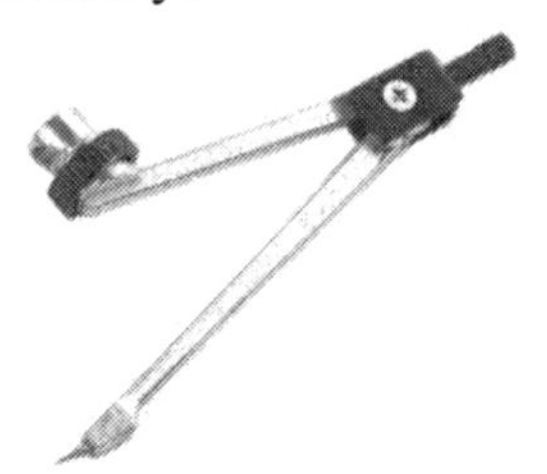

PARTS OF A CIRCLE

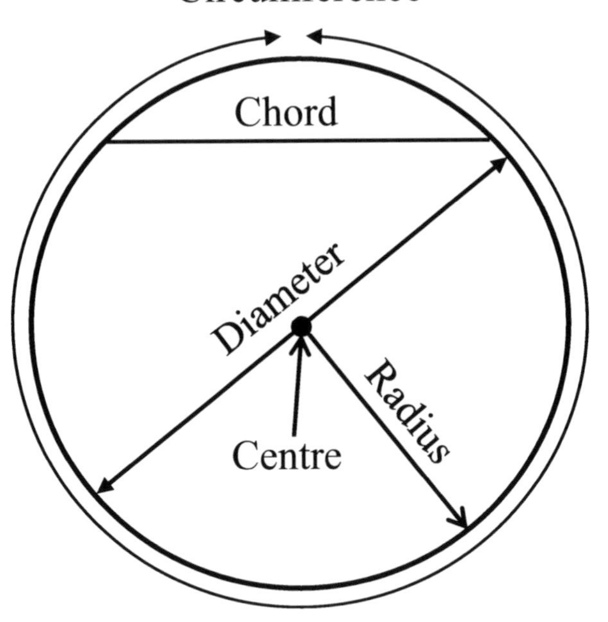

Diameter is a straight line passing through the **centre** and joining two points on the circumference

Radius is a straight line from the centre to the edge of the circle. Twice the radius will equal a diameter.

Circumference is the distance around the circle. It is similar to the perimeter.

Chord is a straight line joining two points on the circumference. If the chord passes through the centre, then it is a diameter.

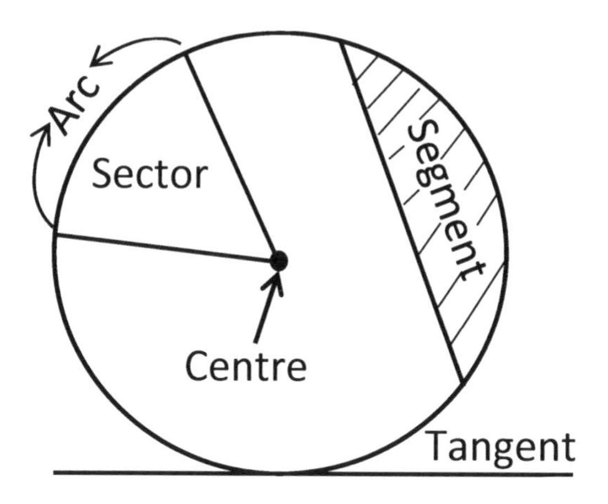

Arc is a part of the circumference of the circle.

The sector is the part of the circle lying between two radii and an arc.

The segment is the part of the circle between a chord and the circumference.

Tangent is a straight line that touches the outside of a circle at one point only.

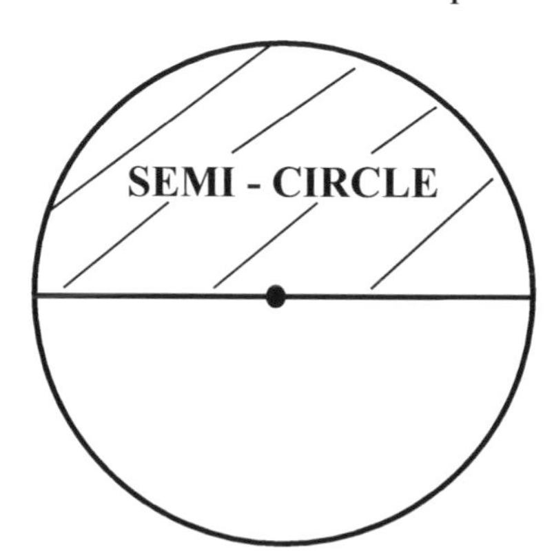

Two semi-circles = one full circle.

EXERCISE 15 E

1) Is a circle a polygon? Explain fully

2) Name any four polygons together with the sum of their interior angles.

3) What would a shape with 12 sides be called? What would the sum of their interior angles add up to?

4) By using a pair of compasses, draw a circle with
a) radius of 4 cm
b) radius of 2.5 cm
c) diameter of 5 cm.

5) a) Draw a circle of diameter 12 cm.
   b) Draw a chord and label it P.
   c) Draw a tangent and label it Q.
   d) Mark a point R on the circumference.

6) a) Draw a circle of radius 5.5 cm.
   b) Draw a diameter of the circle.
   c) Shade the area between the diameter and the circumference.
   d) What is the name of the shaded part?

7) a) Draw a circle of radius 6 cm.
   b) Identify and name any six parts of the circle.

8) Okoro says "A circle with a diameter of 20 cm must have a radius of 40 cm." Is Okoro correct? Explain fully

15.9 ANGLES AND PARALLEL LINES

Parallel lines are lines that will **never** meet, no matter how far they are extended.

Parallel lines exist everywhere in real life including rail tracks and the sides of a piece of paper. Also, in a square or a rectangle, the two opposite sides are parallel.

Small arrows are used to identify two or more parallel lines.

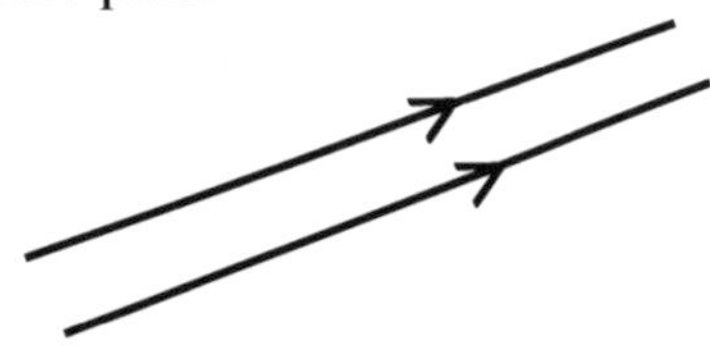

A line that cuts a pair of parallel lines is called a **transversal**.

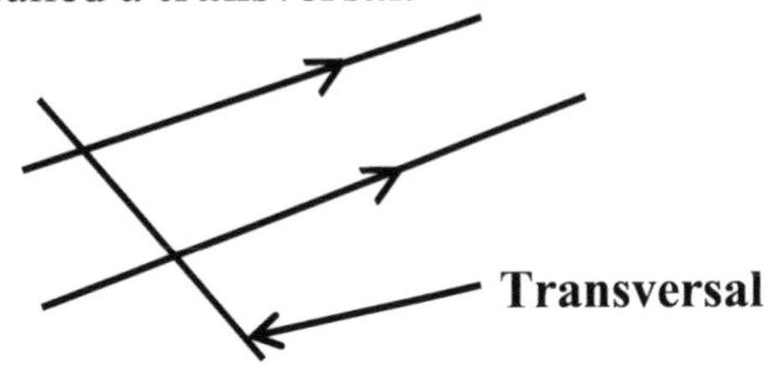

Eight angles are formed when a line cuts through a pair of parallel lines.

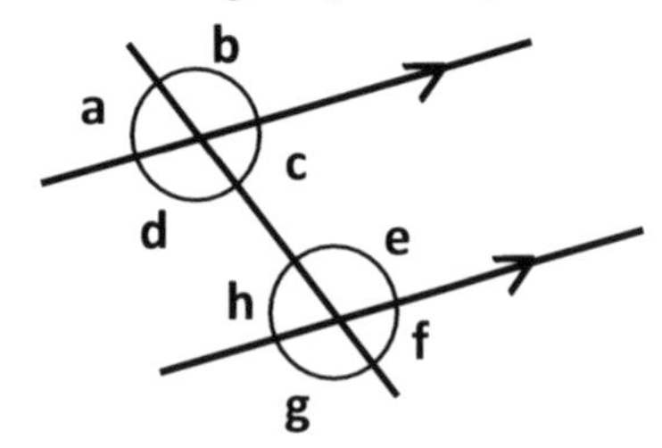

a = c = h = f and b = d = e = g
If the transversal is not perpendicular (at right angles) to the parallel lines, 4 acute and 4 obtuse angles are formed.

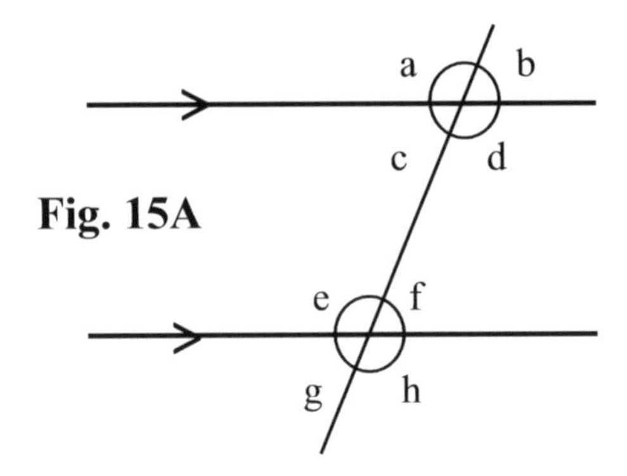

ALTERNATE ANGLES

Alternate angles are equal. They are angles on the opposite sides of the transversal. They form a **Z - SHAPE**.

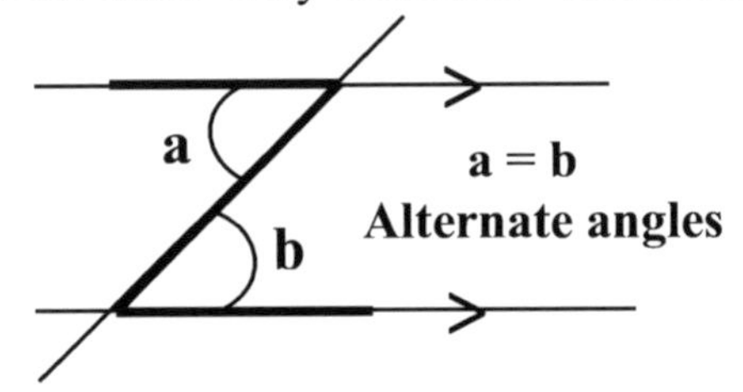

In Fig 15A above, angles **c** and **f** are alternate angles and are equal.
Also, angles **d** and **e** are alternate angles and are equal too.

Example 1: Write down the value of angles x and y. Give a reason for your answer.

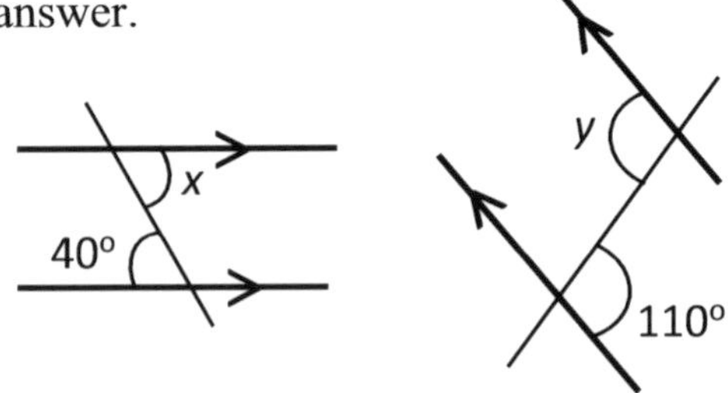

x = **40°** because they are alternate angles. y = **110°** because they are alternate angles.

CORRESPONDING ANGLES

The angles in **matching corners** when a line crosses two parallel lines are called **corresponding angles**.

Also, corresponding angles are **equal**. They also form an F- SHAPE.

In figure 15A above, angles **b** = **f**, **a** = **e**, **d** = **h** and **c** = **g** and all are corresponding angles.

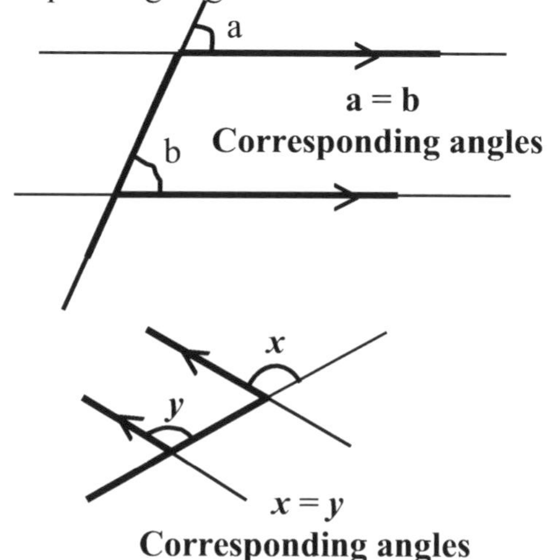

In the above examples, if **a** = 30°, **b** will also be 30° because they are corresponding angles.

Also, if x = 120°, y will be 120° because they are corresponding angles.

Note: Alternate and corresponding angles are always equal when the lines are parallel. If the lines are not parallel, **do not** assume that the angles formed are corresponding or alternate. They may not be equal.

EXERCISE 15 F

1) Calculate the value of the angles marked by letters. Give a reason for your answer.

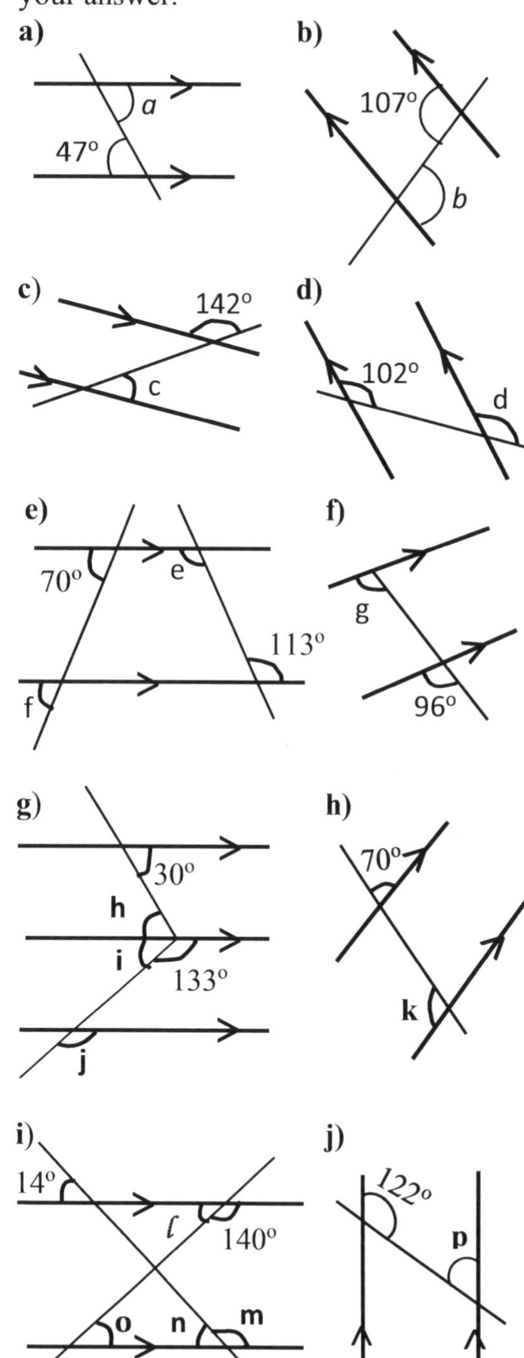

Chapter 15 Review Section

Assessment

1) Work out the missing angles

a)

b)

109°
a b
c d

c)

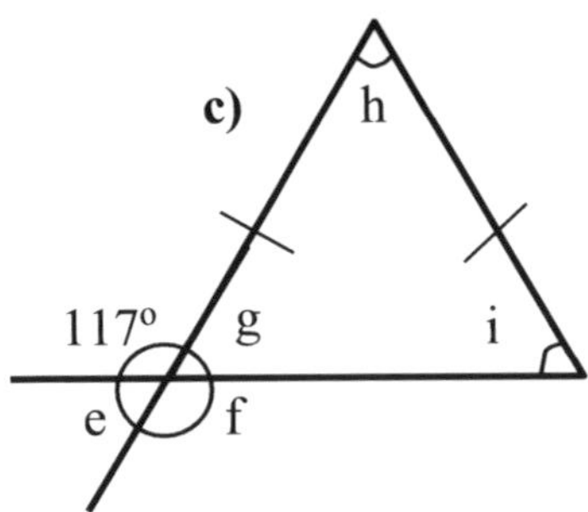

……………………**10 marks**

2) Work out the missing angles.

a)

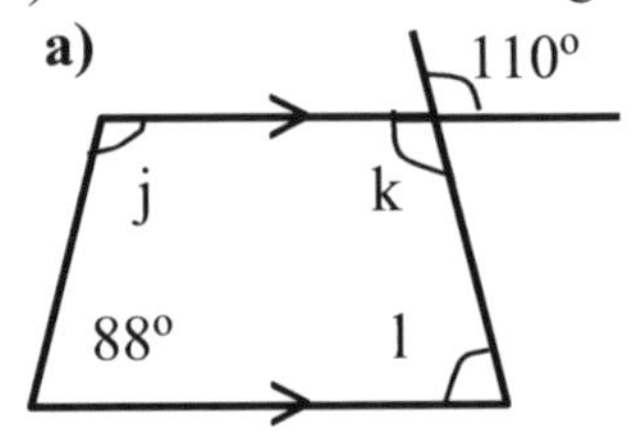

b)

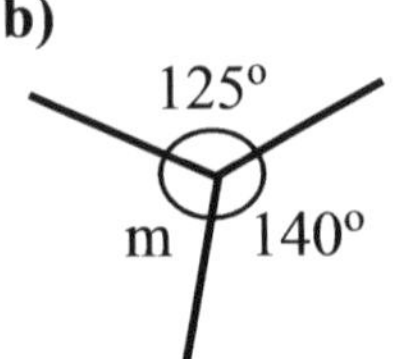

c)

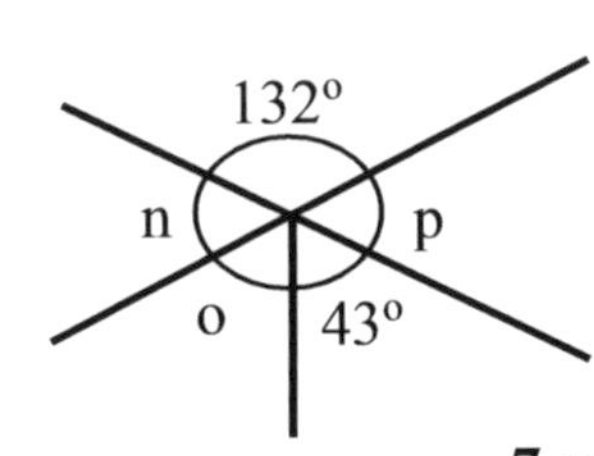

……………………**7 marks**

3)

A **B** **F** **C** **E** **D**

a) How many sets of parallel lines are there in the shape shown?

b) Name the line which is parallel to AF.

c) Name the line which is parallel to ED.

d) What is the mathematical name for shape ABCDEF?

e) How many lines of symmetry does shape ABCDEF have?

f) What type of angle is EDC?

……………….. **6 marks**

4) Work out the missing angles. Give a reason for each answer.

a)

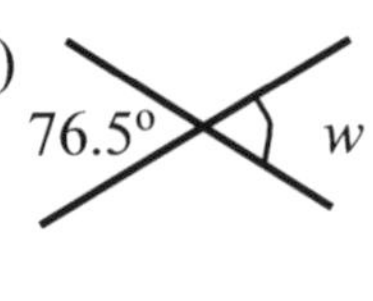

b)

g 65°
e
f
a b
c d

……………….. **8 marks**

16 Arithmetic Skills

This section covers the following topics:

- Function machines
- Magic square

LEARNING OBJECTIVES

By the end of this unit, you should be able to:

a) Explain how to use number machines for basic calculations
b) Reinforce skills in addition, subtraction, multiplication and division
c) Understand magic squares

KEYWORDS

- Number machine
- Input
- Output
- Magic Square

16.1 NUMBER MACHINES

This section explains how to use number machines for basic calculations. Arithmetic skills will be reinforced.

Example 1: What number belongs in the empty boxes?

a) 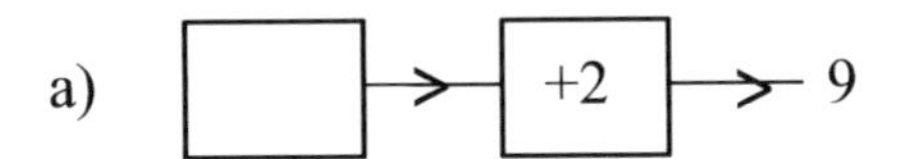

b) □ → - 5 → 15

c) □ → ×4 → 12

d) □ → ÷6 → 7

Answers
a) **7** because 7 + 2 = 9
b) **20** because 20 – 5 = 15
c) **3** because 3 × 4 = 12
d) **42** because 42 ÷ 6 = 7

Alternatively, working backwards would achieve the same goal. However, we do the reverse (opposite) of the signs.

a) 9 – 2 = **7** b) 15 + 5 = **20**
c) 12 ÷ 4 = **3** d) 7 × 6 = **42**

Example 2: What rule belongs in the box?

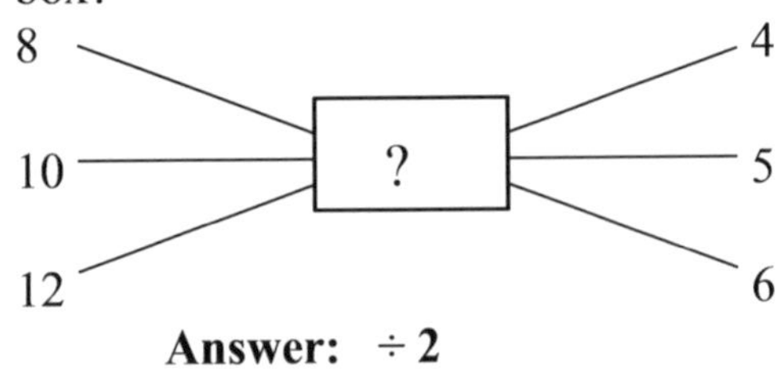

Answer: ÷ **2**

EXERCISE 16 A

1) What number belongs in each box?

a) 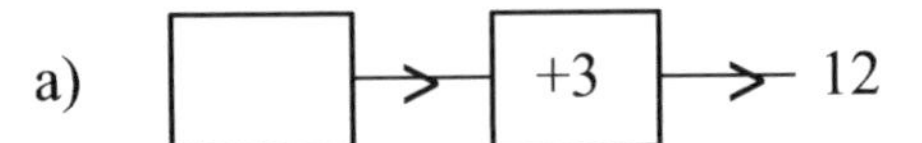

b) 

c) 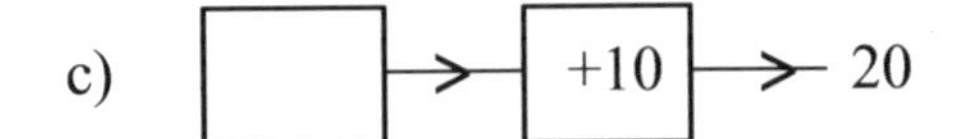

d) 

e) 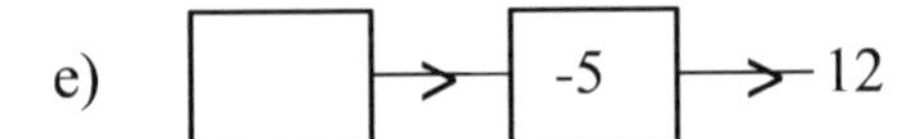

f) 

g) 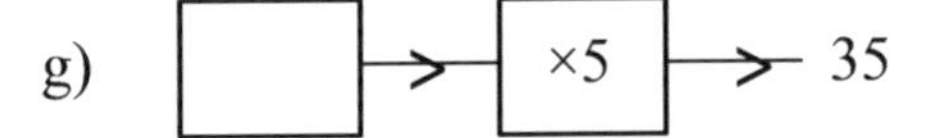

h) 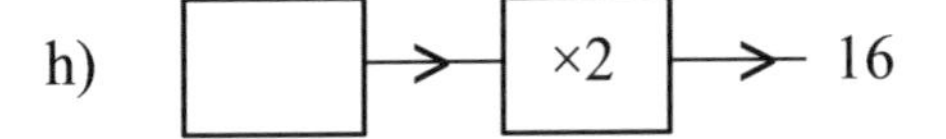

i) 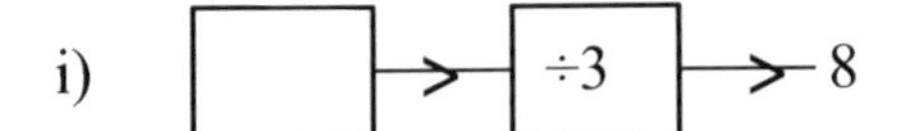

j) □ → ÷4 → 5

k) □ → ÷9 → 4

l) 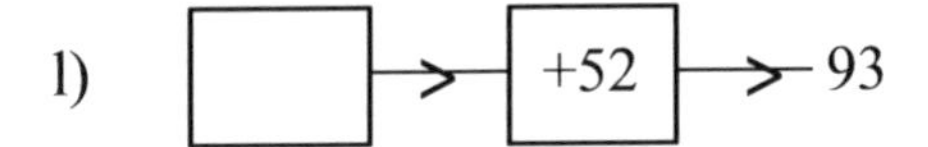

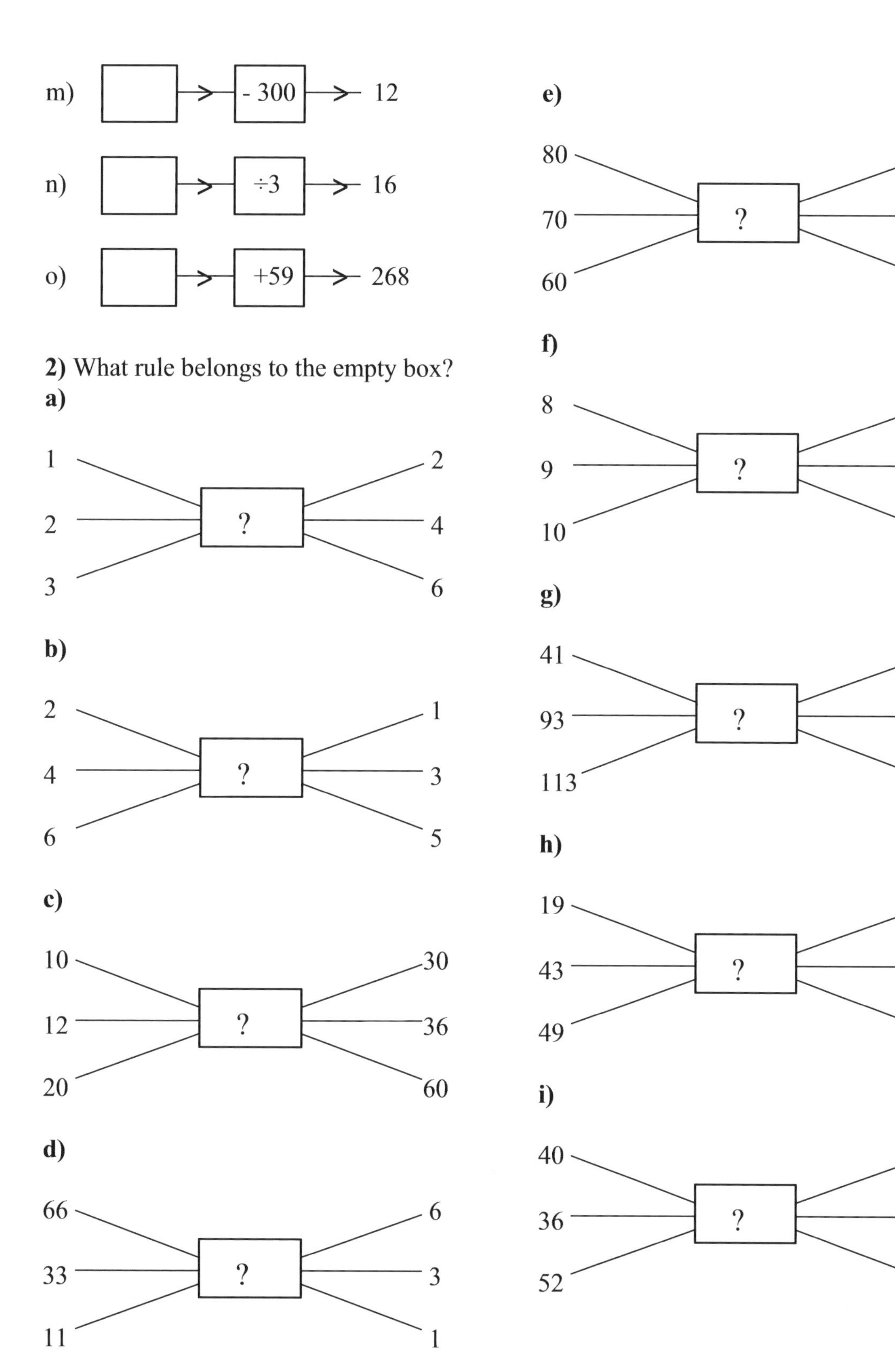

m)
- 300
12
n)
÷3
16
o)
+59
268
2) What rule belongs to the empty box?
a)
1
2
3
?
2
4
6
b)
2
4
6
?
1
3
5
c)
10
12
20
?
30
36
60
d)
66
33
11
?
6
3
1
e)
80
70
60
?
8
7
6
f)
8
9
10
?
80
90
100
g)
41
93
113
?
25
77
97
h)
19
43
49
?
69
93
99
i)
40
36
52
?
10
9
13

16.2 INPUT AND OUTPUT

A function machine could contain an input, instruction box and output.

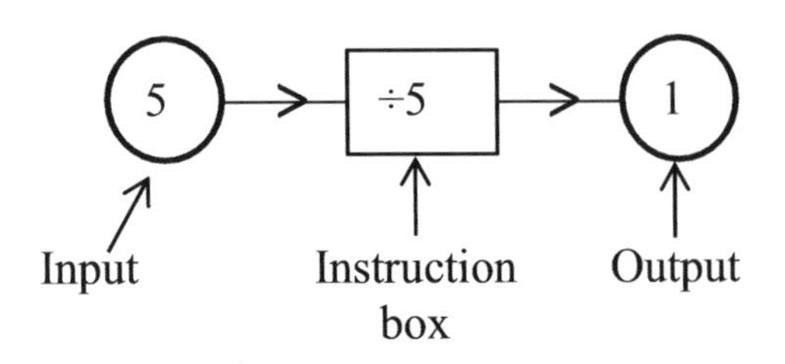

The arrow indicates the direction of the operation. The result of dividing the input value by 5 is **1**.

Number machines can have more than one instruction box.

Example 1

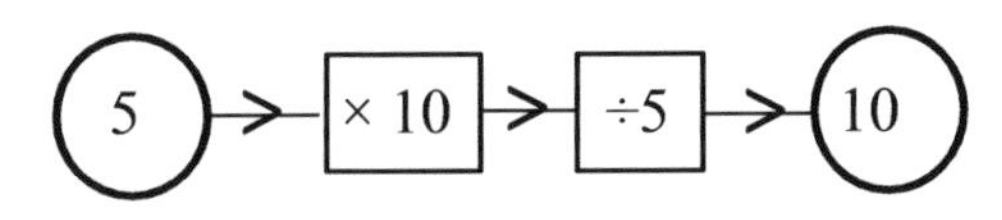

Example 2: Calculate the input value from the number machine.

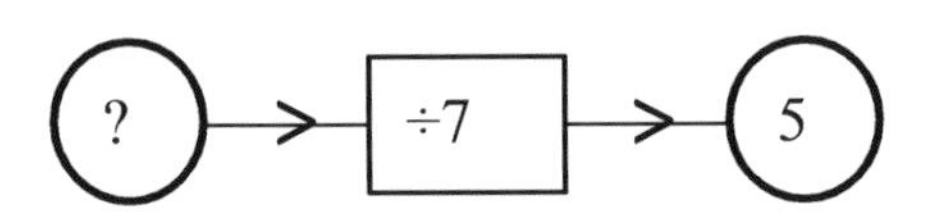

To get the input value, we do the reverse of ÷7 which is ×7. Therefore, $5 \times 7 = 35$. The input value is **35**.

EXERCISE 16 B

1) Calculate the output if the input value is **6** for each question below.

a)

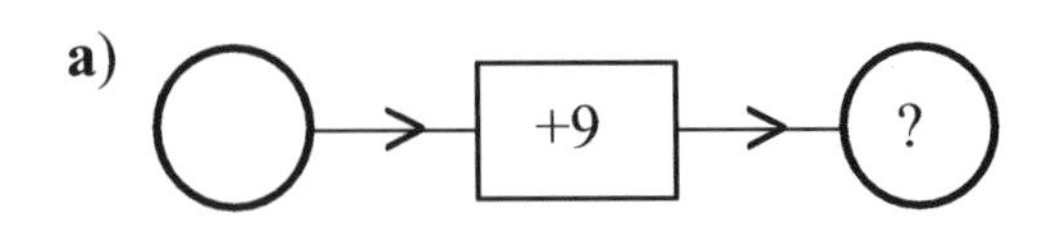

b)

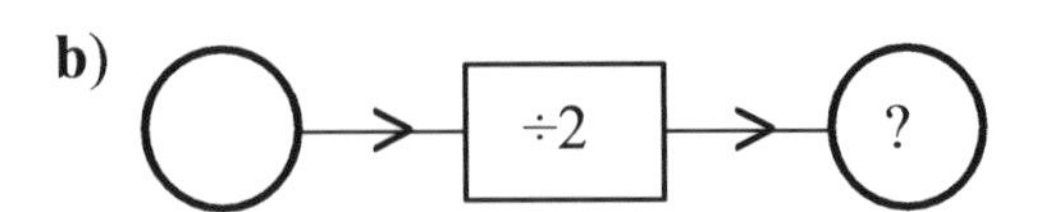

c)

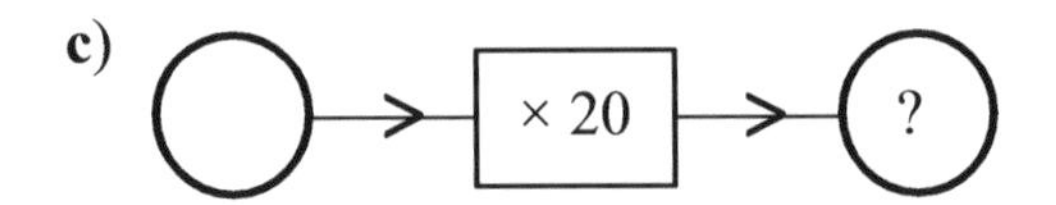

d)

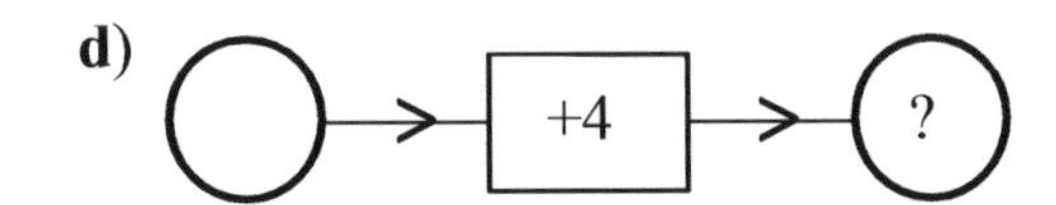

e)

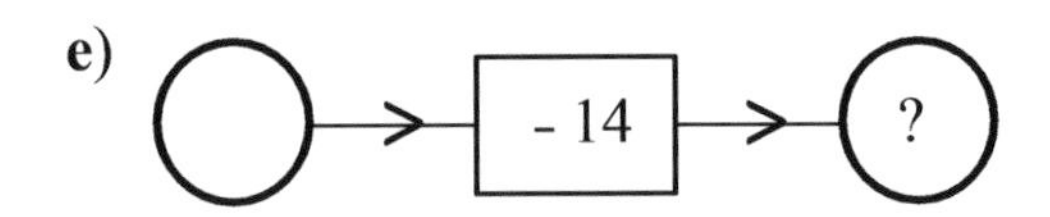

f)

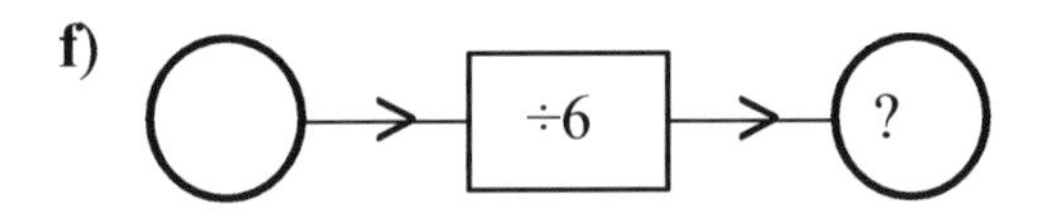

2) Copy and complete the table for the function machine.

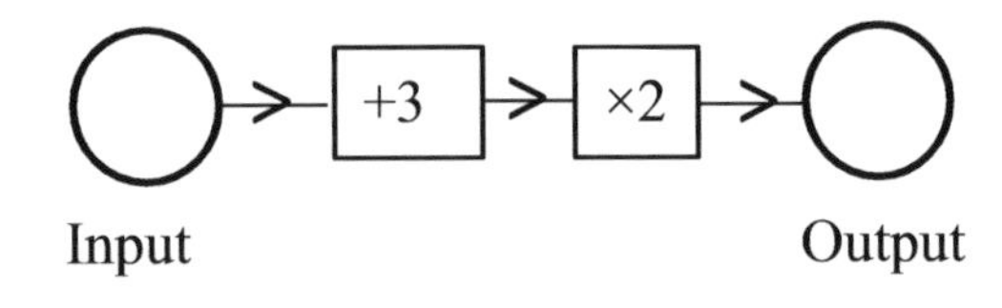

Input	2		0	10	7		5	15
Output		16				12		

3) Copy and complete the table for the function machine.

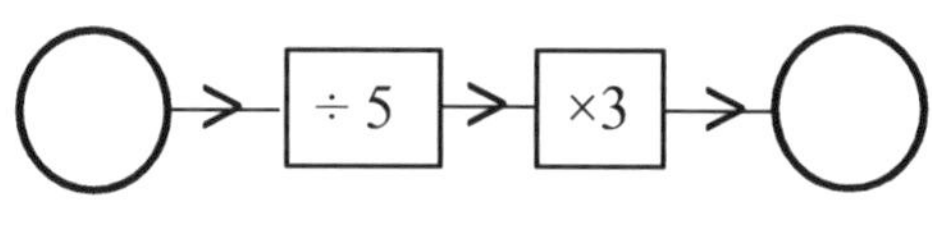

Input Output

Input	10		5	15		100
Output		12			48	

4) For the number machine below, make an input and output table. Use the input 2, 10, 7, 13, 20 and 65.

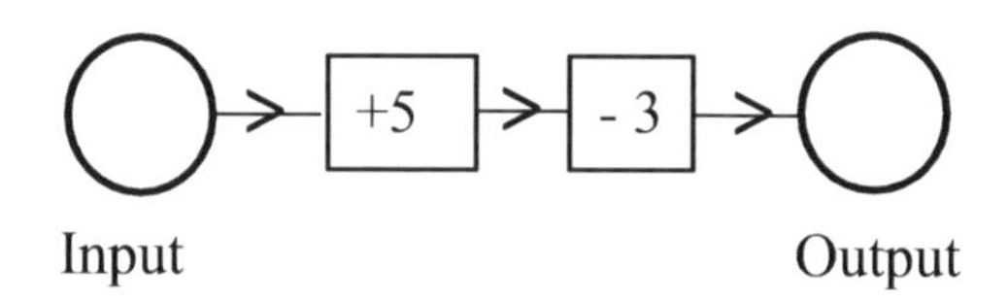

Input Output

5) Work out
a) the input when output is 300
b) the output when the input is 3.

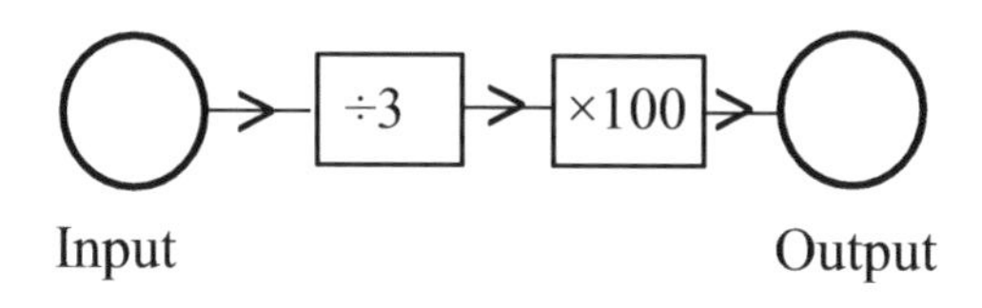

Input Output

6) Copy and complete the table.

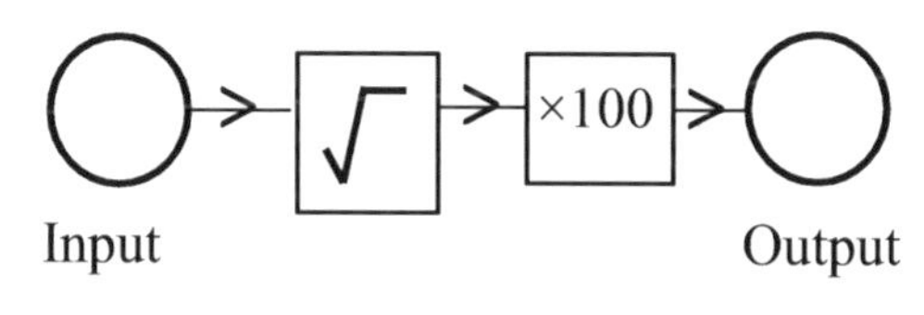

Input Output

Input	16	25	100	64	144
Output					

16.3 MAGIC SQUARES

In a magic square, the sum of each column, row and diagonals are the same.

7	14	3
4	8	12
13	2	9

$7 + 14 + 3 = 24$ $7 + 4 + 13 = 24$
$4 + 8 + 12 = 24$ $14 + 8 + 2 = 24$
$13 + 2 + 9 = 24$ $3 + 12 + 9 = 24$

Also the diagonals, $13 + 8 + 3 = 24$ and $7 + 8 + 9 = 24$

EXERCISE 16 C

1) Complete the magic squares below.

a)

8	3	4
	7	

b)

6		
	7	
4		8

c)

		7
		14
	12	3

d)

	4	11
	8	
5		

e)

5	8	11
		9

f)

	6	
9	10	5

17 Perimeter and Area

This section covers the following topics:

- Perimeter of 2D shapes
- Area of shapes

LEARNING OBJECTIVES

By the end of this unit, you should be able to:

a) Find perimeter of regular shapes
b) Find perimeter of irregular shapes
c) Find perimeter by measurements
d) Find perimeter and area by using a formula
e) Find perimeter of circles and other circular shapes circular shapes
f) Find the area of plane shapes and circles.

KEYWORDS

- Perimeter
- Regular shape
- Irregular shape
- Circles
- Circumference
- Length
- Area
- Pi (π)

17.1 PERIMETER ON CENTIMETRE GRID

The perimeter of any shape is the sum of the lengths of all its sides. It is the outside edge of a plane shape.

Example 1

On a centimetre grid, the perimeter can be worked out by counting the sides (edges) of the squares.

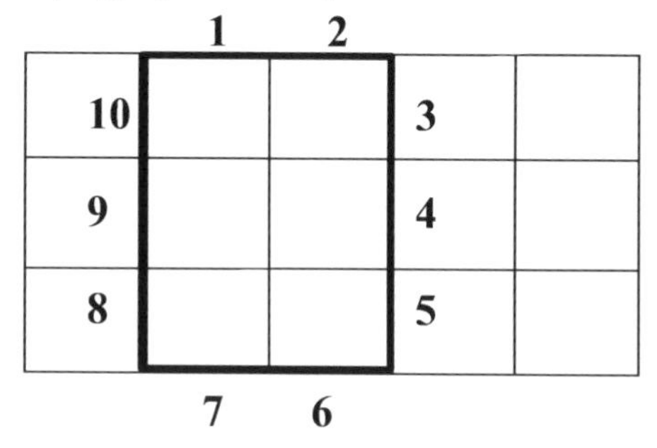

By counting the sides, the perimeter of the above shape is **10 cm**.

Example 2

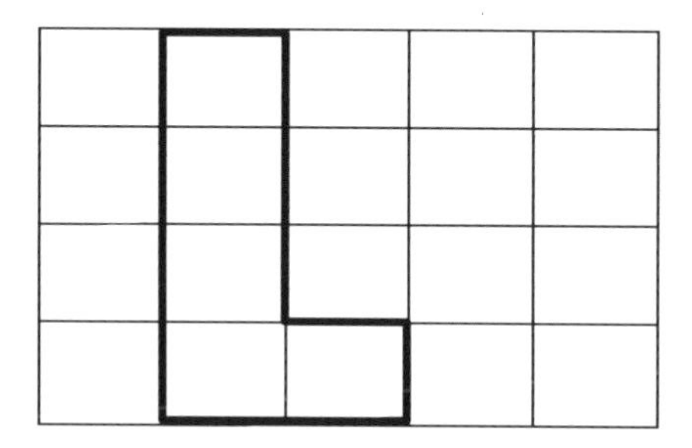

By counting the outside (edges), the perimeter of the above shape is **12 cm**.

Note: Since perimeter is length, the unit will be cm.

Example 3

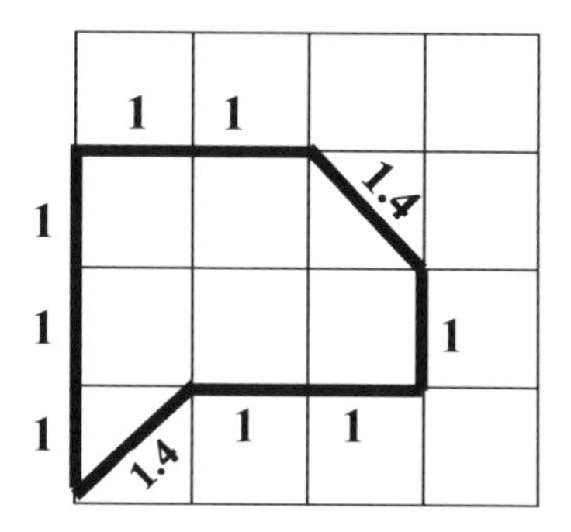

The diagonal of a square is approximately 1.4.

1 + 1 + 1.4 + 1 + 1 + 1 + 1.4 + 1 + 1 + 1 = 10.8

Therefore, the perimeter of the shape above is approximately **10.8 cm**.

17.2 PERIMETER BY MEASUREMENT

To calculate the perimeter of regular or irregular shapes, use a ruler or tape measure if the surface is rounded or curved.

Example 1: Measure the perimeter of the shape below.

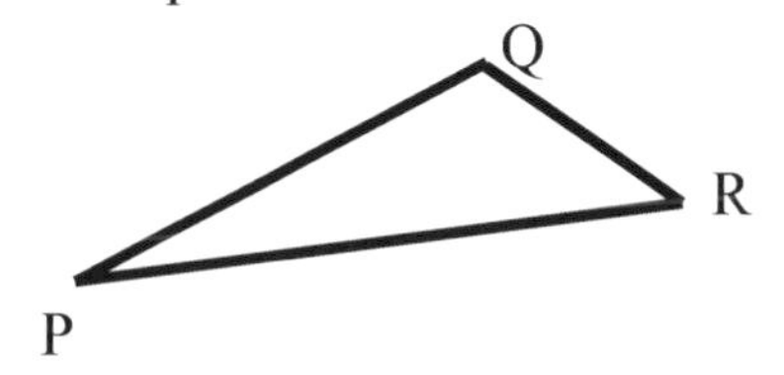

Using a ruler, PQ = 3cm
QR = 1.5 cm
PR = 4 cm

Perimeter = 3 + 1.5 + 4 = **8.5 cm.**

Example 3
Measure the perimeter of the semi-circle below.

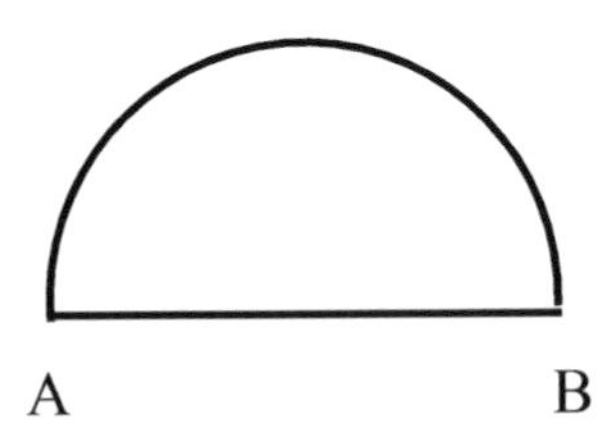

Using a ruler, Length AB = 3.3 cm
Using a rope or tape measure, curved length = 10.4 cm (approximately).

Total perimeter = 3.2 + 10.4 = **13.7 cm**

EXERCISE 17 A

1) By using a ruler and tape or rope where appropriate, measure the perimeter of the shapes in cm.

a) 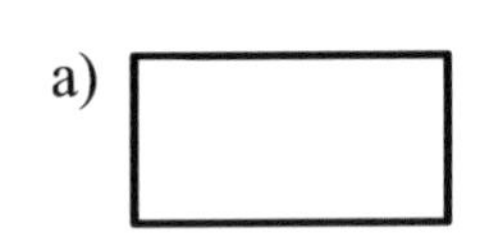

b) 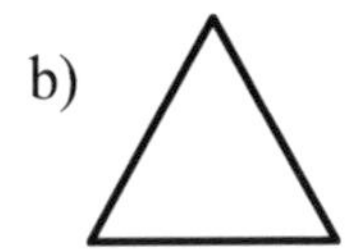

c) 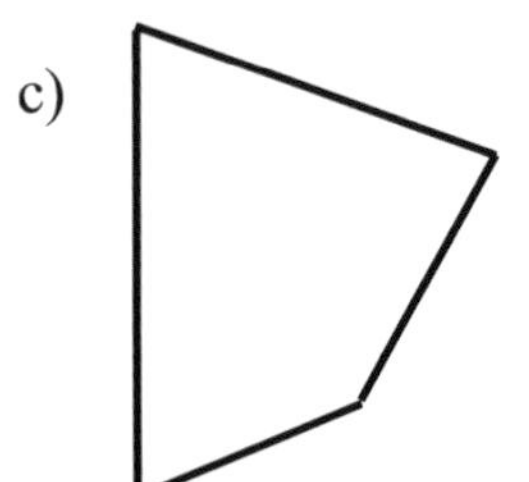

d) 

e) 

f) 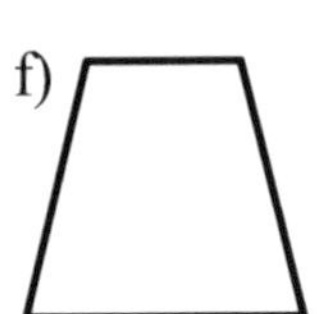

g) 

h) 

2) On the centimetre grid below, calculate the perimeter of the shapes.

a)

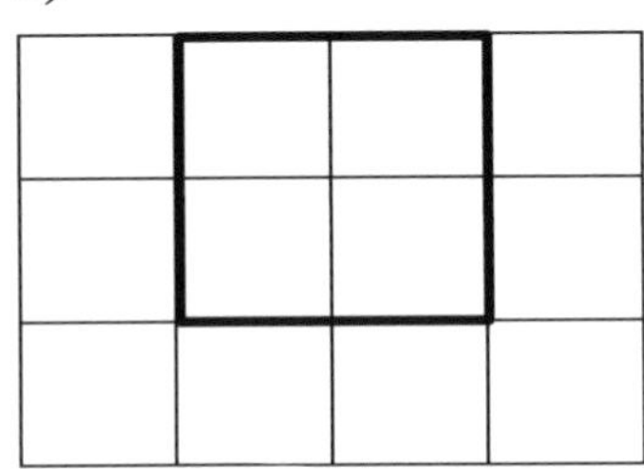

b)

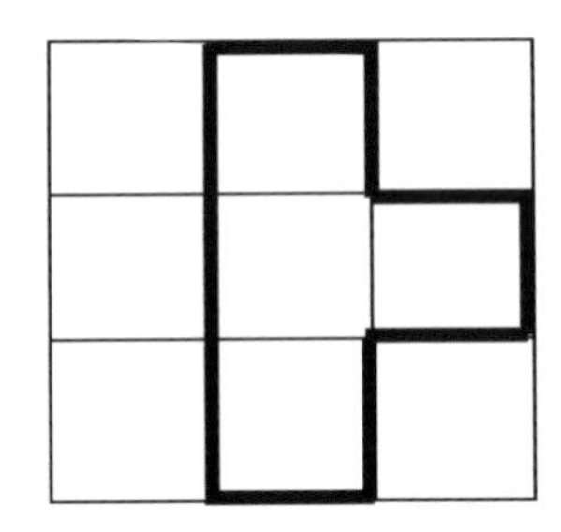

c)

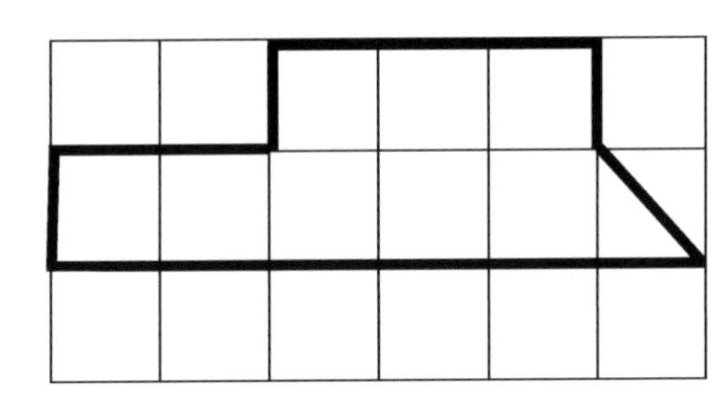

d)

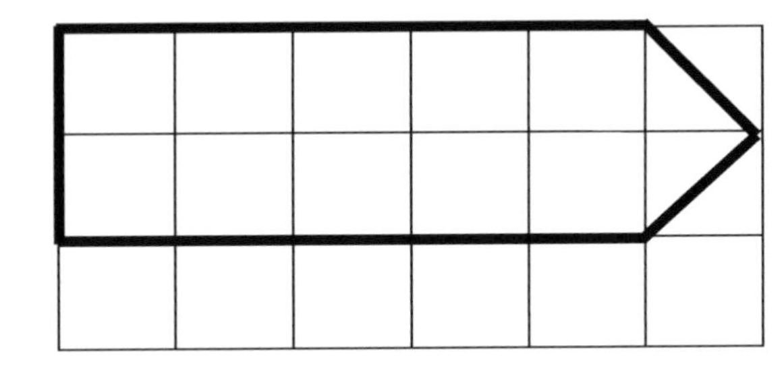

17.3 PERIMETER BY CALCULATIONS

PARALLELOGRAMS

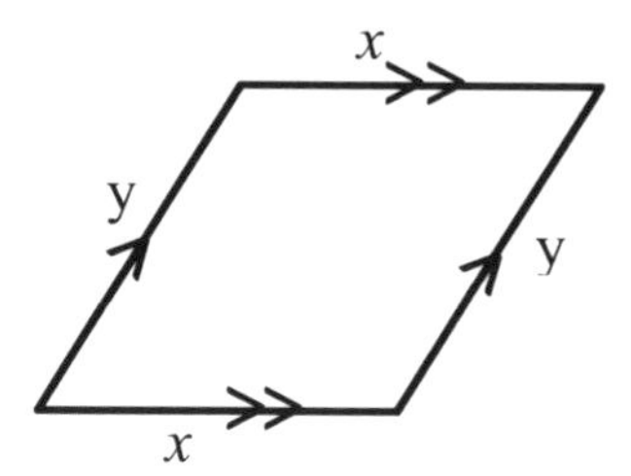

Refer to section Chapter 15, section 15.5 for properties of parallelograms.

Opposite sides are equal with side lengths x and y units. Also, opposite sides are parallel.

The perimeter $= x + x + y + y$
$= 2x + 2y = \mathbf{2(x + y)}$

Example 1: Work out the perimeter of the parallelogram below.

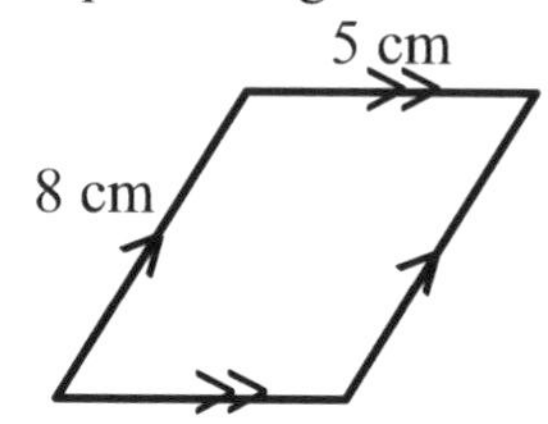

$P = 2(5 + 8)$
$= 2 \times 13$
$=$ **26 cm**

Example 2: Calculate the length of the remaining three sides of a parallelogram if the perimeter is 40 cm and one of its sides is 5 cm long.

$2 \times 5 = 10$ cm therefore, $40 - 10 = 30$cm and $30 \div 2 =$ **15** cm. The three lengths are **15 cm, 15 cm and 5** cm.

RECTANGLES

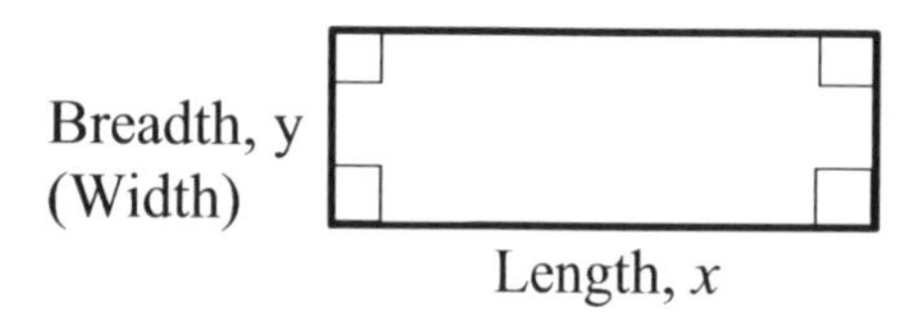

Perimeter of a rectangle
$= x + x + y + y = 2x + 2y = \mathbf{2(x + y)}$

The formula for the perimeter of rectangles and normal parallelograms are the same.

Example 3: Calculate the perimeter of the rectangle below.

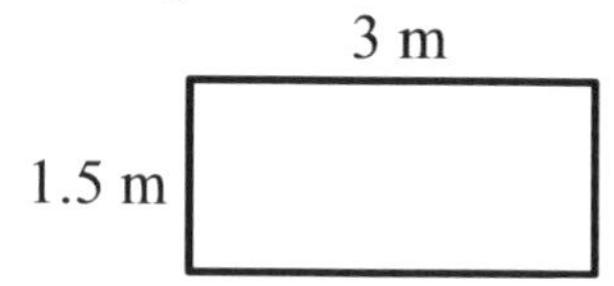

Perimeter $= 2\ (3 + 1.5)$
$= 2 \times 4.5 =$ **9 m**

Or simply adding all the lengths together $3 + 3 + 1.5 + 1.5 = 9$ m

Example 4: Work out the length, w of the rectangle.

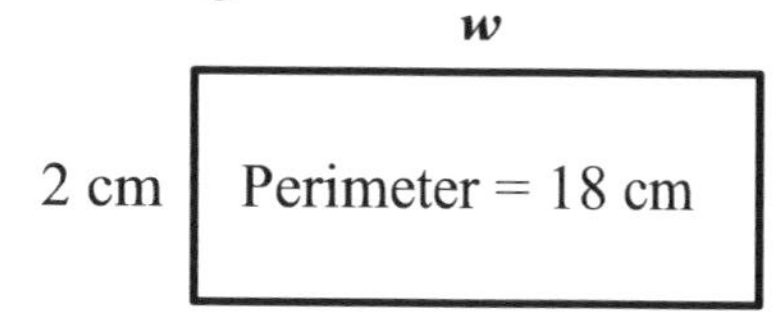

Solution: A rectangle has two opposite sides equal. Therefore, the side opposite 2 cm is also 2cm.
$2 + 2 = 4$
$18 - 4 = 14$
$14 \div 2 = 7$
Therefore, the length $w =$ **7 cm**
Check: $7 + 7 + 2 + 2 = 18$ cm

SQUARE

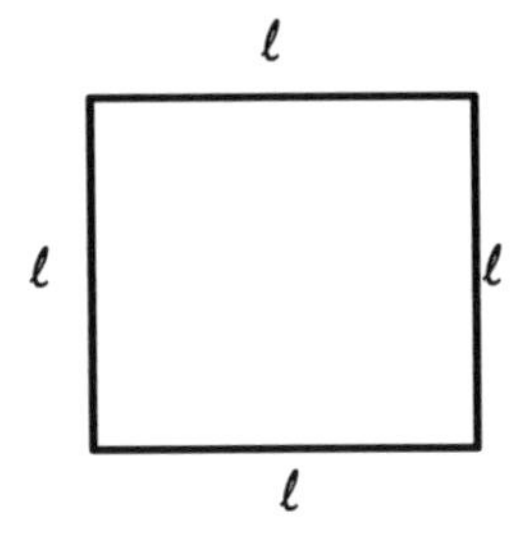

Perimeter = $\ell + \ell + \ell + \ell = 4\ell$

Example 5:

If length, ℓ = 3 cm, the perimeter of the square will be 4 × 3 = **12 cm**

Alternatively, 3 + 3 + 3 + 3 = 12 cm

Example 6: Work out the length of a square with a perimeter of 10 cm.

Perimeter = 4ℓ

10 = 4ℓ

ℓ = 10 ÷ 4 = **2.5 cm**

Example 7: Work out the perimeter of the compound shape below.

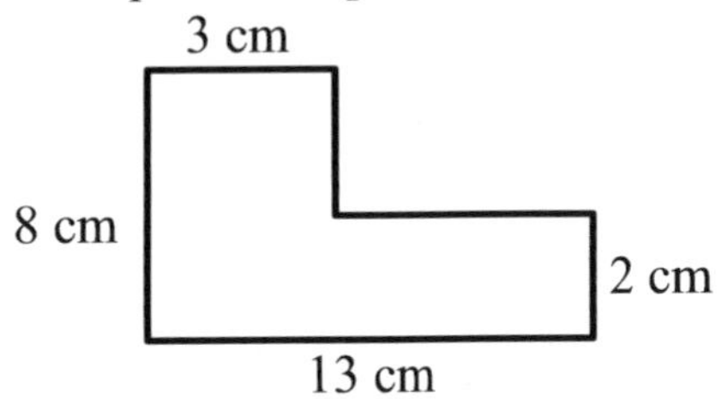

Solution:

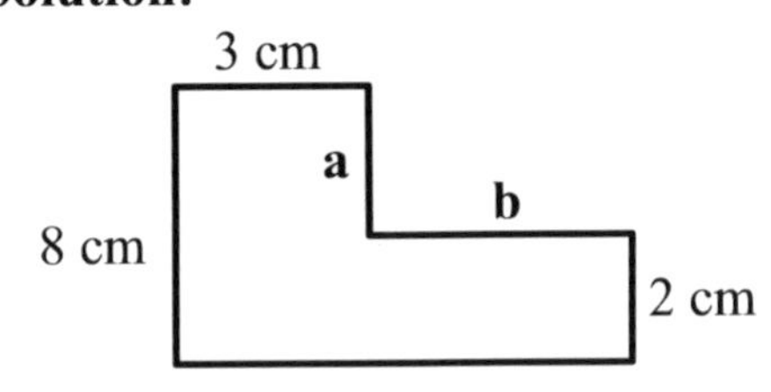

a = 8 - 2 = 6 cm

b = 13 - 3 = 10 cm. Total perimeter = 3 + 8 + 13 + 2 + 10 + 6 = **42 cm**

EXERCISE 17 B

1) Find the perimeter of the shapes below. All lengths are in cm.

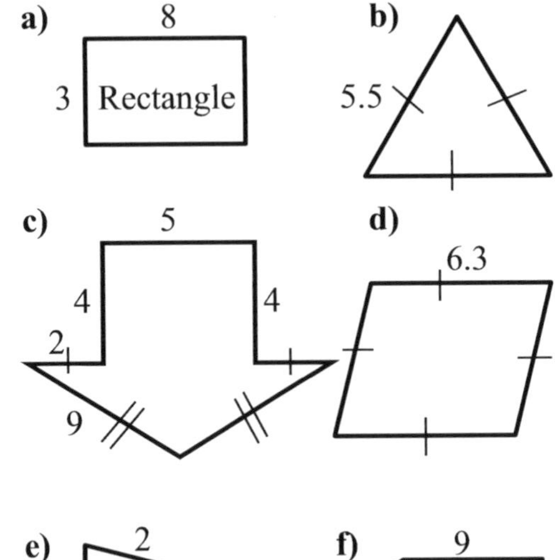

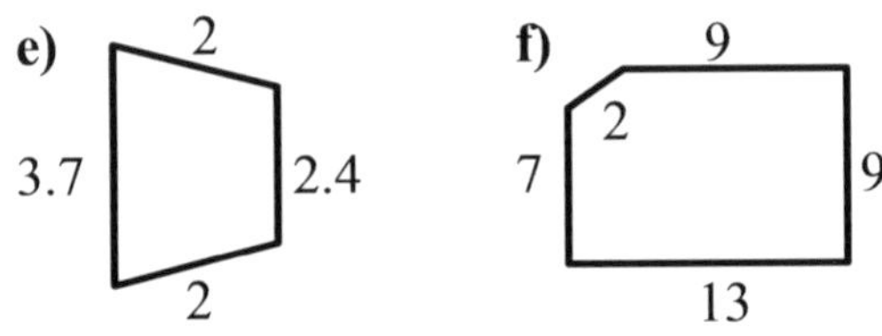

2) Work out the perimeter of the compound shape below.

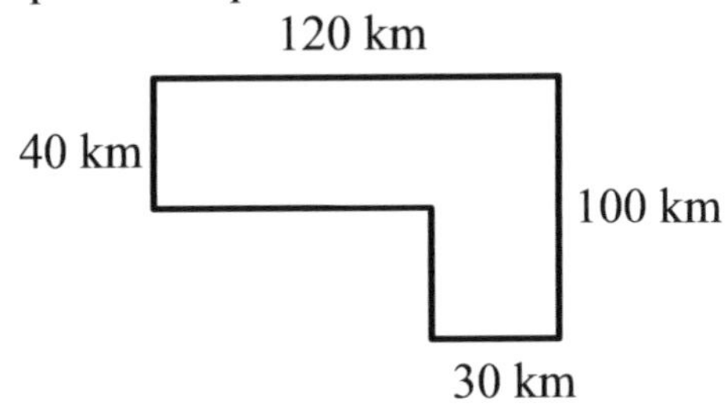

3) Copy and complete for rectangles.

Perimeter	Length	Width
	9 m	2 m
	11 cm	4 cm
22 mm		4 mm
42 m	14m	
	90 cm	60 cm
	7.5 km	4.2 km

17.4 CIRCLES AND PERIMETER

Circumference

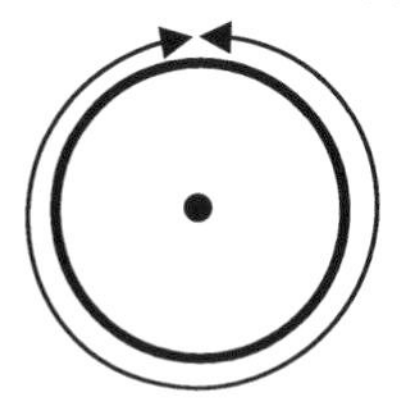

The total distance around the circle is called the circumference. Therefore, the perimeter of a circle is known as its circumference.

Since the outer surface of a circle is circular, it is difficult to use a ruler for measuring the circumference.

A more simplistic way of finding the circumference of a circle is by using a thread.

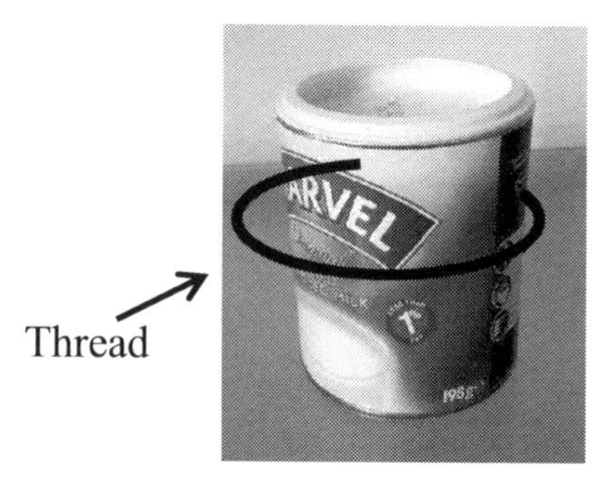

To find the circumference of the circular base of the cylindrical container, wind a piece of thread around the container once. Mark the thread where they cross each other. Pull the thread straight and measure against a ruler.

The length of the thread is the circumference of the circle.

Known fact: If you measure the distance around a circle (circumference) and then divide it by the distance across the circle through the centre (diameter), your answer will always come close to a particular value. The reliability depends on the accuracy of the measuring instrument(s).

That particular value is approximately 3.141592653…….

In mathematics, we use the Greek letter π (pronounced as ***pi***) to represent this number. In some textbooks, you might see π as 3.14, 3.141, $\frac{22}{7}$ or $3\frac{1}{7}$. All these numbers are not even accurate as it is impossible to express the number as an exact fraction or a decimal number.

Therefore,

Circumference (c) = $\pi \times$ diameter (d)

$\mathbf{C = \pi d}$

We also know that twice the radius is the diameter of a circle. Therefore, the above formula could also be written as

$\mathbf{C = 2\pi r}$,
where r is the radius of the circle.

Example 1
Calculate the circumference of a circle with a diameter of 14 cm. Use π as $\pi\frac{22}{7}$.

Solution: Using the formula, $C = \pi d$
$C = \frac{22}{7} \times 14$ cm
$C = 22 \times 2 =$ **44 cm**

Example 2: Calculate the circumference of the circle below. Use π as 3.14.

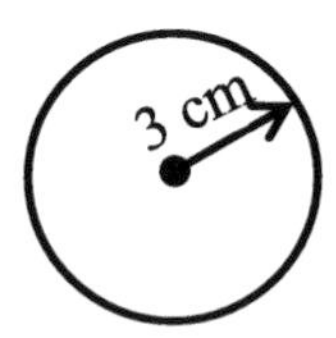

Solution: $C = 2\pi r$
$C = 2 \times 3.14 \times 3$
$C =$ **18.84 cm**

Example 3:

The front wheel of the bicycle has a radius of 35 cm.

a) What is the circumference of the front wheel?
b) How many complete revolutions does the wheel make when the bicycle travels 500 **metres**?

a) $C = \pi d$ or $C = 2\pi r = 2 \times \frac{22}{7} \times 35$
$C =$ **220 cm**

b) 100 cm = 1 m and
220 cm = 2.2 m

Complete revolution will be
$500 \div 2.2 =$ **227**

Example 4: Calculate the perimeter of the semi-circle below. Use π as 3.14.

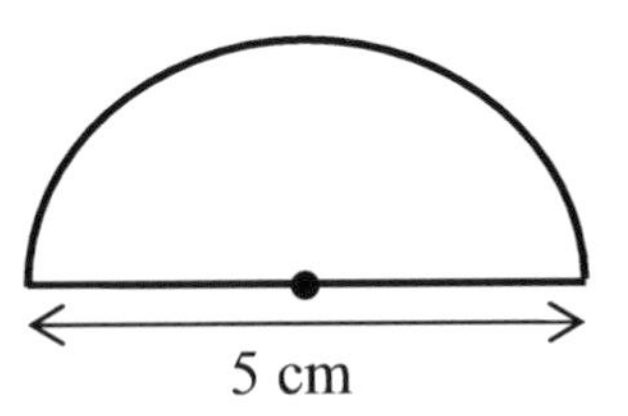

The perimeter of the semi-circle is the length of the circular face (circumference) plus the diameter.

The length of the circular face is the circumference ÷ 2.

$\frac{\pi \times d}{2} = \frac{3.14 \times 5}{2} = \frac{15.7}{2} = 7.85$ cm
Perimeter of the semi-circle
$= 7.85 + 5 =$ **12.85 cm**

Example 5: Calculate the perimeter of the object below. Use π as 3.14.

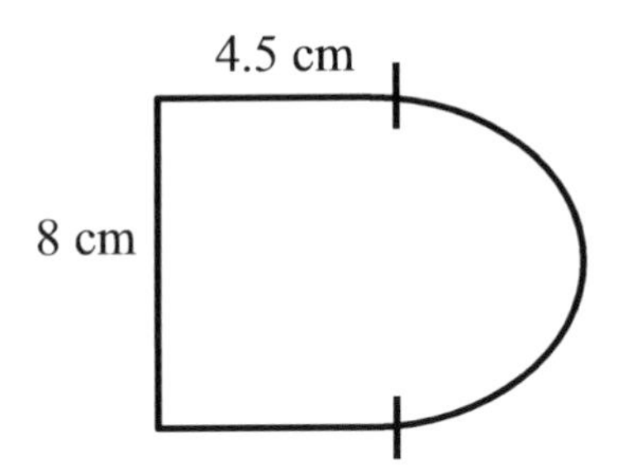

The total perimeter will be 8 cm + 4.5 cm + 4.5 cm + length of the circular end (semi-circle).

Length of the semi-circle
= circumference ÷ 2
$= (\pi \times 8) \div 2 = (3.14 \times 8) \div 2$
= 12.56 cm

Perimeter = 8 + 4.5 + 4.5 + 12.56
= **29.56 cm**

EXERCISE 17 C

1) **Group work:** Look around and find three circular objects. Measure the circular ends and record your answers. Remember to include the unit(s) you may have used.

In this exercise, you require a measuring instrument like a thread or a tape.

2) By measurements, work out the perimeter of the shapes below.

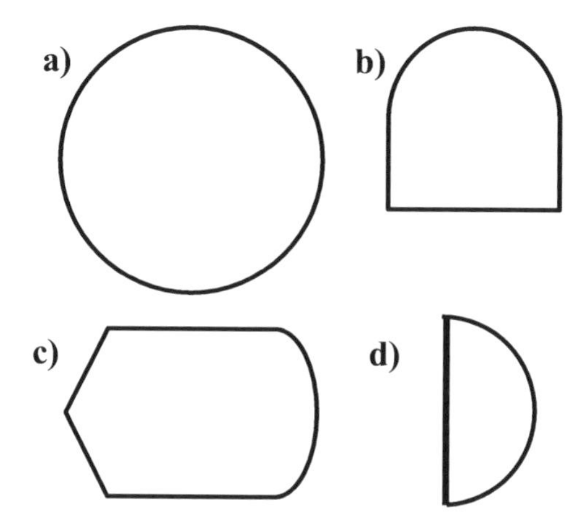

3) Calculate the circumference of the circles below. Use π as $\frac{22}{7}$.

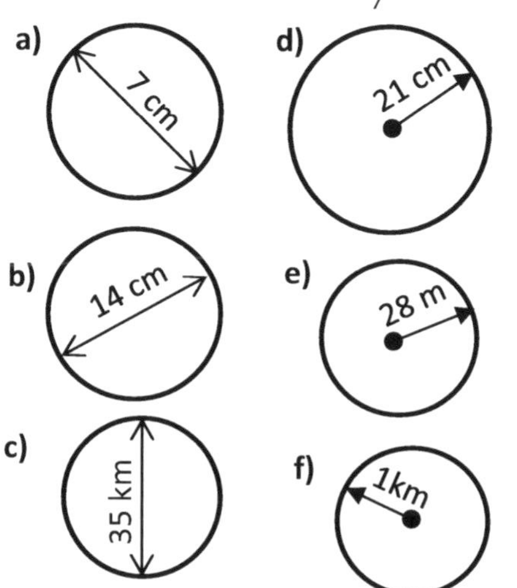

4) Calculate the perimeter of these shapes to one decimal place where possible and use π as $3\frac{1}{7}$.

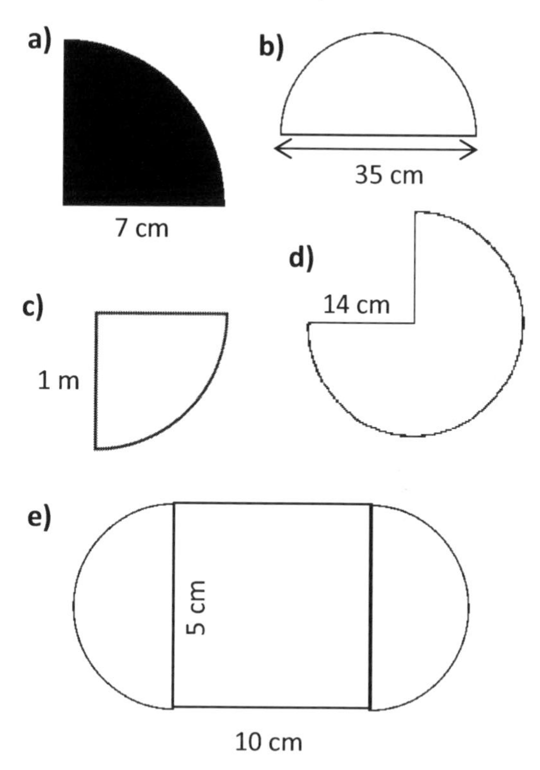

5) A circular piece of string has a radius of 21 cm. What is the circumference of the string? Use π as 3.14.

6) A duct tape is wound 20 times round a cylinder of diameter 8 cm. How long is the duct tape? Use π as 3.14

7) Calculate the diameter of a circle with a circumference of 7 cm. Take π as $\frac{22}{7}$.

8) A bicycle wheel has a diameter of 40 cm. How many complete revolutions does the wheel make when the bicycle travels 200 metres? Take π as 3.14.

17.5 AREA OF PLANE SHAPES

In simple terms, the **area** of a shape is the amount of **space** inside it.

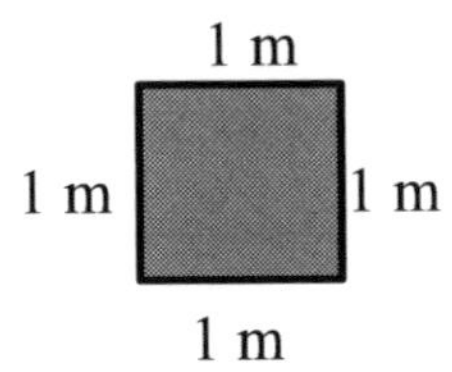

The square above has a length of 1metre and covers an area of $1 \times 1 =$ **1 m²**.

Conventionally, the square is used as the shape for the unit of area. Similarly, squares will have units depending on the unit of its length. If a length is in centimetres, the unit of the area will be **cm²**.

AREA BY COUNTING SQUARES

When shapes are drawn on a centimetre square grid, the area can be worked out by simply **counting** the squares.

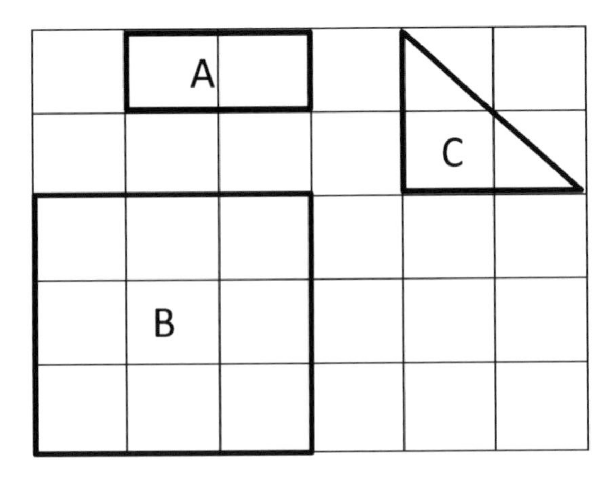

The area by counting the squares will be A = 2 cm², B = 9 cm² and C = 2 cm².

AREA BY ESTIMATE

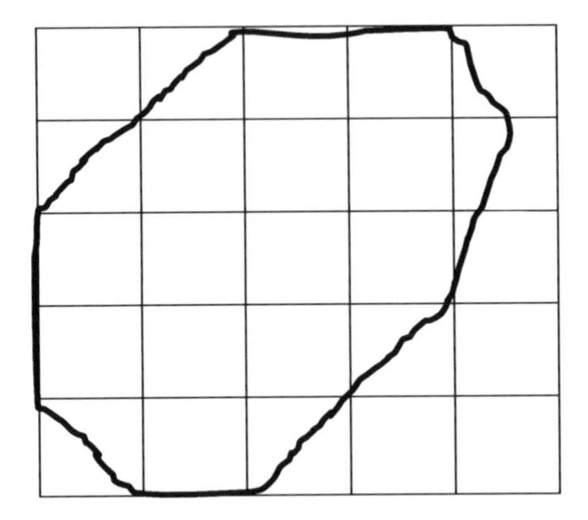

It will be almost impossible to find the exact area of the above shape. We estimate by counting the full squares and add up smaller parts to make up.

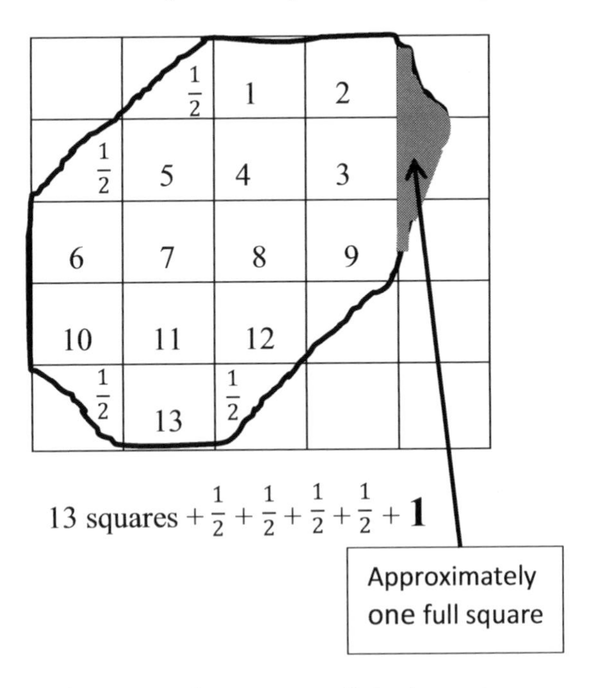

$13 \text{ squares} + \frac{1}{2} + \frac{1}{2} + \frac{1}{2} + \frac{1}{2} + \mathbf{1}$

The approximate area of the irregular shape above is **16 square units**.

AREA BY CALCULATIONS

AREA OF A RECTANGLE

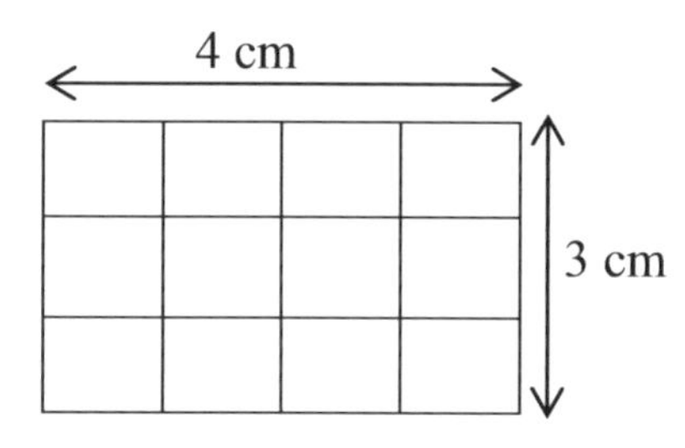

Area = length × breadth (width)
= $4 \times 3 =$ **12 cm²**

AREA OF A SQUARE

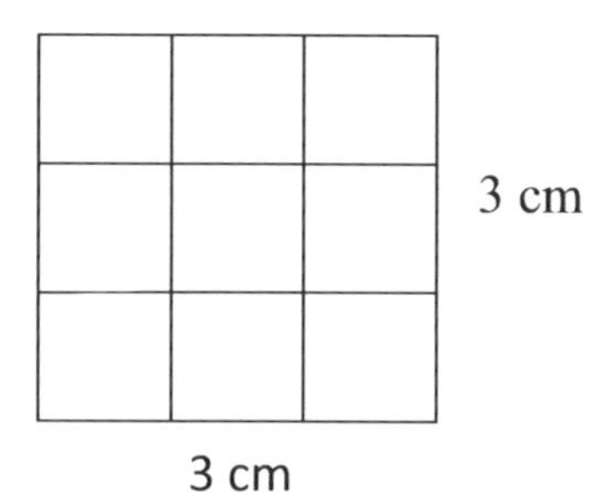

Area of a square = Length × length
= $(\text{Length})^2$
= $3 \times 3 = 9$ cm²

Example 1: The area of a rectangle is 42 cm². What is the width of the rectangle is the length is 7 cm?

Area = Length × breadth (width)
42 = 7 × width
Width = $42 \div 7 =$ **6 cm**

Example 2: Work out the area of a square with length 2 cm.
Area of a square = length × length
= 2×2
= **4 cm²**

Example 3: Work out the area of the shaded part.

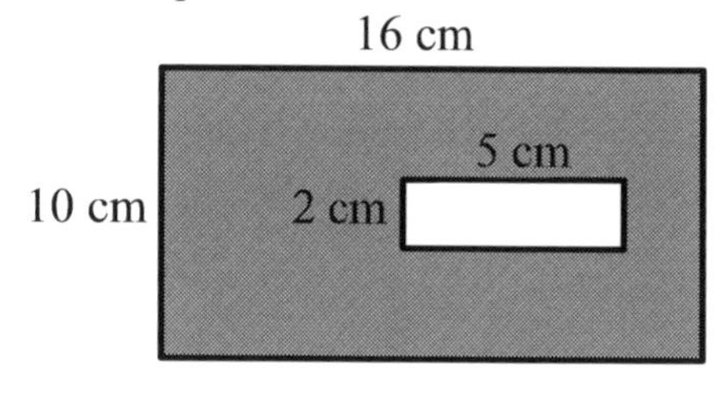

Area of big rectangle = 16×10
= 160 cm²
Area of small rectangle = 2×5
= 10 cm²
Area of shaded part = $160 - 10$
= **150 cm²**

Example 4: Calculate the length of a side of a square with an area of 49 km².

The length = $\sqrt{49} =$ **7 km**

Check: $7 \times 7 = 49$

Example 5: Calculate the area of the shape below.

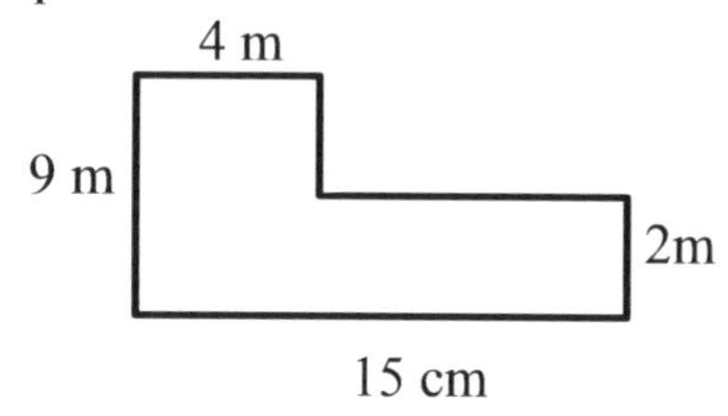

Solution
Split the shape into two rectangles.

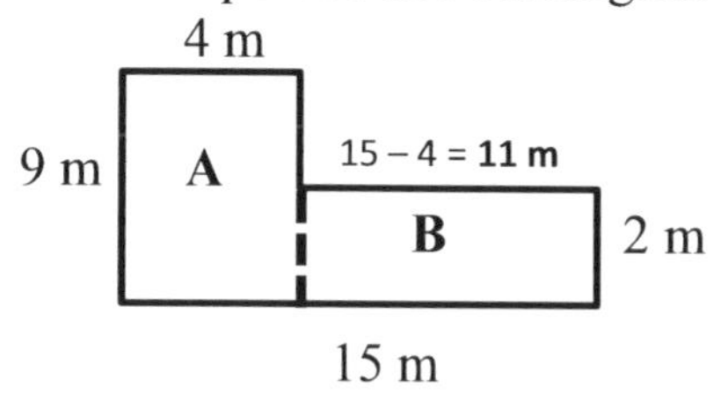

Area of A = $4 \times 9 = 36$ m²
Area of B = $11 \times 2 = 22$ m²
Total area = 36 m² + 22 m² = **58 m²**

EXERCISE 17 D

1) Calculate the area of the rectangles and square below. All lengths in cm.

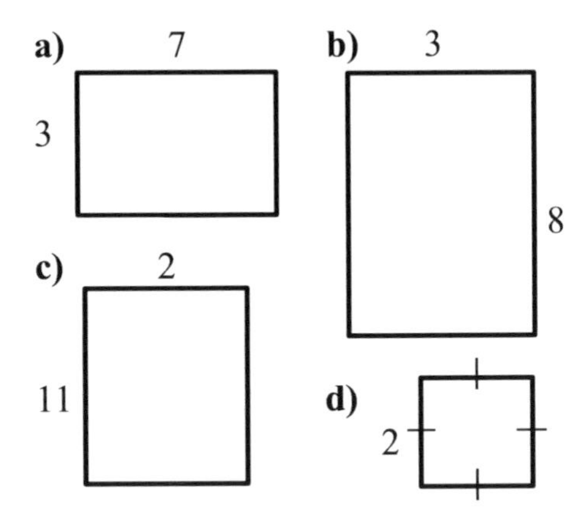

2) Below is a centimetre square grid. Work out the area of the shapes.

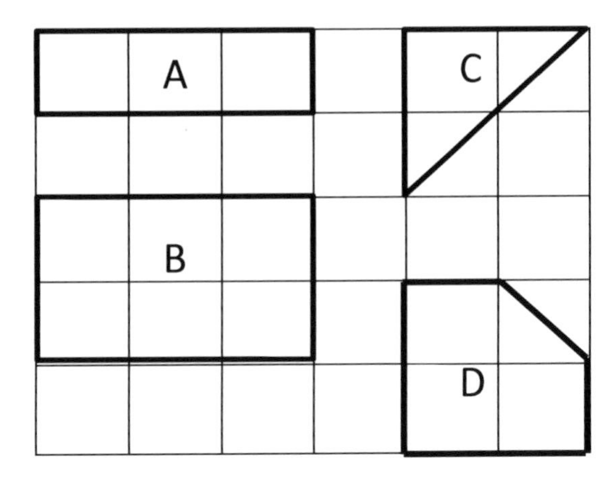

3) Estimate the area of the shapes drawn on a centimetre square grid below.

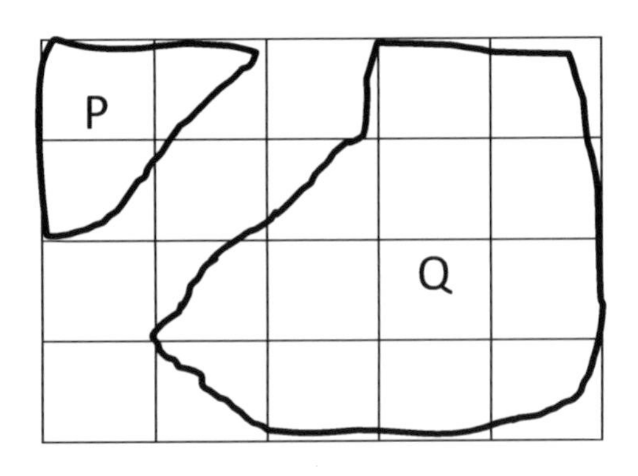

4) The area of a rectangle is 8 cm^2. What could be the length and breadth (width) of the rectangle?

5) A square has an area of 36 m^2. A length is 6 m, calculate the size of the other length.

6) Work out the area of the shaded region of the two rectangles.

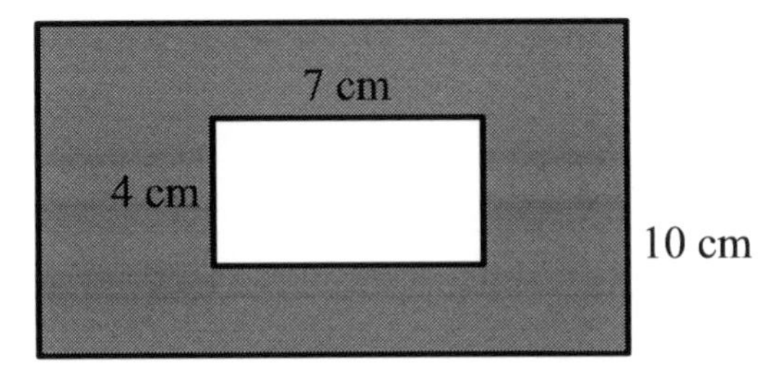

7) The floor below is to be carpeted.

5.5 m

FLOOR

4 m

a) Work out the area of the floor.
b) A vinyl carpet costs ₦1 500 per square metre. The cost of labour to lay the carpet is ₦4 300.
What is the total cost to successfully lay the carpet on the floor?

8) Complete the table of squares below.

Area	**Length of side**
1 cm^2	
9 m^2	
	8 cm
	4.3 cm
81 cm^2	

AREA OF A PARALLELOGRAM

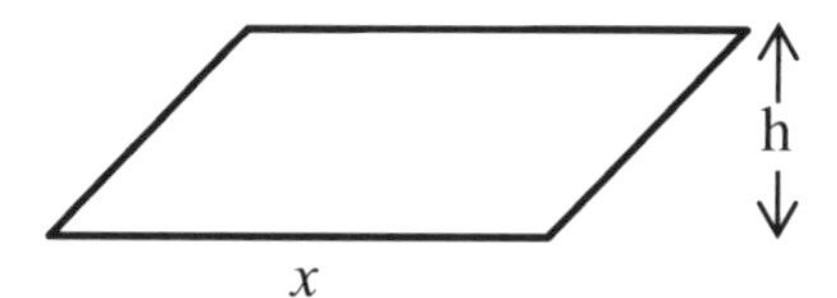

The area of a parallelogram is **base × perpendicular height**

Perpendicular means at 90°.

Example 1: Calculate the area of the parallelogram below.

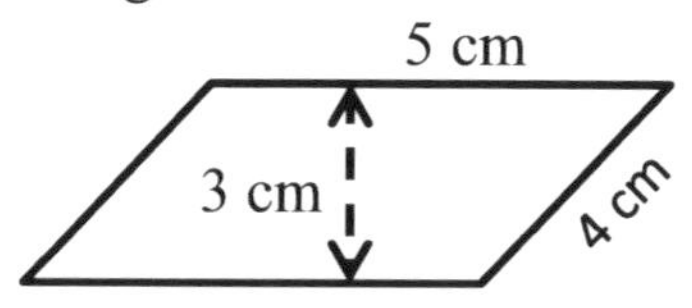

Area = 5 × 3 = **15 cm²**

Notice that 4 cm was not used. It is not perpendicular (at 90°) with the base.

Do **not use** the slant height for calculating the area.

Example 2: Calculate the area of a parallelogram with a base of 10 cm and perpendicular height of 5 cm.

Area = 10 × 5 = **50 cm²**

Example 3: Calculate the height of the parallelogram below.

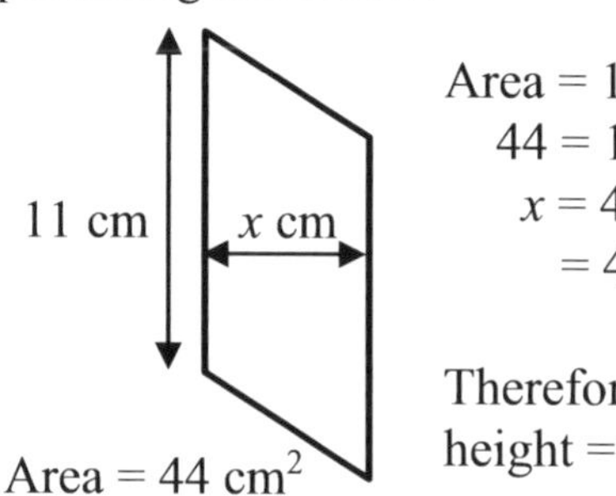

Area = 11 × x
44 = 11 × x
x = 44 ÷ 11
= 4

Therefore, the height = **4 cm**

Useful formula:

Height of a Parallelogram = area ÷ its base

Base of a Parallelogram = area ÷ perpendicular height

AREA OF A TRIANGLE

Two identical triangles will always join to produce a parallelogram.

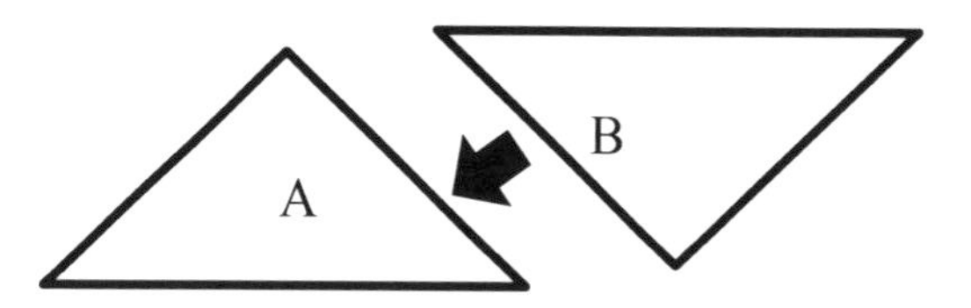

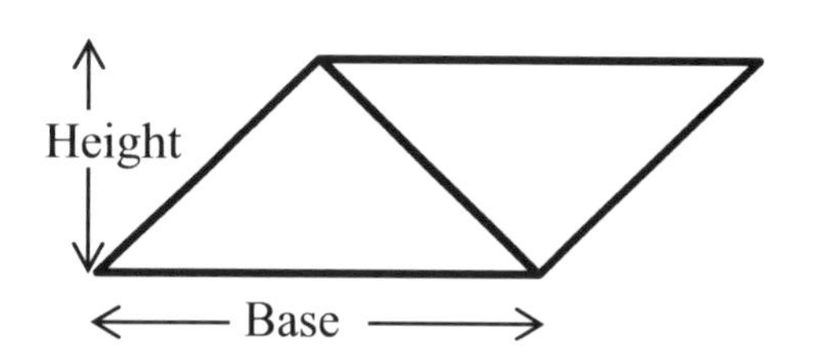

It goes to say that two triangles make up one parallelogram.

The area of a parallelogram is base × perpendicular height; therefore, the area of a triangle must be the area of a parallelogram ÷ 2.

Area of a triangle $= \frac{1}{2}$ **base × perpendicular height**

Example 4: Calculate the area of the area of the triangle below.

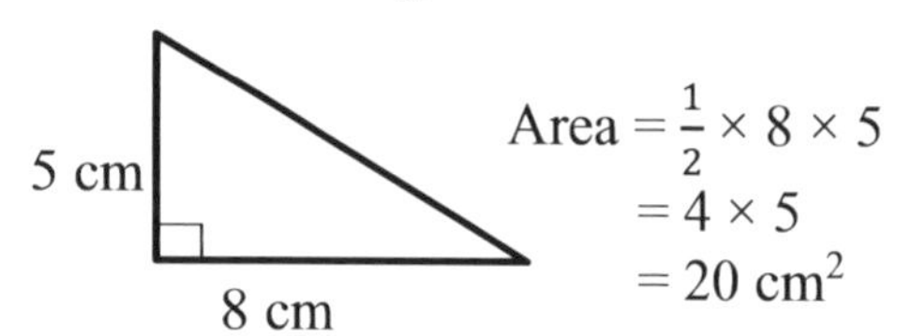

Area = $\frac{1}{2}$ × 8 × 5
= 4 × 5
= 20 cm²

Example 5: Calculate the area of triangle PQR.

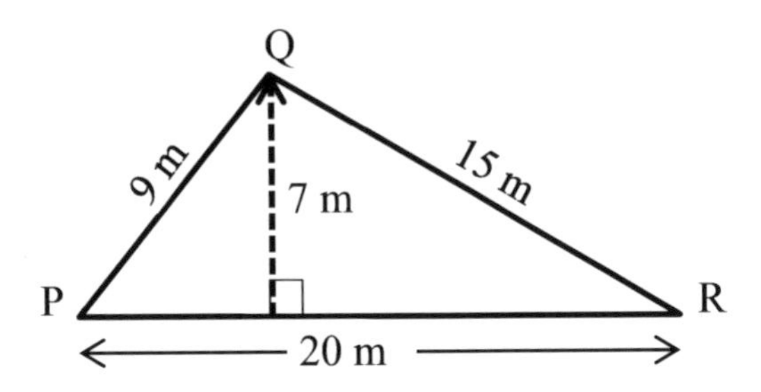

Area of PQR $= \frac{1}{2} \times 20 \times 7$
$= \mathbf{70\ m^2}$

Notice that 9 m and 15 m were not used. Only the lengths that are perpendicular to each other are used for calculating the area of a triangle.

In this case, 7 m height is perpendicular to the base of 20 m.

Example 6: Work out the area of triangle STU

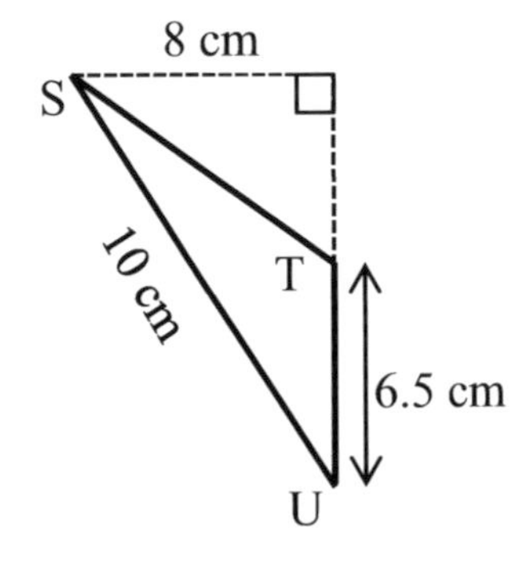

Area $= \frac{1}{2} \times 6.5 \times 8$
$= \mathbf{26\ cm^2}$

EXERCISE 17 E

1) All lengths are in cm. Work out the area of the triangles below.

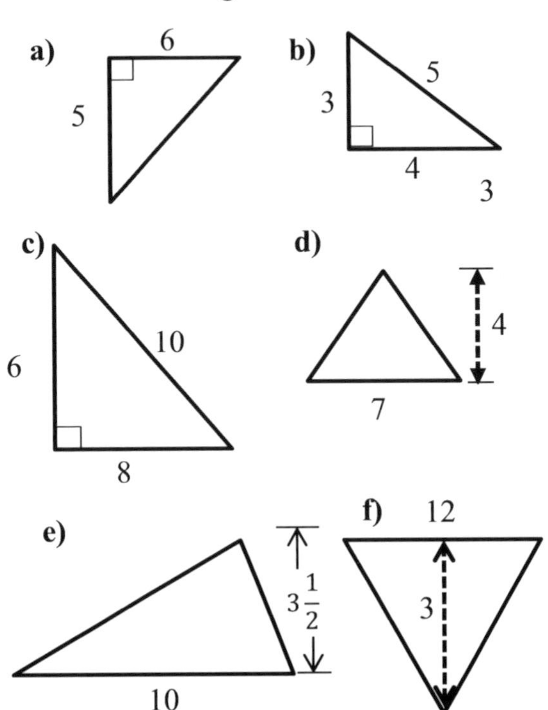

2) Calculate the total area of the quadrilateral ABCD. All lengths are in metres.

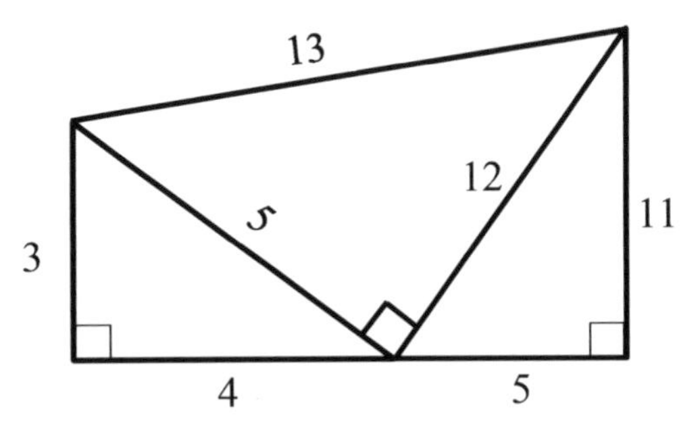

3) A triangle has an area of 30 cm^2 with a base length of 10 cm. Calculate the perpendicular height of the triangle.

4) List a possible base length and height of a triangle with area 28 cm^2.

5) All shapes below are parallelograms. Calculate the area of each shape. All lengths are in cm.

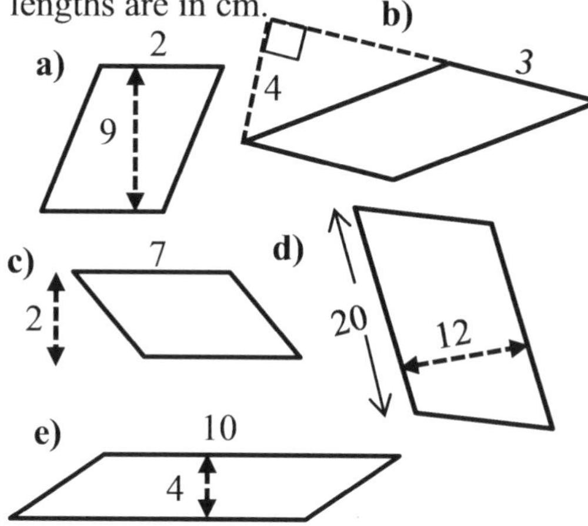

6) All lengths are in cm. Work out the area of the parallelograms and triangles below.

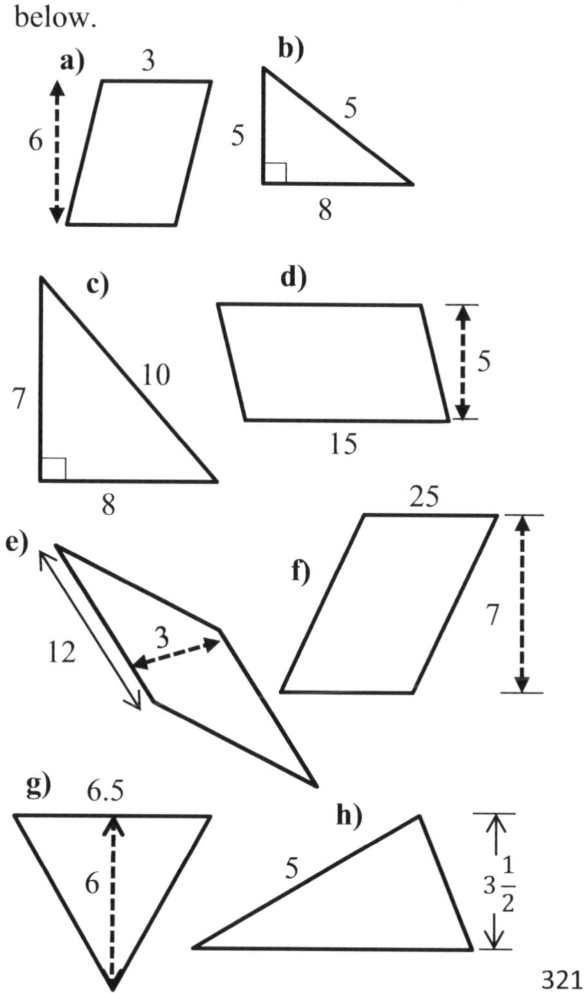

7) In questions 7a - d, calculate the base, *y*.

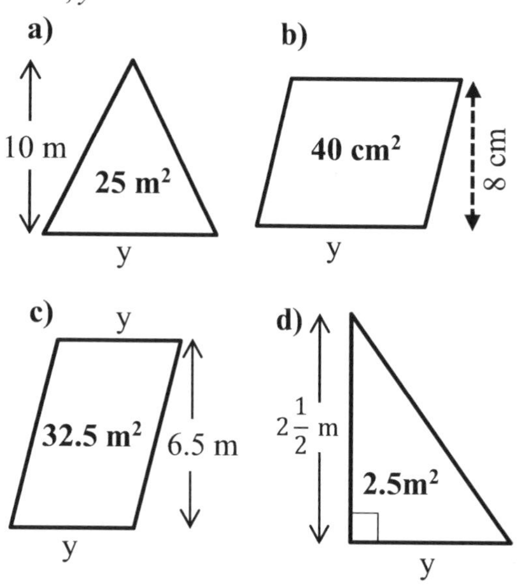

8) Calculate the area of the plane shapes below. All lengths are in cm.

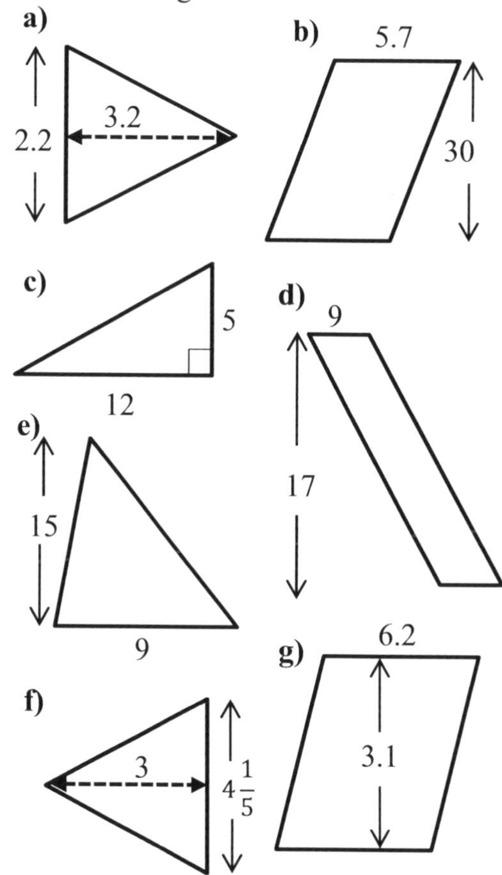

AREA OF A TRAPEZIUM

A trapezium is a four-sided straight shape with **one pair** of parallel sides.

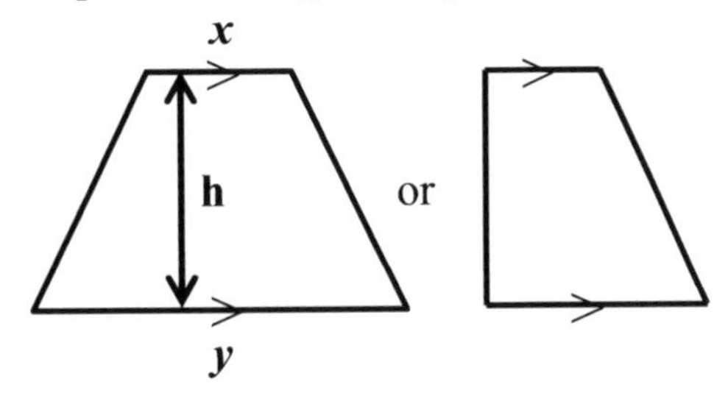

If two identical trapezia are put together, they form a parallelogram.

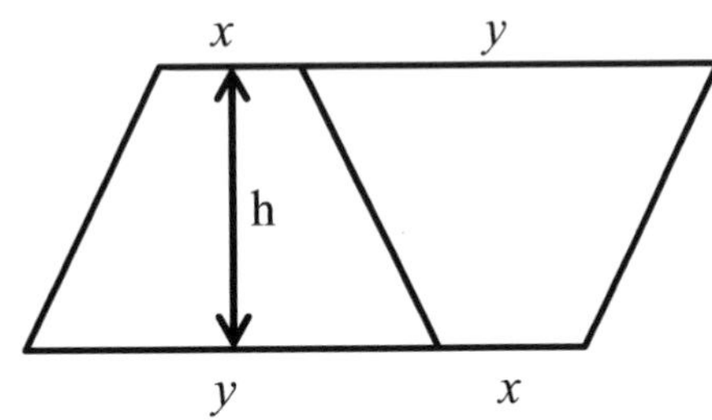

Remember:
Area of a parallelogram = base × perpendicular height $(x + y) \times$ h

Since two trapezia make up one parallelogram, the area of a trapezium
$= \frac{1}{2} \times$ sum of parallel sides × height
$= \frac{1}{2} \times (x + y) \times \mathbf{h}$

Example 1: Calculate the area of the trapezium below.

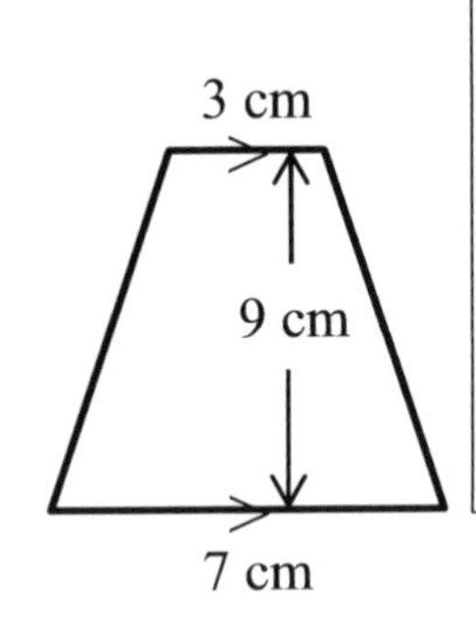

Add the parallel sides $3 + 7 = 10$
Divide by 2
$10 \div 2 = 5$
Multiply by perpendicular height
$5 \times 9 =$ **45 cm²**

Example 2: Calculate the area of the trapezium below.

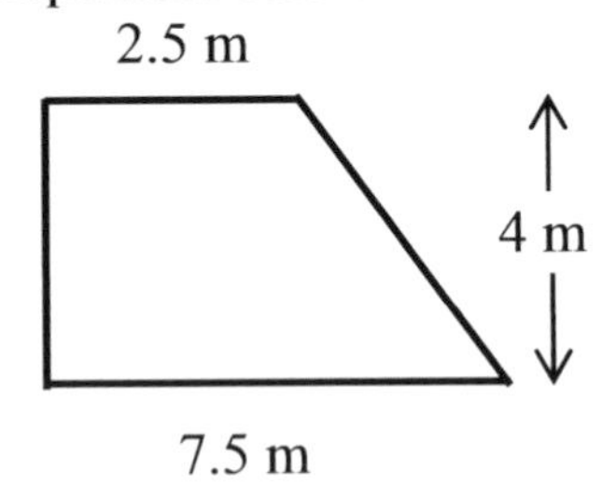

Area $= \frac{1}{2} \times (2.5 + 7.5) \times 4$
$= \frac{1}{2} \times 10 \times 4$
$=$ **20 m²**

EXERCISE 17 F

1) All the lengths are in cm. Work out the area of the trapezia below.

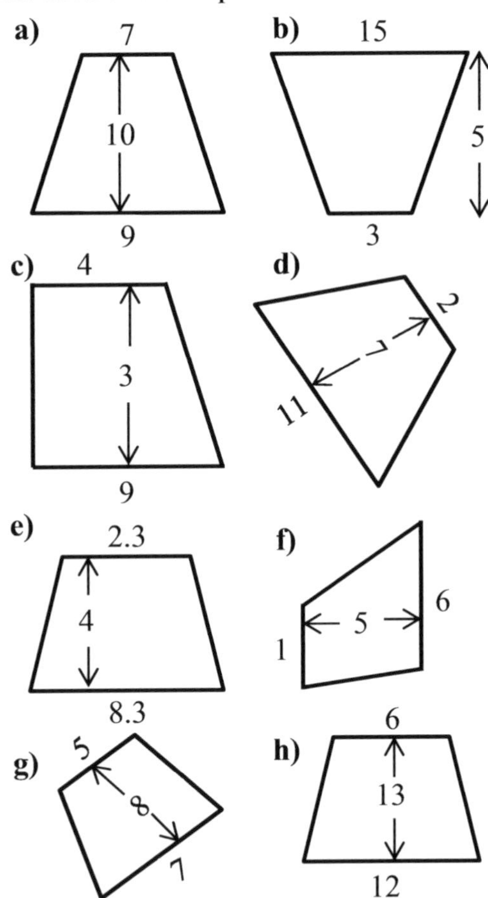

2) Calculate the length of the missing sides in the trapezia below. All the lengths are in *m*.

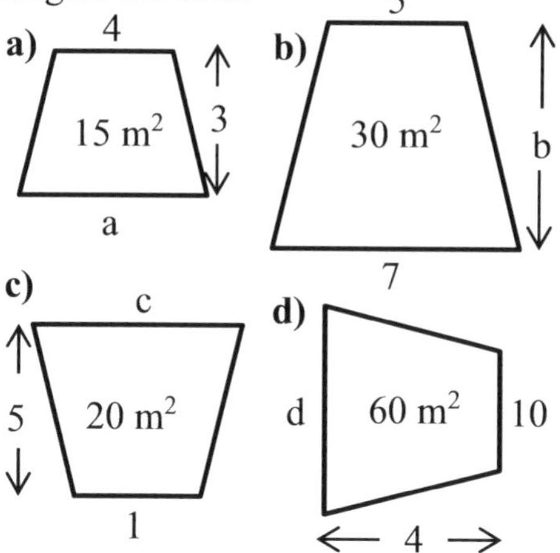

17.6 COMPOUND SHAPES

Any shape made up of more than one basic figure is called a **compound** or **composite** shape.

Areas of compound shapes are calculated by splitting them into its individual shapes and then add together.

Example 1: Calculate the area of the compound shape below.

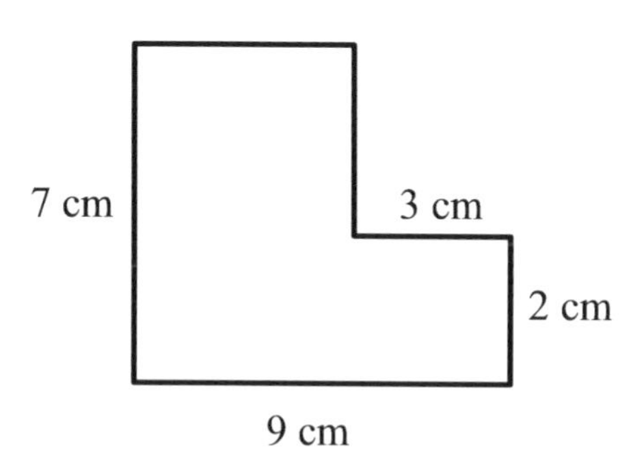

Solution: Split the shape into two rectangles.

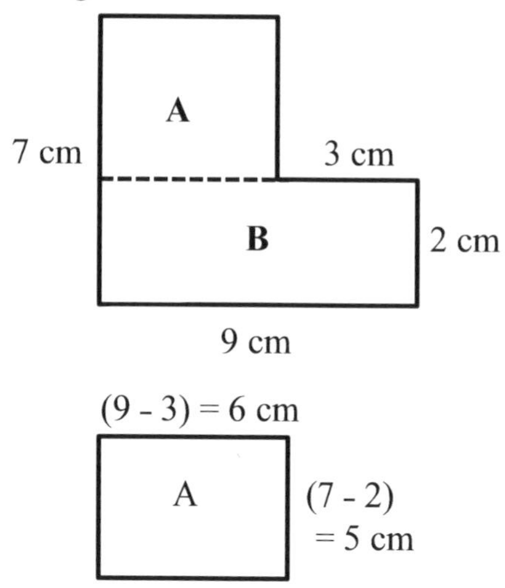

Area of A = $6 \times 5 = 30$ cm^2
Area of B = $9 \times 2 = 18$ cm^2
Total area = $30 + 18 =$ **48 cm²**

Example 2: Work out the area of the shape.

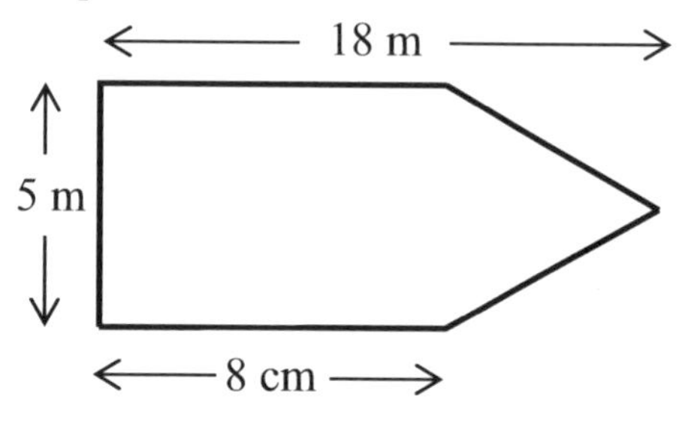

Solution: Split the shape into a rectangle and triangle.

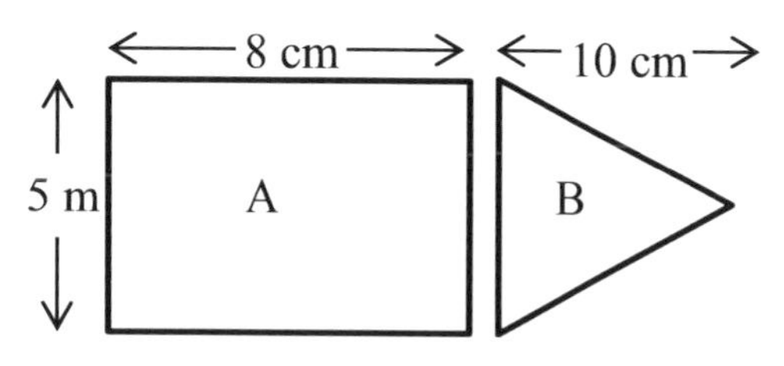

Area of A = $8 \times 5 = 40$ cm^2
Area of B = $\frac{1}{2} \times 5 \times 10 = 25$ cm^2
Total area = $40 + 25 =$ **65 cm²**

EXERCISE 17 G

1) Calculate the area of the compound shapes. All lengths are in cm.

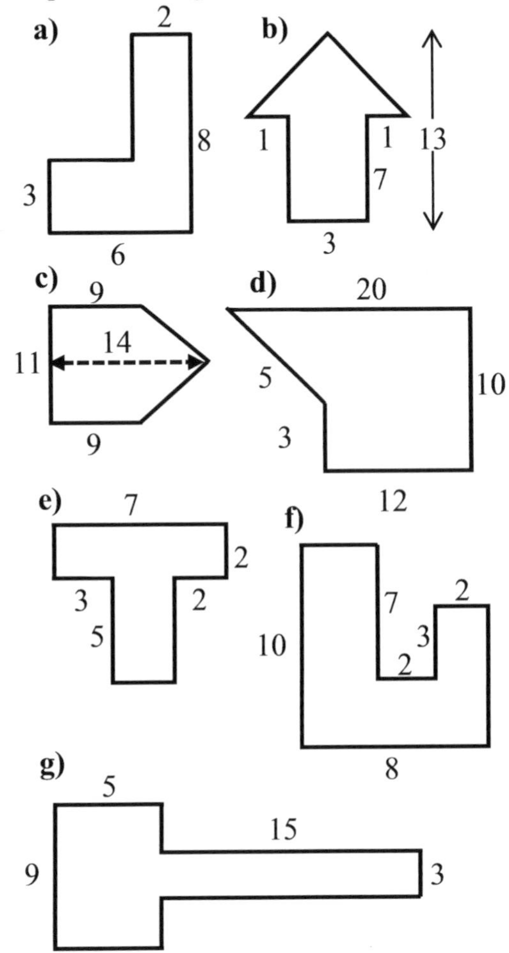

2) The shape below contains two identical triangles at both ends and a rectangle. Calculate the area of the shape. All lengths are in centimetres.

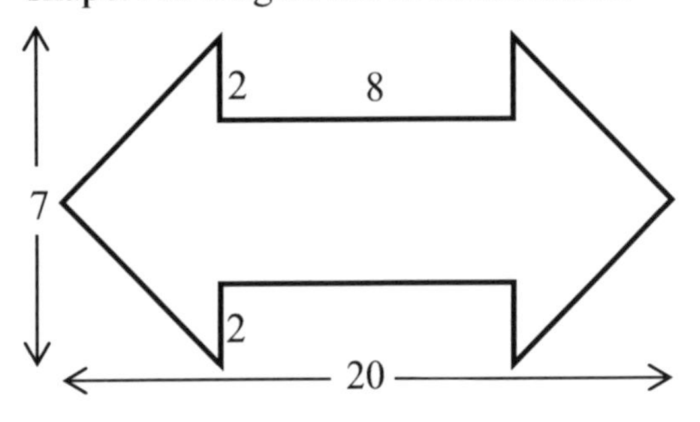

3) Obiora showed his working out for the area of this shape.

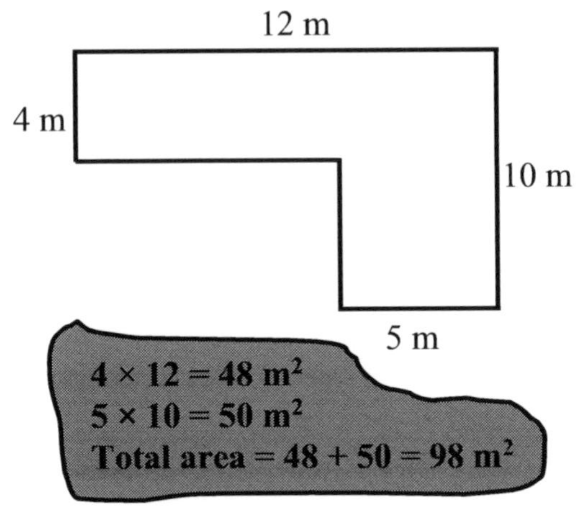

Show that Obiora is wrong.

4) Calculate the area of the shapes below. All lengths are in metres.

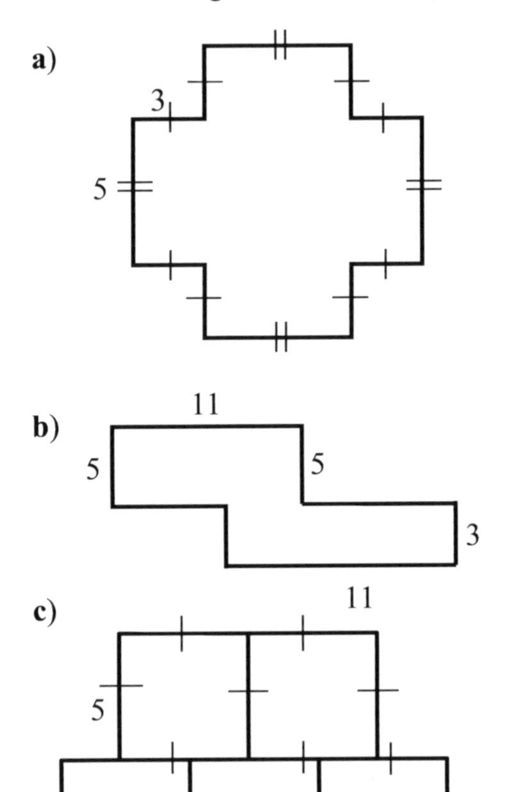

17.7 SHADED AREA

Example 1: Calculate the shaded area.

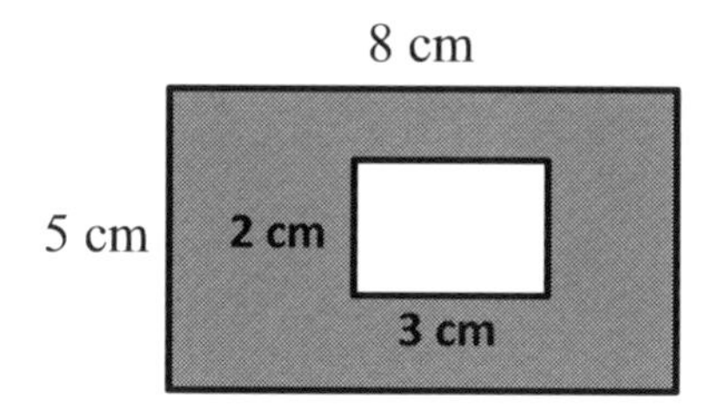

Area of big rectangle = $8 \times 5 = 40$ cm^2
Area of small rectangle = $3 \times 2 = 6$ cm^2
So, are of shaded part = 40 - 6 = 34 **cm^2**

Example 2: Calculate the shaded area

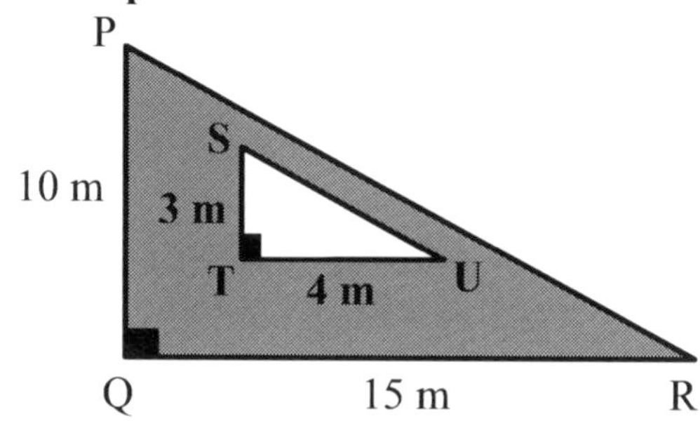

Area of ΔPQR = $\frac{1}{2} \times 10 \times 15 = 75$ m^2
Area of ΔSTU = $\frac{1}{2} \times 4 \times 3 = 6$ m^2
Shaded area = 75 - 6 = **69 m^2**

Example 3: Work out the shaded area.

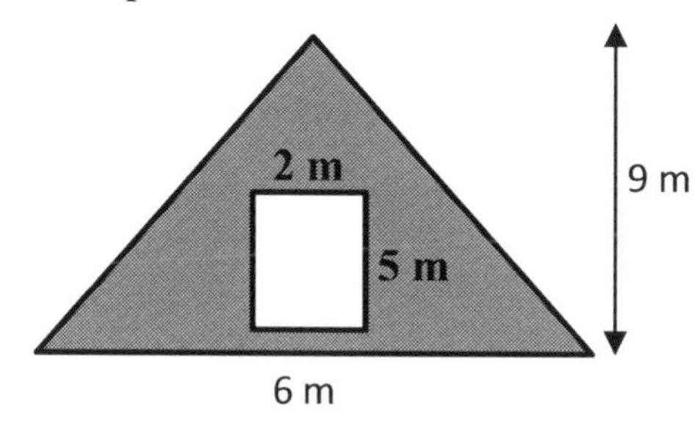

Area of triangle = $\frac{1}{2} \times 6 \times 9 = 27$ m^2
Area of rectangle = $2 \times 5 = 10$ m^2
Shaded area = 27 - 10 = **17 m^2**

EXERCISE 17 H

1) Calculate the shaded area of the shapes below. All lengths are in metres.

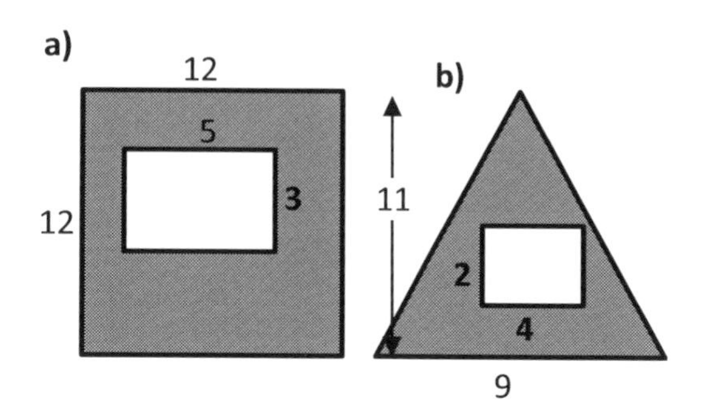

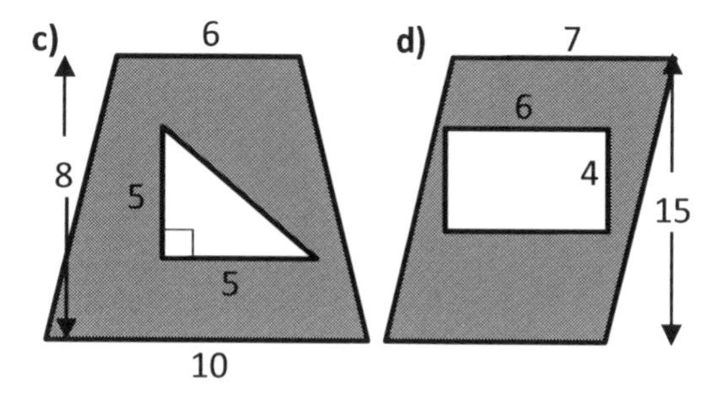

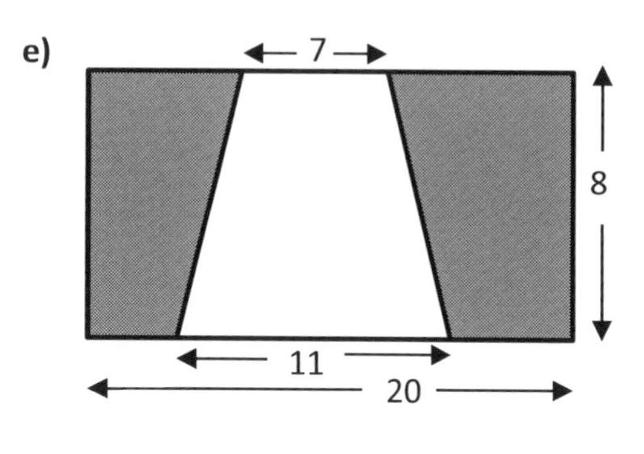

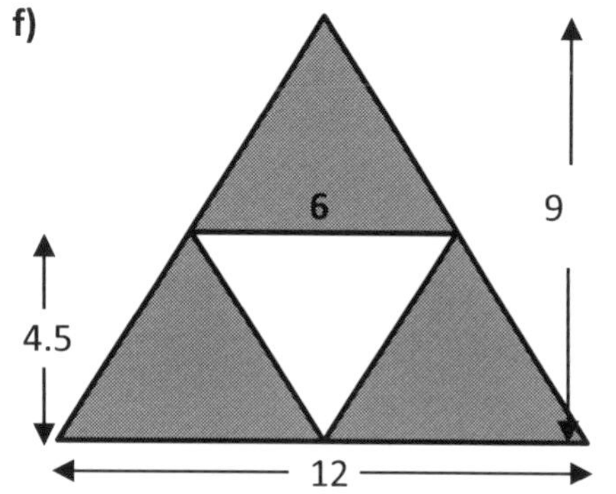

17.8 AREA OF A CIRCLE

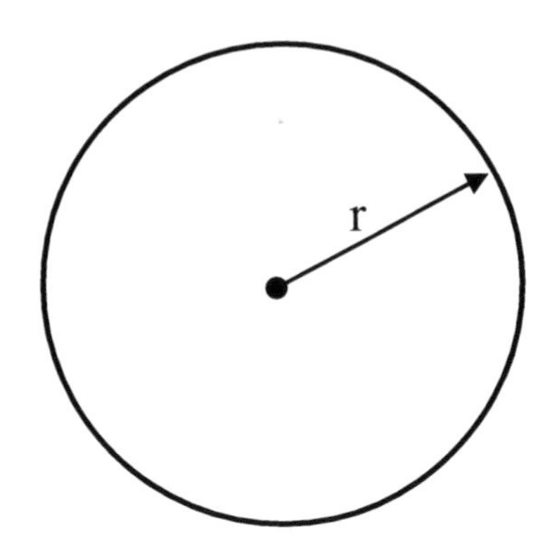

Divide the circle above into different sectors, cut them out and rearrange to form a rectangle. The length of the rectangle will be half the circumference, that is $\frac{1}{2} \times \pi \times$ diameter.
However, diameter = 2 × radius(r)

Length = $\frac{1}{2} \times \pi \times 2 \times r = \pi r$

The shape formed is **close to** a rectangle as shown below.

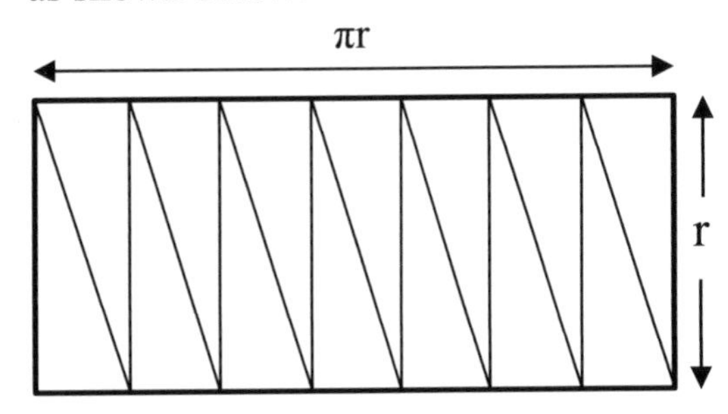

Therefore, the **area of the circle** which is also the area of the rectangle will equal $\pi \times r \times r = \pi r^2$

Area of circle = $\boldsymbol{\pi r^2}$
Always remember to use the radius instead of the diameter when working out the area of a circle.

Example 1: A circle has a radius of 7 cm. Calculate the area of the circle. Use π as $\frac{22}{7}$.

Area = $\pi r^2 = \frac{22}{7} \times 7^2$
$= \frac{22}{7} \times 7 \times 7 = \mathbf{154\ cm^2}$

Example 2: Calculate the area of the circle. Use π as $\frac{22}{7}$ and round to two decimal places.

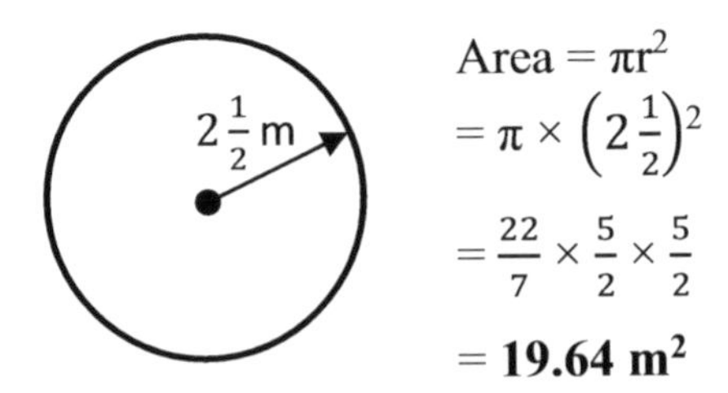

Area = πr^2
$= \pi \times \left(2\frac{1}{2}\right)^2$
$= \frac{22}{7} \times \frac{5}{2} \times \frac{5}{2}$
$= \mathbf{19.64\ m^2}$

Example 3: Work out the area of the circle. Use π as 3.14 and round to one decimal place.

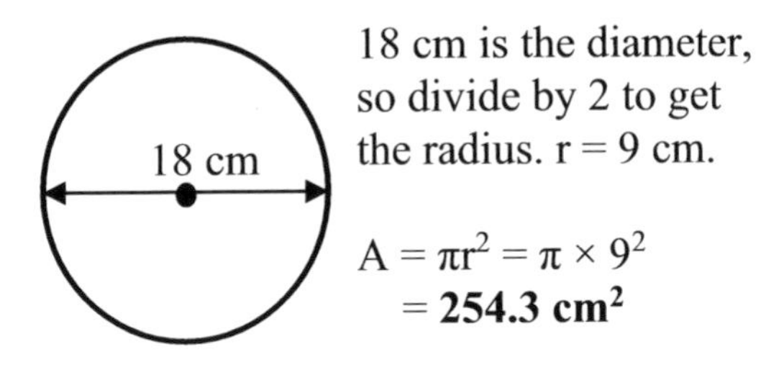

18 cm is the diameter, so divide by 2 to get the radius. r = 9 cm.

$A = \pi r^2 = \pi \times 9^2$
$= \mathbf{254.3\ cm^2}$

Example 4: Calculate the area of the semicircle. Use π as $\frac{22}{7}$.

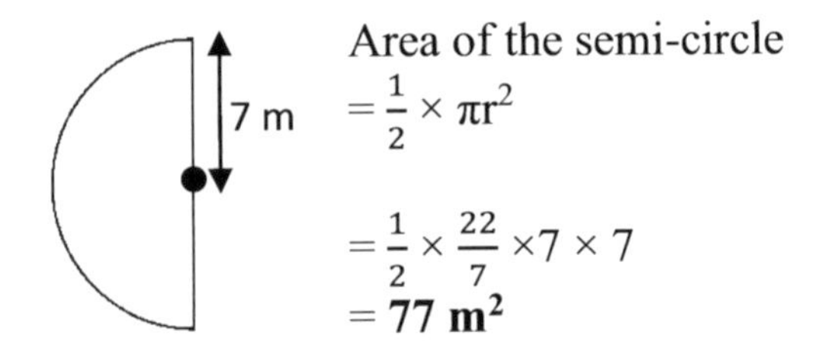

Area of the semi-circle
$= \frac{1}{2} \times \pi r^2$

$= \frac{1}{2} \times \frac{22}{7} \times 7 \times 7$
$= \mathbf{77\ m^2}$

EXERCISE 17 I

1) Calculate the area of the shapes below. Use π as $\frac{22}{7}$ and round to one decimal place where possible.

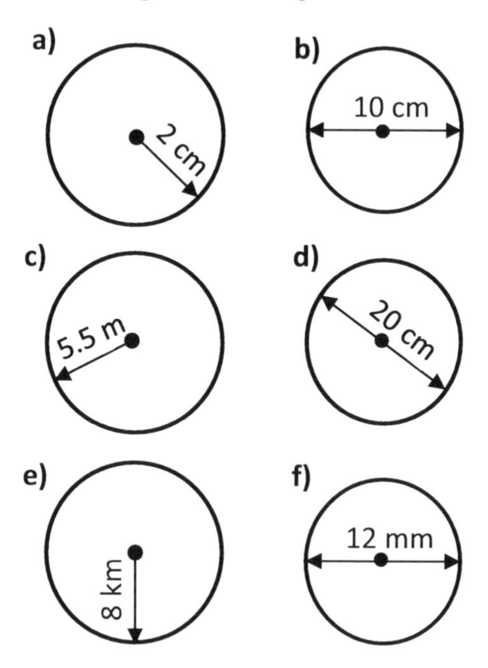

2) A circular track has a radius of 70 m.
a) Calculate the diameter of the track.
b) Calculate the area of the track.
Use π as 3.14.

3) Copy and complete the table below. Use π as $\frac{22}{7}$.

Diameter(m)	Radius(m)	Area(m^2)
14		
42		
	21	
	14	

4) Calculate the area of the semi-circles and quadrants. Give your answers to one decimal place where possible. All lengths are in metres. Use π as 3.14.

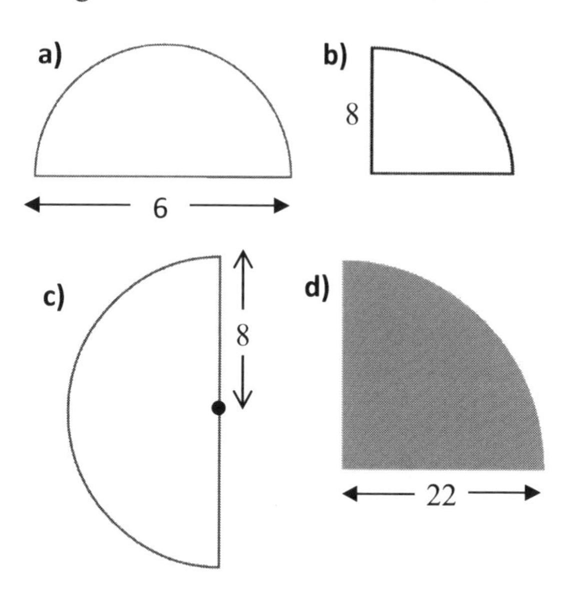

5) Calculate the area of the compound shapes made up of semicircles and a rectangle. Use π as 3.14.

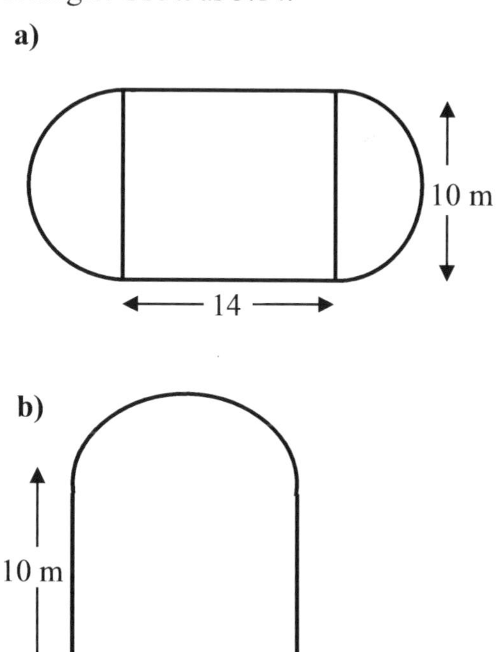

Example 5: Calculate the area of the shaded part. Use π as 3.14.

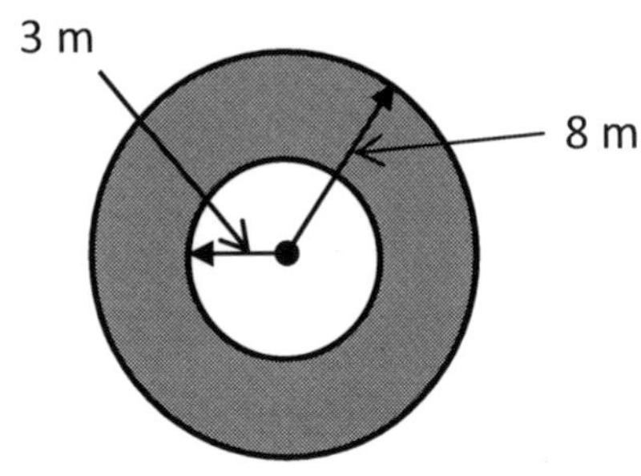

Area of the big circle = πr^2
$= 3.14 \times 8^2$
$= 200.96 \text{ m}^2$

Area of the small circle = πr^2
$= 3.14 \times 3^2$
$= 28.26 \text{ m}^2$

Area of shaded part = 200.96 - 28.26
= **172.7 m²**

Example 6: A rectangle is inscribed in a circle shown below. Calculate the shaded area to one decimal place. Take π as $\frac{22}{7}$.

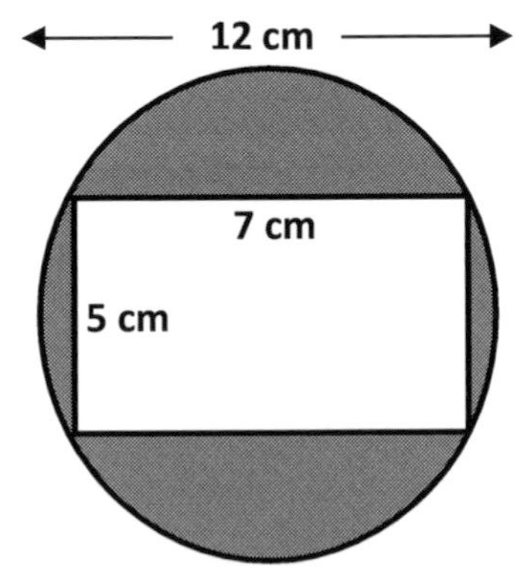

Area of rectangle = $7 \times 5 = 35 \text{ cm}^2$

Area of circle = πr^2
...But diameter = 12 cm, therefore radius = 12 ÷ 2 = 6 cm.

Area = $\frac{22}{7} \times 6^2 = \frac{22}{7} \times 6 \times 6$
$= 113.1428571 \text{ cm}^2$

Shaded area = 113.1428571 - 35
= **78.1 cm²**

EXERCISE 17 J

1) Calculate the area of the shaded part of each of the diagrams below. Take π as 3.14 and round your answers to one decimal place where possible.

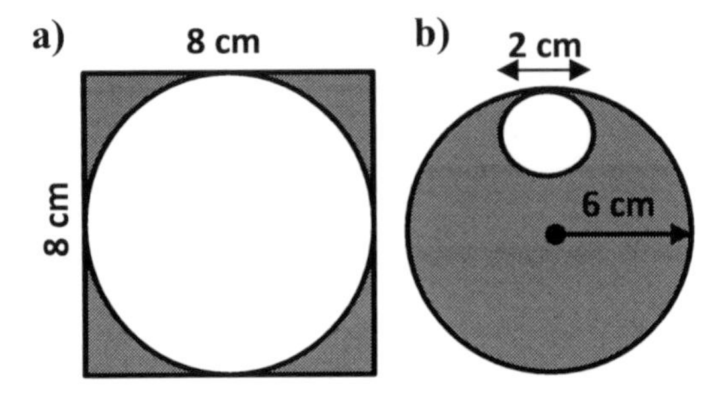

2) Three identical circles are placed in a rectangular box as shown below.

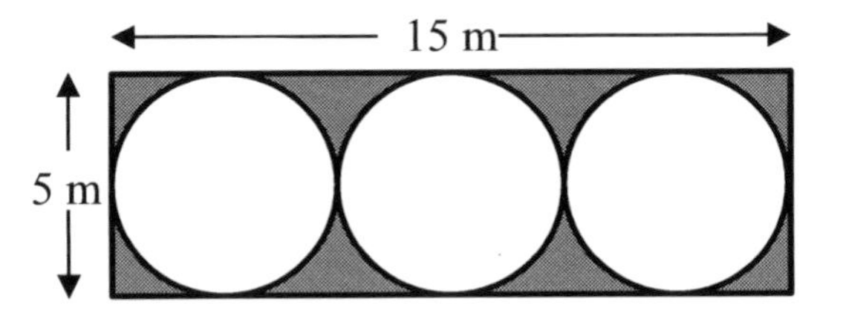

a) Work out the area of a circle.
b) Work out the total area of the three circles.

c) Work out the shaded area.
(Take π as 3.14 and round to two decimal places where possible)

3) Calculate the percentage of the shaded area. Take π as 3.14.

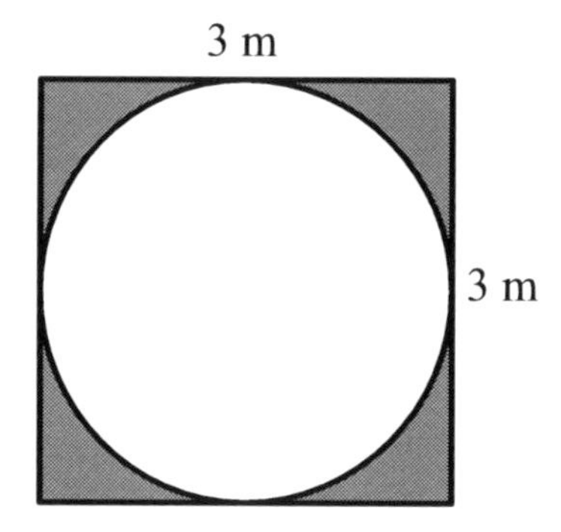

4) A circle of diameter 20 cm fits inside a semi-circle. Calculate the shaded area. Take π as 3.14 and round to one decimal place.

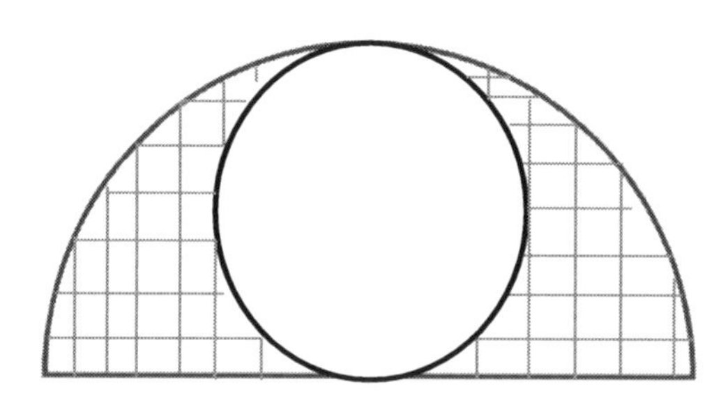

5) Three identical circles of diameter 10 cm each are placed inside a large circle of diameter 30 cm.

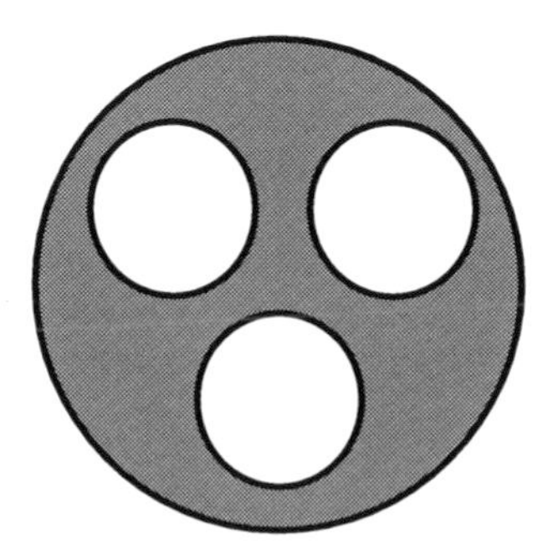

a) Calculate the area of the shaded part.
b) Work out the percentage of the shaded part. Use π as 3.14.

6) Calculate the area of the shaded part. Use π as 3.142.

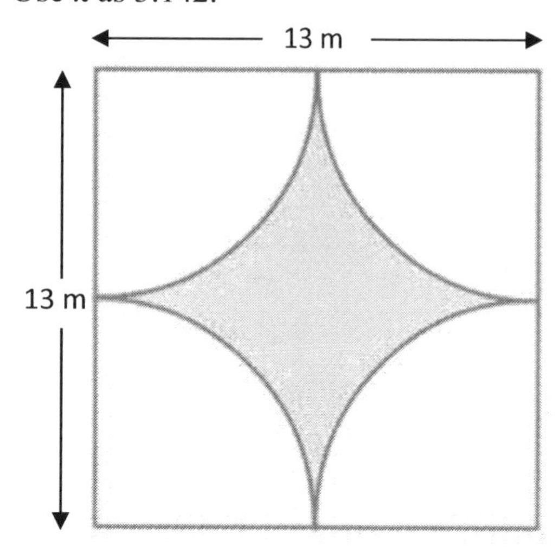

7) You have three semi-circles. The diameter of the large semicircle is 30 m. Work out the area of the shaded part and round your answer to one decimal place.
Use π as 3.142.

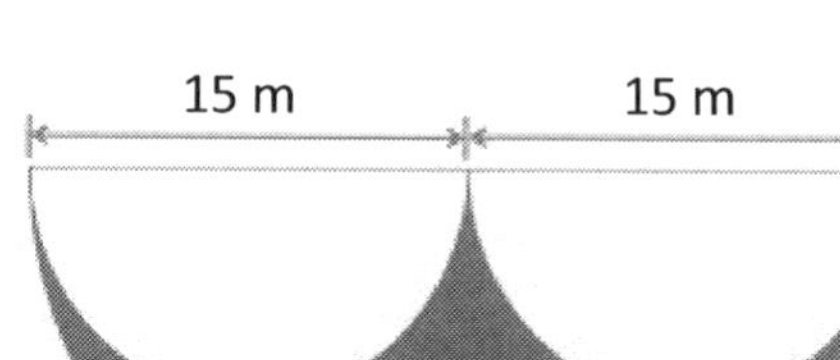

Chapter's 16 & 17 Review Section's
Assessment

1) What number belongs in each box?

a) 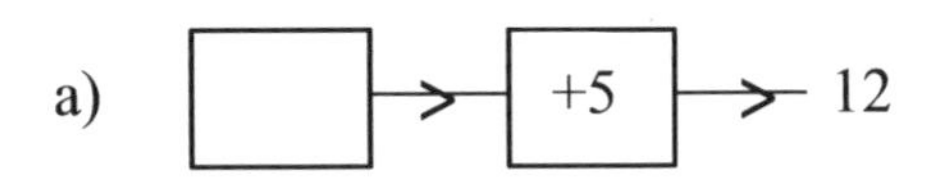

b) 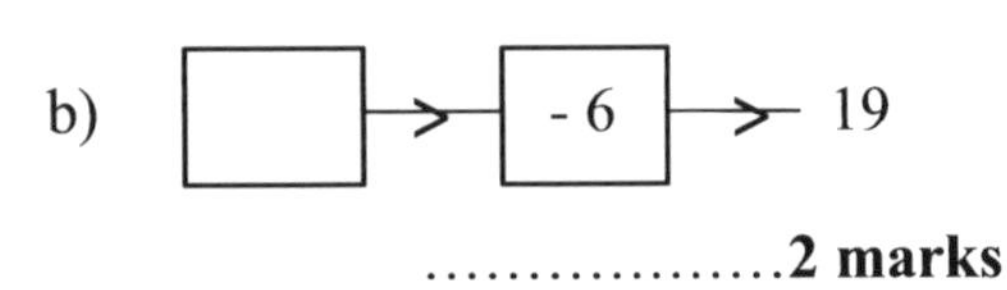

..................**2 marks**

2) What rule goes in the box?

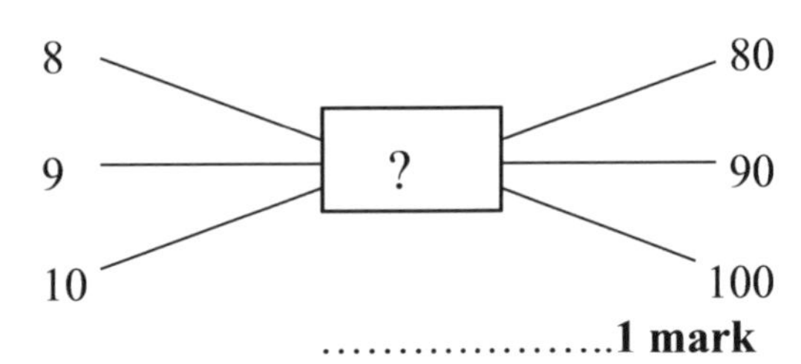

....................**1 mark**

3) Copy and complete the table for the function machine.

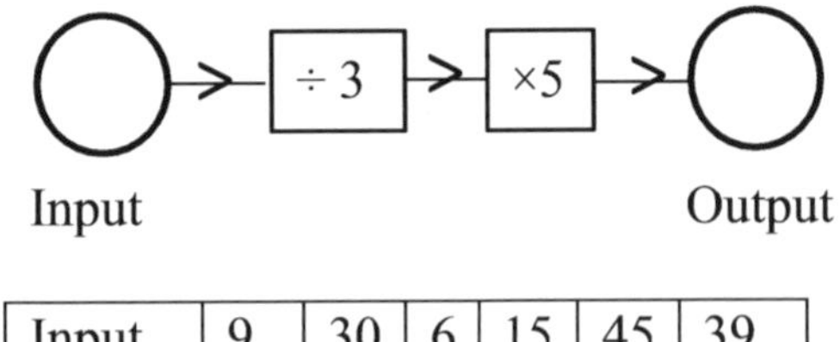

Input	9	30	6	15	45	39
Output						

..................**6 marks**

4) Complete the table for the magic squares below.

a)

8		10
	7	
4		

b)

9		
	8	
5		7

.... **4 marks**

5) On the centimetre grid below, calculate the perimeter of the shapes.

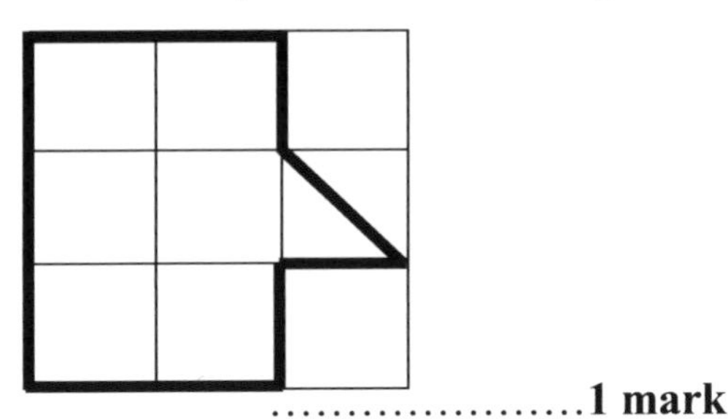

.....................**1 mark**

6) Calculate the perimeter of the object below. Use π as 3.14.

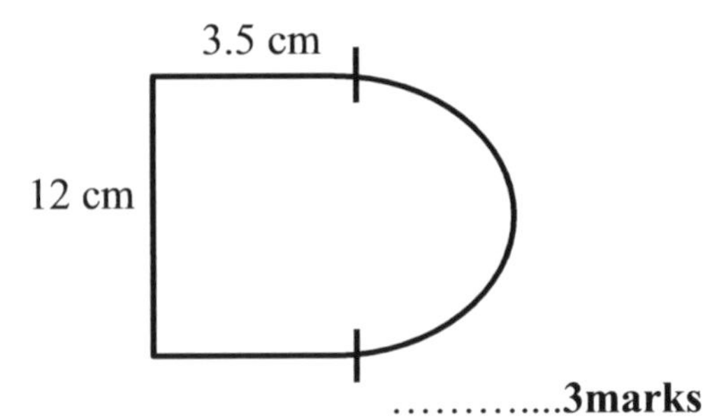

..............**3marks**

7) Calculate the circumference of the circles below. Use π as $\frac{22}{7}$.

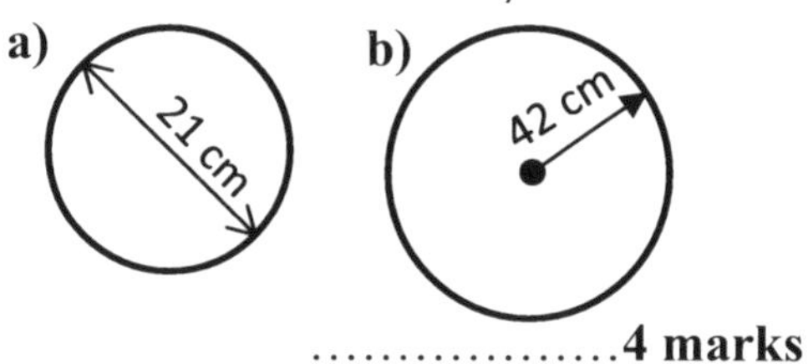

………………**4 marks**

8) Work out the area of the shaded part.

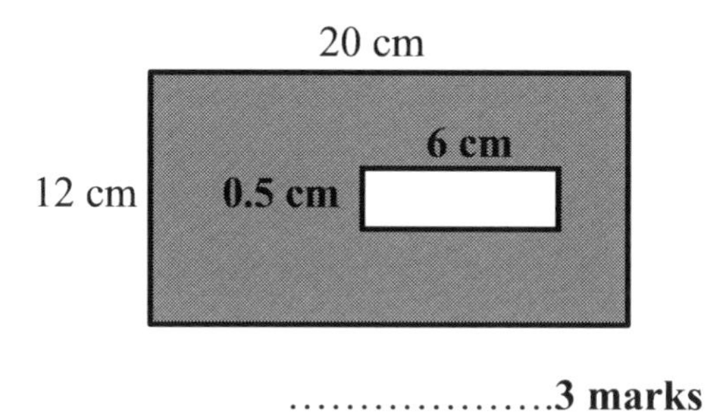

………………**3 marks**

9) All lengths are in cm. Work out the area of the triangles below.

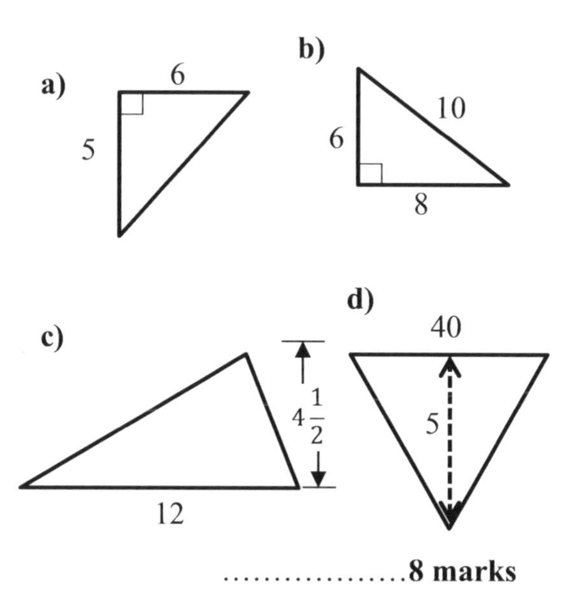

………………**8 marks**

10) A triangle has an area of 21 cm^2 with a base length of 7 cm. Calculate the perpendicular height of the triangle.

………………**2 marks**

11) All lengths are in cm. Work out the area of the parallelograms and triangles below.

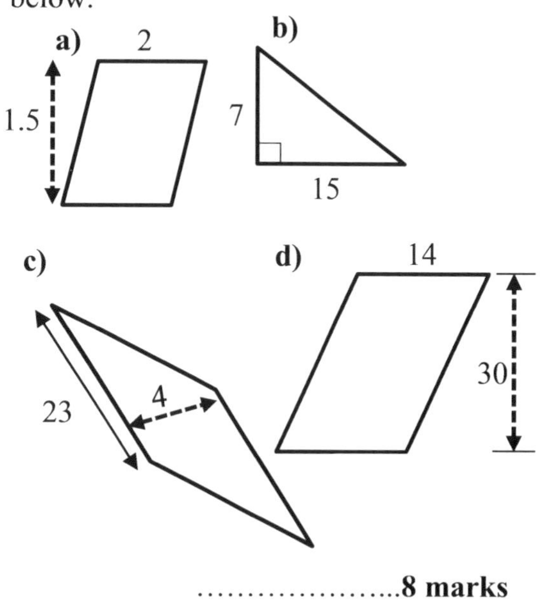

………………**8 marks**

12) Calculate the length of the missing sides in the trapezia below.

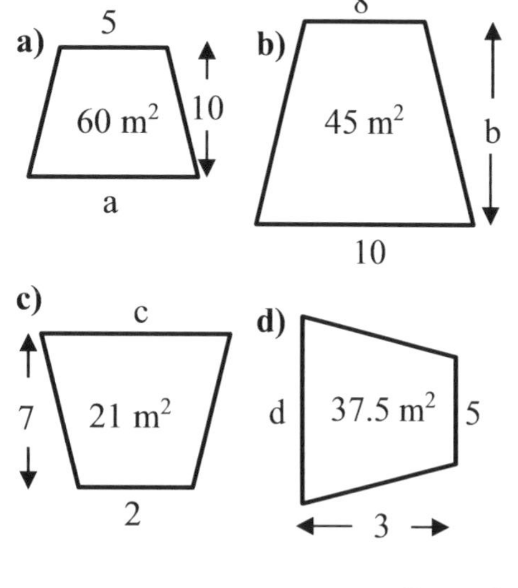

………………**8 marks**

13) Calculate the area of the compound shapes. All lengths are in cm.

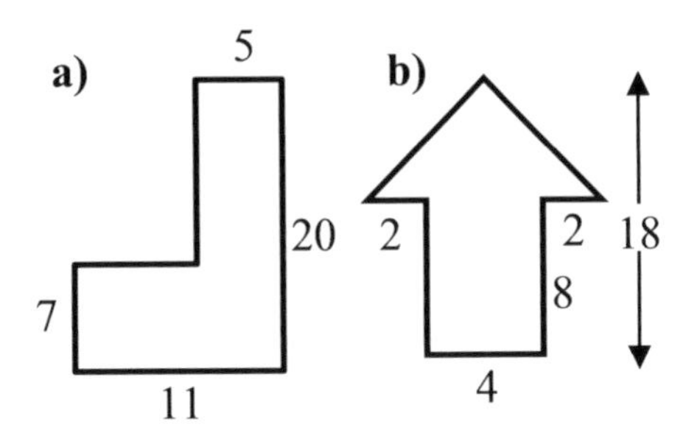

.....................**4 marks**

14) Calculate the area of the shapes below. Use π as $\frac{22}{7}$ and round to one decimal place where possible.

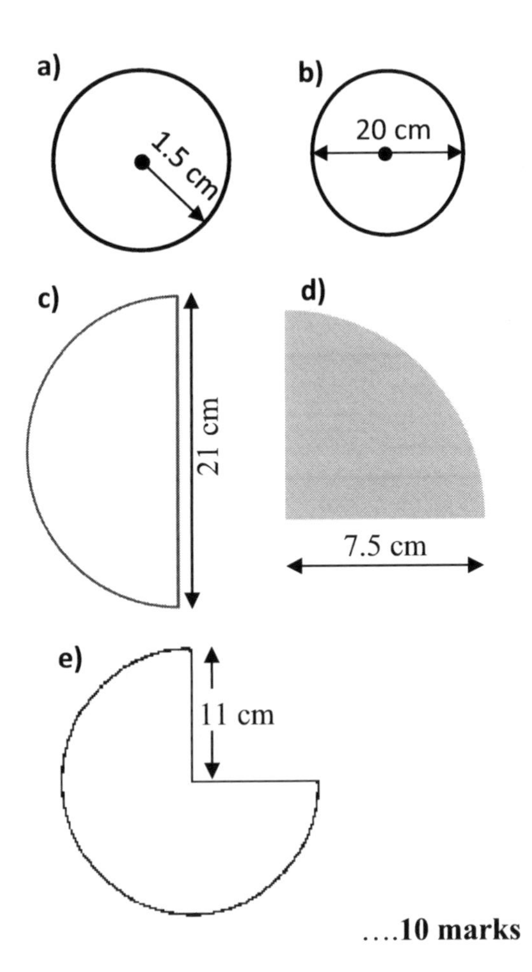

....**10 marks**

15) Four identical white squares of length 7 cm each are placed inside a large circle of diameter 25 cm.

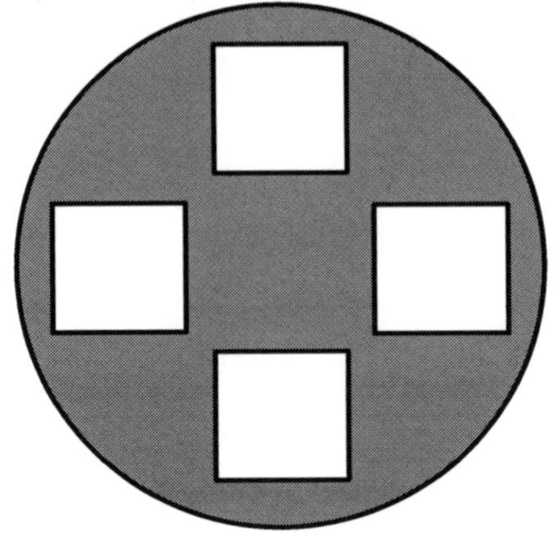

a) Calculate the area of the shaded part.

.... **3 marks**

b) Work out the percentage of the shaded part. Use π as 3.14.

....................**2 marks**

16) An arts theatre has a circular floor as the base. The diameter is 40 metres. A tile specialist charges ₦2 500 per square metre to tile the whole floor. Use π as $\frac{22}{7}$ and work out the cost of tiling the entire floor of the art theatre.

.... **4 marks**

17) A school track is made up of two straights and two semi-circular ends. Calculate the enclosed area within the track. Take π as 3.14 and round to 2 decimal places.

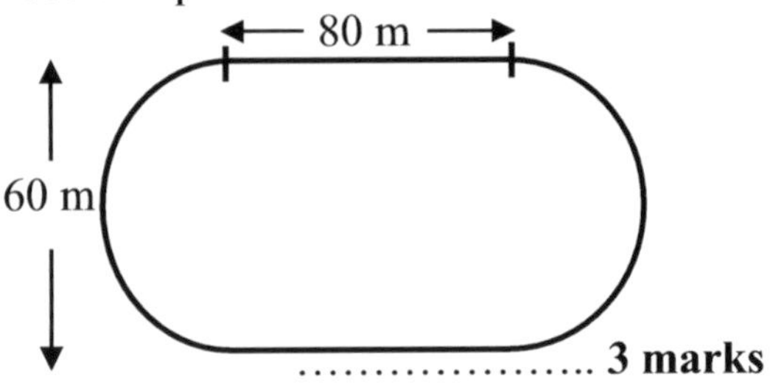

.... **3 marks**

18 Data Collection

This section covers the following topics:

- Understanding statistical data
- Collecting data

LEARNING OBJECTIVES

By the end of this unit, you should be able to:

a) Draw up statistical data
b) Understand why data is collected
c) Record a tally chart
d) Carry out surveys
e) Write questionnaires

KEYWORDS

- Data
- Collection
- Statistics
- Record
- Survey
- Questionnaire
- Tally chart
- Frequency table

18.1 PURPOSE OF COLLECTING DATA

Information could be obtained by asking questions to friends, families or even strangers and then record the details. The information obtained is turned to **data** when we use numbers to represent the details. The practice of collecting and analysing numerical data is known as **statistics**. The information from statistics is called **statistical data.**

Statistics allow sets of numerical data to be analysed and compared so that significant trends and changes could be determined and possible action taken. Statistical data about consumption of soft drinks according to age group in one year is shown below.

Age

Preferred Cola	18 to 24	25 to 29	30 to 39	40 to 49	50 to 54	55 to 64	65 or more
	%	%	%	%	%	%	%
Coca Cola	65	41	55	28	46	36	36
Diet Coke	2	10	13	15	8	12	23
Coke Zero	9	23	19	22	28	16	14
Pepsi Light	0	3	0	3	3	6	9
Pepsi Max	16	18	6	10	13	24	14
Pepsi	7	5	7	22	3	6	5
NET	**100**	**100**	**100**	**100**	**100**	**100**	**100**

Table 18.1

The statistical data like in table 18.1 above could help inform the Health Authorities in any country or State about the rate of consumption of soft drinks according to age range. Other conclusions could be reached from such data.

EXERCISE 18A

1) All questions refer to table 18.1.
a) What age group consumed the most Diet Coke?
b) What age group consumed the least Pepsi Max?

c) What age group consumed the most Coca Cola?

d) What can you say about the intake of Pepsi Light across the different age groups?

18.2 TYPES OF DATA

There are two types of data source.

Data you collect yourself is called **primary data**. These are information obtained by experimenting or by asking people questions.

Carrying out a survey falls under primary data.

Data that has already been collected by a third party (someone else) is known as **secondary data**. Sources of secondary data could be from the newspapers, the internet, magazines and any indirect means.

When data can only be *described in words*, it may be referred to as **qualitative data**. It falls into categories such as gender, place of birth, colour, names of people and so on.

When data can be given *numerical values*, they are classified as **quantitative data**. It falls into categories of the mass of an object, shoe size, temperature, time and so on.

Quantitative data can be **discrete** or **continuous**. Discrete data only take certain values. It is usually whole numbers but may include data such as shoe size. We may have size 6 or $6\frac{1}{2}$ in shoe size and not $6\frac{1}{4}$.
Continuous data are mostly obtained from measurements and can take any value in a range of numbers.

EXERCISE 18 B

1) For each statement below, say whether the data is **qualitative** or **quantitative**.

a) The softness of a dog
b) The age of a car
c) The shoe size of an adult
d) The colour of a car
e) The time to complete 100 m sprint.
f) The colour of the sky
g) The number of sweets in your pocket
h) The weight of eggs in a crate
i) The length of a perimeter fence
j) The heights of students

2) To test whether these statements are true, state whether you need primary or secondary data.

a) "Boys in class 5 at Williams's school prefer Geography to Chemistry."

b) "It rained in Abuja last summer."

c) "London has more illiterates than literates."

d) "70% of pupils in year 9 would prefer to visit the zoo than to go shopping."

e) "More women read *Hello Magazine* than men."

f) Tertiary education is better than secondary education."

18.3 COLLECTING DATA

To organise data, you need data collection table or usually called **frequency table or tally chart.**

You may need three columns for a frequency table. The first column is for listing the items you are collecting or going to count, the middle column for tally marks and the last column for recording the frequency of each unit/article.

Example 1: Here are some examination grades recorded by Tunde for some Senior Secondary School students.

B2 A1 C4 C4 P7 F9 C4 C6
A1 B2 B2 B2 C4 C4 C6 F9
F9 F9 F9 F9 F9 C4 C5 C6

a) Draw a frequency table (Tally Chart) to record the grades
b) How many students got grades F9?
c) How many students are represented in Tunde's survey?

a)

Grades	Tally	Frequency
A1	//	2
B2	////	4
C4	~~////~~ /	6
C5	/	1
C6	///	3
P7	/	1
F9	~~////~~ //	7

b) 7 students
c) 24 (Add the frequency column)

EXERCISE 18 C

1) A spinner has five sides with the colours Blue (B), Red (R), Yellow (Y), Green (G) and White (W). It is spun many times and the result recorded.

W R Y G B W W Y B
W R G B Y G Y B W
Y W B W R G Y Y W

a) Record the information on a frequency table (Tally Chart).
b) What colour was the least recorded?
c) What colour occurred the most?
d) How many times was the spinner spun?
e) What is the probability of the spinner landing on red?

2) A secondary school class was asked how many pairs of shoes they had. Some of the results were recorded below.

Number of shoes	Tally	Frequency
1	//	2
2	~~////~~ //	
3		11
4+		3

a) Complete the table.
b) How many people were in the class?
c) Joe says "The number of people who said they own two pairs of shoes is less than those that stated that they own more than four pairs." Is Joe correct? Explain fully.

3) The marks out of 50 of some students in a history examination are recorded below.

12	13	15	20	25	42
15	45	13	15	18	15
40	13	15	14	15	20
42	42	15	12	42	14

a) Draw a tally chart to show the data.
b) How many students scored 40 marks or less?
c) How many students scored the highest mark?
d) How many students were surveyed?
e) What was the most popular mark?
f) What percentage is 13 marks?

4) Adetola surveyed to find out how many Year 8 pupils travelled to school on a Friday. Adetola took a sample and recorded his answers as shown below.

Travel method	Tally	Frequency
Cycle		2
Bus	~~////~~ ///	
Taxi	////	
Walk		17

a) Copy and complete the table.
b) How many pupils were surveyed?
c) What is the most common mode of transport?
d) How many more pupils walked to school than used the bus?

QUESTIONNAIRES

A survey collects primary data or information. A **questionnaire** is only one way of collecting data.

It is a *form* that asks people some questions, and it is important to write it correctly.

CRITERIA FOR A GOOD QUESTIONNAIRE

When writing a questionnaire, use simple language. Ask questions that can be answered with a 'Yes' or 'No' and provide tick boxes where possible. Avoid overlap of data and make sure there is an introduction/title for the questionnaire. Personal and leading questions should not be encouraged.

USE SIMPLE LANGUAGE

"People say that an average home has more than two brothers. We have three brothers in my family. How many do you have?" ✖

The above question is not an acceptable format for a good questionnaire because it is too wordy. Always keep it short and simple. A better question could be:

How many brothers do you have?
Tick one box.

☐ 0 ☐ 1-2 ☐ 3-4 ☐ More than 4 ✓

Tick boxes are provided to make it easier to respond. Also, it is shorter and has time frames.

CHOICES MUST NOT OVERLAP

"How many pets do you have?

☐ 0 ☐ 1-2 ☐ 2 - 4 ☐ 5 - 6

The choices overlap. ✖

Someone with two pets will not know what box to tick as there is an overlap of the number 2 in the second and third boxes.

Instead write:

☐ 0 ☐ 1-2 ☐ 3 - 4 ☐ More than 4 ✓

AVOID LEADING QUESTIONS

"Do you agree that watching musical channels on TV is bad for your eyesight?" ✖

That is a leading question and must be avoided to reduce bias. Instead, write:

Watching musical TV is bad for your eyesight.

☐	☐	☐	☐	
Strongly Disagree	Disagree	Agree	Strongly Agree	✓

AVOID VAGUE QUESTIONS/ANSWERS

How much money do you spend on breakfast each month?

☐	☐	
A little	A lot	✖

The answer options are too vague. 'A lot' and 'A little' does not say much.

Instead write:

☐	☐	☐	
Less than ₦500	₦500 - ₦2 000	More than ₦2 000	✓

AVOID PERSONAL QUESTIONS

How old are you? ________ ✖

The above question is too personal. Instead, write:

What age group are you? Please tick one box.

☐	☐	☐	☐	☐	☐	
Less than 10	10 – 20	21 – 30	31 – 40	41 – 50	More than 50	✓

EXERCISE 18 D

1) i)Write one thing that is not right about each question below.
   ii) Design a better questionnaire.

a) How much pocket money do you receive from your parents each week?

☐	☐	☐	☐
₦100 - ₦200	₦200 – 300	₦300 – 400	₦400 – 500

b) Have you ever been imprisoned?

☐ Yes No ☐

c) How much time do you spend on homework each week?

d) Your favourite food?

☐	☐	☐
Beans	Fish and chips	Rice

e) Do you eat breakfast every day?

☐	☐
Not really	At times

f) How many uncles do you have?

☐	☐	☐
0	1 – 3	3 – 5

g) Don't you agree that smoking is bad for your health?

☐ Yes ☐ No

19 Presenting Data

This section covers the following topics:

- Representing data
- Interpreting data

LEARNING OBJECTIVES

By the end of this unit, you should be able to:

a) Draw a pictogram
b) Interpret a pictogram
c) Draw a bar chart
d) Interpret a bar chart
e) Interpret data numerically and graphically
f) Dra and interpret a pictogram
g) Draw and interpret a pie chart

KEYWORDS

- Numerical data
- Bar chart
- Pictogram
- Pie chart
- Interpret
- Represent
- Frequency table

19.1 BAR CHARTS

A **bar chart** is a statistical way of presenting data. They show patterns in data, and the bars could be vertical or horizontal. Also, the width of the bars must be equal.

24 pupils are given a mark out of 4 for a short physics test, and results were:

1 4 3 3 2 4 2 1 1 1 2 4
0 4 4 4 3 3 3 4 2 3 3 3

The numbers above are **raw** data and must now be organised and structured statistically. We organise the raw data by drawing a tally chart (refer to section 18.3) shown below.

Mark	Tally	Frequency
0	/	1
1	////	4
2	////	4
3	~~////~~ ///	8
4	~~////~~ //	7
Total		24

A **bar chat** can now be drawn as shown below.

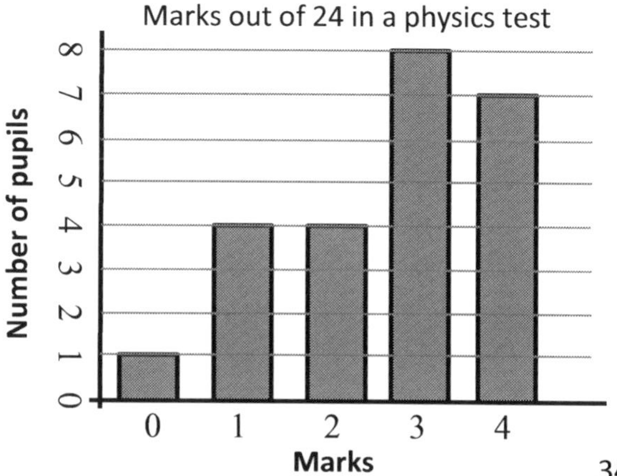

EXERCISE 19 A

1) The frequency table shows the type of TV programmes that Chika's schoolmates liked the best.

TV Programme	Frequency
Sport	5
Cartoon	8
Horror	3
Documentary	2

a) Draw a bar chart for this data.
b) What programme is the most popular?
c) How many of Ibrahim's schoolmates liked Horror and Sports?

2) The bar chart below shows the hair colour of some hairdressers.

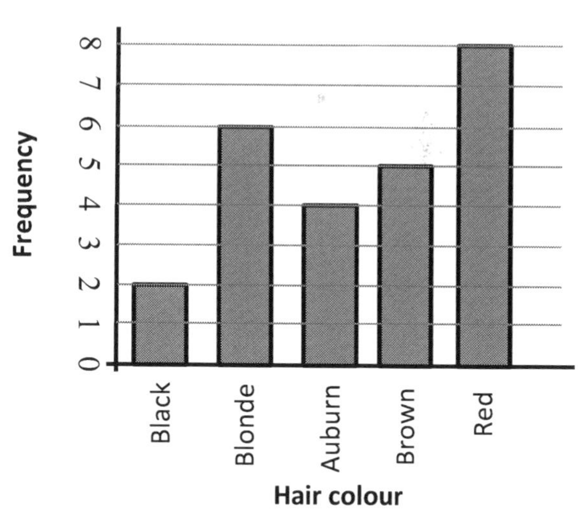

a) How many hairdressers have red hair?
b) How many hairdressers were surveyed?
c) Which hair colour is the least common?

3) A farm attendant checked the number of eggs in different nets and recorded his results.

1	0	3	6	4	6
2	5	4	2	2	4
4	0	2	4	5	6

a) By using tally table/chart, organise the results above.
b) Draw a bar chart to represent the information.

4) In a colour test, 22 people were asked to identify the colours (Yellow Y, Red R, Pink P and Blue B) they see from images as displayed on a computer. These are the results:

Y Y R P B Y R P B Y Y
B B P Y B Y R B P P B

a) Draw a frequency table.
b) Draw a pie chart to represent the information above.
c) What colour was the most common?
d) What colour was the least common?

5) Ngozi recorded the number of soft drinks sold over five days and drew a bar chart as shown below.

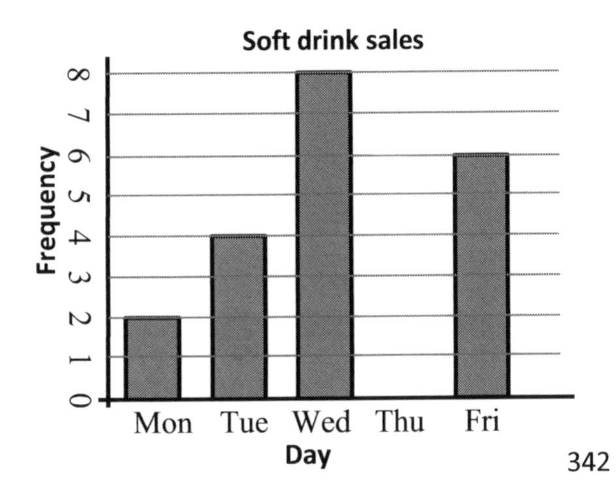

a) How many soft drinks did Ngozi sell altogether?
b) What might have happened on Thursday?
c) How many more soft drinks were sold on Wednesday than on Monday?
d) Each soft drink was sold for ₦250. What profit did Ngozi make if she bought the soft drinks for ₦200 each?

19.2 DUAL BAR CHARTS

It is used to compare two or more sets of data. Multiple bars are drawn alongside each other. It may also be called **compound bar chart.** A key must be used to identify the different colour bars.

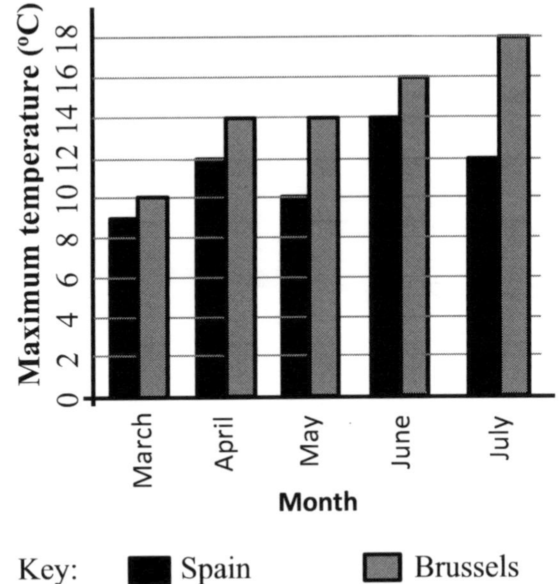

The maximum temperature in Brussel in June was 16°C.
The maximum temperature in Spain in March was 9°C.

19.3 PICTOGRAMS

A **pictogram** is a pictorial representation of statistical data on a chart or graph. It is often used in mathematics and shows the clear distinction of data in picture form.

Each of the pictures represents a number or items. Just as in dual bar charts, a **key** is used in pictograms to show how many items are represented by one image.

Example 1: Draw a pictogram to represent the data below.
Use ◆ to represents 4 people

Sport	Frequency
Football	20
Tennis	8
Boxing	6
Hockey	2
Basket ball	12
Golf	16
Rugby	8

Sport	
Football	◆ ◆ ◆ ◆ ◆
Tennis	◆ ◆
Boxing	◆ ▲
Hockey	▲
Basket ball	◆ ◆ ◆
Golf	◆ ◆ ◆ ◆
Rugby	◆ ◆

EXERCISE 19 B

1) Draw a pictogram to represent the information below. Remember to include a key.

Colour of shirts	Frequency
Yellow	3
Red	15
Blue	12
Pink	6

2) The pictogram below shows the number of envelopes Emma bought in six days. Key: ☺ = 2 envelopes.

Monday	☺ ☺
Tuesday	☺ ☺ ☺ ☺ (
Wednesday	☺ (
Thursday	☺ ☺ ☺
Friday	☺ ☺
Saturday	☺ ☺ ☺ ☺ ☺

a) How many envelopes did Emma receive on Thursday?
b) How many envelopes did Emma receive on Tuesday?
c) How many more envelopes did Emma receive on Friday than on Wednesday?
d) How many envelopes altogether did Emma receive in six days?

3) Draw a pictogram for the data below.

Favourite State	Frequency
Anambra State	12
Ondo State	8
Rivers State	24

19.4 PIE CHARTS

When something is shared out, a **pie chart** is a good way of representing the information. A pie chart is a circle split up into sectors and each of the sectors representing a proportion of the data, usually discrete.

DRAWING A PIE CHART

Pie charts are drawn by first drawing a circle of any reasonable radius. The angle of **each sector** is calculated before using a protractor for measuring the angles.

Example 1: Draw a pie chart to represent the information below.

Season	Tourists
Summer	20
Autumn	15
Winter	5
Spring	5

Solution: First convert the frequencies to angles by adding up all the frequencies. 20 + 15 + 5 + 5 = 45.
One degree/angle = 360 ÷ 45 = **8**

Each Angle is calculated as follows:

Season	Tourists	Angles
Summer	20	20 × 8 = **160°**
Autumn	15	15 × 8 = **120°**
Winter	5	5 × 8 = **40°**
Spring	5	5 × 8 = **40°**

Check that the angles add up to 360°.
160 + 120 + 40 + 40 = 360
Now, draw a circle of a reasonable radius and mark the centre. Starting from the centre and using the radius as a starting point, draw the angles. Label the pie chart.

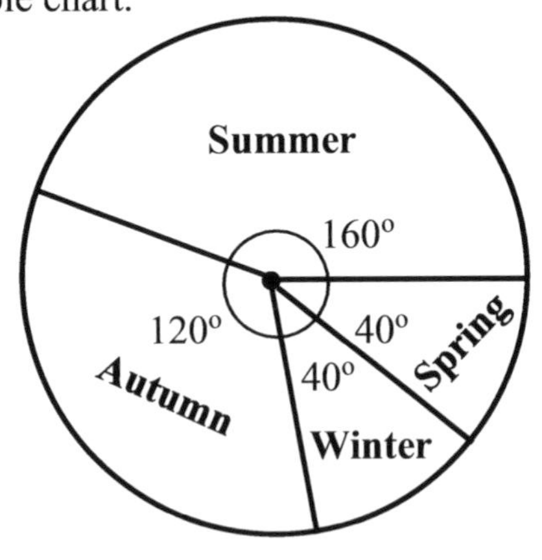

INTERPRETING PIE CHARTS

Example 2: 80 children were surveyed about their favourite subject and the results shown in the pie chart below. Work out the number of children who liked each subject.

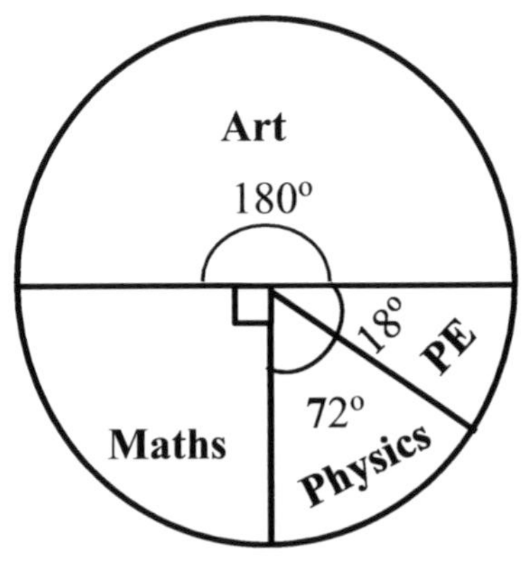

Art: 80 ÷ 2 = 40 children
Maths: 80 ÷ 4 = 20 children
Physics: $\frac{72}{360} \times 80$ = 16 Children
PE: $\frac{18}{360} \times 80$ = 4 Children

EXERCISE 19 C

1) A class completed a survey on how 20 pupils travel to school. To draw a pie chart, the pupils calculated what each pupil would represent by $360 \div 20 = 18^o$. Part of the table is shown below.

Mode of transport	**Number of pupils**	**Angle**
Private car	4	$4 \times 18 = 72^o$
Walk	10	
Cycle	1	
Taxi	2	
Other	3	
Total	20	

a) Why was 4 multiplied by 18?
b) Copy and complete the table.
c) Draw a pie chart to illustrate the data.

2) The table below shows the number of pets owned by a Grade 9 class in the United States of America.

Pets	**Frequency**
Cat	15
Rabbit	5
Dog	20
Hamster	4
Guinea pig	10
Gold fish	6

Draw a bar chart to show the information.

3) In an experiment, 78 people were asked to choose their favourite colour. They did and the information recorded as shown in the pie chart below.

Favourite colour

Blue
150°
Yellow
Pink

Show that the above information is incorrect.

4) 120 students went on a science trip to different parts of Africa and their destinations shown in the pie chart below.

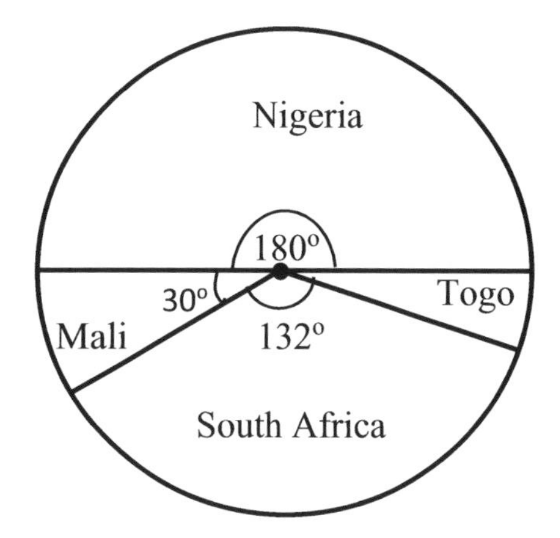

a) What fraction of the students went to Mali?
b) Which country was the most popular destination for the science trip?
c) How many students went to South Africa?
d) How many students went to Togo?
e) What fraction of the students went to Mali?

Chapters 18 & 19 Review Sections
Assessment

1) As part of a survey on TV viewing, Chika wants to find out the following:
a) Whether men watch more TV than women
b) The number of hours of TV people watches in a day
Design at least two questions that Chika should include in her survey (questionnaire) .... **4 marks**

2) 32 amateur drivers were given marks out of 7 in a driving test. The results are shown below:
1 3 5 2 1 6 7 1 5 3 1 1 2 6 1 3
4 4 4 7 6 6 7 4 7 5 5 5 5 2 6 7

a) Draw a bar chart to illustrate the data.
b) What was the most common test score? .... **4 marks**

3) Ibrahim went to a cultural event in Brixton, London. The table below shows how much he spent.

Products	Cost (£)
Hats	12
Sausages	3
Drinks	10
T-shirts	15
Souvenirs	20

Draw a pie chart to show these costs. .... **4 marks**

4) The pictogram below shows the number of late comers in a school.

Week	
Monday	♥ ♥ ♥ ◗
Tuesday	♥ ♥ ♥
Wednesday	
Thursday	♥ ♥
Friday	♥ ♥ ◗

Key: ♥ represents 2 students

a) What do you think happened on Wednesday? .... **1 mark**
b) How many students were absent for the whole week? ....**2 marks**

5) Temperatures in Spain and Brussels

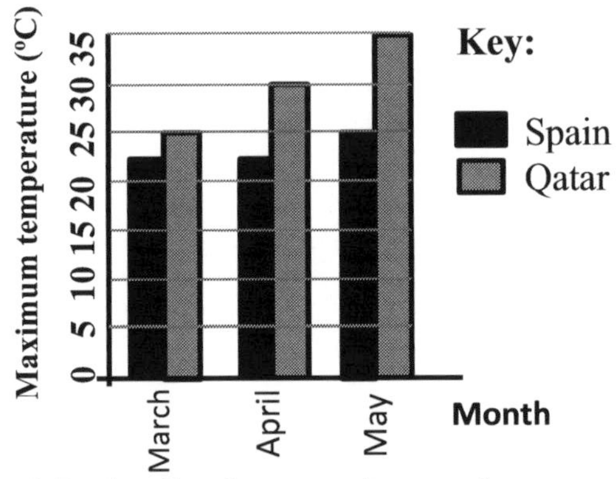

a) In April, what was the maximum temperature in Spain? .... **1 mark**
b) The difference between the average temperatures in Spain and Qatar was maximum in which month? ...**1 mark**

20 Averages and Range

This section covers the following topics:

- Mean
- Median
- Mode
- Range

LEARNING OBJECTIVES

By the end of this unit, you should be able to:

a) Calculate the mean of a set of numbers
b) Find the median of a set of numbers
c) Identify the mode from a given set of data
d) Write down mode from a bar chart
e) Find the range from a bar chart
f) Work out the mean from a frequency table
g) Find the mode from a frequency table
h) Find the median from a frequency table
i) Find the range from a frequency table

KEYWORDS

- Averages
- Mean
- Median
- Mode
- Range

20.1 AVERAGES

An **average** is a central number that is representative of all the numbers in a set. The **mean**, **median** and **mode** are the three most commonly used averages.

There is also a concept known as the ***range,*** which **is not** an average, but the difference between the highest and lowest numbers.

Example 1:
Joe completed seven mathematics homework and his marks out of 20 recorded as shown below.

9 11 7 8 15 14 6
A set of values like the one above is called a **distribution**.

MEAN

The mean of a probability distribution is obtained by adding all the numbers in the distribution and then divide by the number of items.

The mean for Joe's marks would be
$= \frac{9+11+7+8+15+14+6}{7}$
$= \frac{70}{7} = \mathbf{10}$ ✓

Example 2: Calculate the mean of 6, 7, 3, 4, 1, 9.
Mean $= \frac{6+7+3+4+1+9}{6} = \frac{30}{6} = \mathbf{5}$ ✓

Example 3:
Calculate the mean of 4, 7, 4, 2, 7.
Mean $= \frac{4+7+4+2+7}{5} = \frac{24}{5} = \mathbf{4.8}$ ✓

EXERCISE 20 A

1) Calculate the mean of each distribution.
a) 3, 7, 2
b) 10, 5, 12, 6, 12
c) 2, 4, 6, 6, 8, 10
d) 7, 9, 2, 4, 2, 1, 3
e) 5, 8, 4, 9, 10, 12, 23, 35

2) In a sale, Eric bought: three books at ₦250 each, two books at ₦450 each, six books at ₦125 each and four books at ₦300 each.
a) How much did Eric pay for all the books he bought?
b) What was the mean cost of the books?

3)

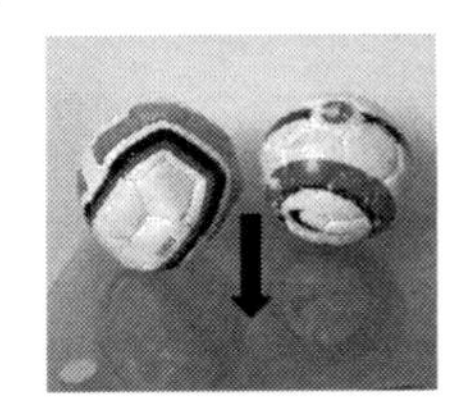

410 g each

450 g

a) Calculate the overall mass of the three balls.
b) Calculate the mean mass of the three footballs.

4)

Two baby pandas weigh 225 g and 350g respectively. What is the average weight of the two pandas?

5) Work out the mean of the following distributions. Leave some answers as ***mixed numbers*** in their simplest form.

a) 7, 5, 9, 9, 5, 6, 4, 3
b) 10, 7, 12, 7, 24, 20
c) 9, 7, 4, 3
d) 12, 3, 4, 5, 2, 5, 4
e) 7 m, 2 m, 10 m, 3 m, 8m
f) 3.75, 0.5, 0.8, 0.95

6) Five students took a geography examination and scored the following marks: 12%, 33%, 50%, 30%, and 60%. Calculate the mean percentage mark.

7) The mean of three numbers is 9.
a) What is the sum of the three numbers?
b) If two of the numbers are 15 and 4, what is the third number?

8) Ifeoma covered a number from the set of numbers below and said: "The mean of the numbers is 6.34."

3.5 4 9.2 [covered] 8

What is the value of the covered number?

9) To one decimal place, work out the mean of these numbers.
a) 8, 11, 3, 4, 3, 2
b) 10, 3.3, 0.7, 6.5, 9.5, 3, 6, 7
c) 6, 5, 7, 4, 3, 2, 1, 6, 7, 8, 9
d) $\frac{1}{2}, 2\frac{1}{4}, 5\frac{1}{5}$

10) Enugu Rangers football team's goal average was 2.3 after 30 matches. How many goals did Enugu Rangers football team score?

MEDIAN

The **median** is the middle value in the list of numbers when arranged in **order of size**. The order could be ascending or descending.

However, if there are two middle numbers, find the average of the two numbers (add them and divide by 2). That would give the median of the numbers.

Example 1: Ebuka sometimes cycles to school. He recorded the number of times he cycled to school in the last seven weeks, as shown below.

1, 2, 5, 0, 3, 2, 4

To find the median, first, arrange in order of size.

It becomes: 0, 1, 2, (2), 3, 4, 5

There are seven numbers in the list, the 4th number ($\frac{7+1}{2}$) is the median. So, the 4th number which is also the middle value is 2; therefore the median number of times Ebuka cycled to school is **2**. ✓

Notice that you would still get the same answer if you had arranged from highest to lowest. 5, 4, 3, (2), 2, 1, 0

Example 2:
Find the median of 5, 4, 7, 2, 3, 1

First arrange in order: 1, 2, (3, 4) 5, 7

Notice that 3 and 4 are the middle numbers. $(3 + 4) \div 2 = 7 \div 2 = 3.5$
Therefore, **3.5** is the median. ✓

EXERCISE 20 B

1) Write down the median value of each set of numbers.

a) 3, 6, 5, 2, 9, 8, 7
b) 9, 7, 1, 0, 3, 2
c) 12, 8, 4, 3, 1, 7, 6, 4, 2
d) 8, 1, 6, 4
e) 10, 20, 80, 40, 50
f) 4, 3, 2, 8, 7, 6, 5, 4

2) Two students, Chuba and Tochukwu, had five maths tests over a term. The tests were marked out of 30, and the results are shown below:

Chuba: 19, 23, 17, 16, 20
Tochukwu: 6, 21, 27, 3, 20

a) Find the median mark for Chuba.
b) Find the median mark for Tochukwu.
c) Comment on the marks of both students.

3) The morning temperatures (°C) in Lagos for a week in January were:

27, 19, 30, 16, 21, 15, 20
Find the median temperature in Lagos.

4) The show sizes of eight sailors are:
12, $10\frac{1}{2}$, 10, 11, 8, 10, 9, 7
Find the median shoe size for the data.

5) The height of some tables in metres was recorded as 0.9, 1, 1.2, 0.8, 0.85
Work out the median of the heights.

6) Find the median value of
a) 8, 9, 12.5, 4.5, 3, 2.5, 7, 10.5, 4, 3
b) 60, 70, 45, 35, 75

MODE

The most occurring number or value in a distribution is called the **mode** or **modal value**.

You may also have two modes which are explained in Example 2 below. The distribution with two modes is sometimes said to be **bimodal**. Other distributions with more than two modes are said to be **multimodal**. The mode can also represent qualitative data.

Example 1:
Find the mode of the set of numbers.
5, 3, 4, 5, 6, 5, 5, 7, 2

It is always advisable to arrange the distribution in order of size (Not a must). It helps in identifying the mode or modes quicker.

In order of size, the above distribution is
2, 3, 4, 5, 5, 5, 5, 6, 7

It is very clear from the numbers above that the most occurring number is 5, as it appeared four times and no other number or numbers appeared four times. Therefore, **5** is the mode. ✓

Example 2:
Find the mode of the numbers below.
6, 9, 10, 15, 6, 3, 10, 7

In order of size: 3, **6, 6**, 7, 9, **10, 10**, 15
Two numbers appeared twice.
Therefore, the modes are **6** and **10**. ✓

Example 3: Work out the mode for 1, 4, 3, 5. There is **no mode.** ✓

EXERCISE 20 C

1) From the list of numbers below, work out the mode.

a) 2, 5, 4, 1, 2, 1, 1, 7,
b) 3, 5, 8, 9, 3,
c) 4, 2, 12, 7, 12, 7, 5
d) 1, 9, 8, 15, 9, 3, 9, 4, 9
e) 20, 30, 40, 45, 58

2) Kolade asked six people what size of shoes they wear, and he recorded their sizes. 7, $8\frac{1}{2}$, 9, 8, $10\frac{1}{2}$, 8. What is the mode of their shoe sizes?

3) In a school survey, six students said their favourite colours. The results are given below.
Blue Red Blue Yellow Pink Blue
Work out the mode of their colours.

4) A pyramid was designed with different colours as shown. What is the mode of the colours?

5) Write down the mode for each set of data below.

a) 13, 13, 43, 34, 53, 13, 50, 34
b) 7, -8, 6, -3, -8, 2, 3, 8
c) 3.4, 3.5, 1.2, 3.5, 6.7, 2.8
d) $\frac{2}{5}, \frac{1}{5}, \frac{2}{5}, \frac{3}{5}, \frac{3}{6}, \frac{2}{5}, \frac{2}{4}, \frac{2}{5}, \frac{2}{4}$

6) Write down the modal type for each set of data below.
a) Blue, white, yellow, black, blue, red,
b) π, £, √, ₦, π, Θ, π, Θ
c) Dog, cat, rabbit, cat, mouse, dog

RANGE

The **range** is not an average, but the difference between the highest and lowest values in the **distribution**.

Example 1: Work out the range of the set of numbers: 4, 7, 2, 9, 13, 4, 2

Solution:
You may decide to first arrange the numbers in order of size before looking for the highest and lowest numbers.
In order of size: 2, 2, 4, 4, 7, 9, 13

Highest number = 13,
Lowest number = 2
Therefore, the range is 13 - 2 = **11** ✓

EXERCISE 20 D

1) Work out the range of the set of numbers below.

a) 6, 11, 8, 2, 10
b) 4, 5, 3, 12, 15
c) 19, 23, 35, 14, 8, 5
d) 90, 40, 30, 85, 20, 10, 60

2) The number of goals scored in seven football matches in one state is shown.
2 1 3 1 2 2 1
Work out the range.

3) In a small class of five pupils, their heights are 140, 156, 134, 170, and 165. Work out the range of their heights in centimetres.

4) In exercise 20 C, work out the range of the data in question 5.

20.2 AVERAGES AND RANGE FROM FREQUENCY TABLES AND DIAGRAMS

AVERAGES AND RANGE FROM A FREQUENCY TABLE

Example 1: The frequency table below shows the marks obtained by a class in a history test.

Marks	Frequency
5	1
8	2
12	5
15	3

The **modal mark** is the mark with the highest frequency. The highest frequency is 5, therefore the modal mark is **12.**

The **range** is
(Highest mark - the lowest mark) 15 – 5 = **10**

For the **median mark**, first, add up the frequencies. 1 + 2 + 5 + 3 = 11.
The median is $\left(\frac{n+1}{2}\right) th$ value. (11 + 1) ÷ 2 = 6.

It means that the median lies on the **6th value.** Go back to the frequency and start from the top, keep adding the numbers until you get to 6 or more and read off the corresponding ***Mark's*** value.

1 + 2= 3……..But 3 is not up to 6, so consider adding the next number which is 5.
3 + 5 = 8 this is more than 6 but still within limits. Read across to give 12. Therefore, the **median mark** is **12.**

To calculate the mean, extend the column as shown below.

Mark	Frequency	Mark × Frequency
5	1	5 × 1 = 5
8	2	8 × 2 = 16
12	5	12 × 5 = 60
15	3	15 × 3 = 45
Total	11	126

Mean $= \frac{\text{total of (mark} \times \text{frequency)}}{\text{total number of frquency}} = \frac{126}{11} =$ **11.45** to 2 decimal places.

Note: A common mistake is to divide by 4. Always divide by the total of all the frequencies and in the above example, 11.

EXERCISE 20 E

1) The table below shows the weights of the apples in a bag.

Weight (g)	Frequency
100	5
110	7
115	10

Work out the
a) modal weight b) median weight
c) range d) mean weight.

2) A six-sided dice is rolled in an experiment and the results shown in the frequency table below.

Score	Frequency
1	10
2	20
3	6
4	25
5	12
6	15

Work out the following:
a) the modal score b) the range
c) the median score d) the mean score.

3) The table below shows the number of goals scored in each football match by Enyimba FC for a season in Nigeria.

Goals	Frequency
0	8
1	19
2	11
3	2

a) Work out the mean number of goal scored per match.

b) How many matches did Enyimba FC play in the season?

4) In a primary school, pupils are awarded merits for 90% attendance or more. The table shows the number of merits awarded to pupils in two different year groups.

Number of merits	Year 3 frequency	Year 4 frequency
0	5	3
1	15	27
2	20	12
3	14	8
4	7	6

a) How many students are in Year 3?
b) How many students are in Year 4?
c) How many merits were awarded in total to pupils in Year 4?
d) Work out the mean number of merits given per pupil for pupils in year 3?
e) How many more pupils received three merits in Year 3 than in year 4?

5) The table below shows the number of off days taken by some employees at a company in August 2017.

Number of off days	0	1	2	3	4	5
Frequency	5	4	1	0	3	2

a) Calculate the modal number of off days.
b) Calculate the mean number of off days.
c) Calculate the range
d) Work out the median number of off days.

AVERAGES AND RANGE FROM DIAGRAMS

Example 1: Some adults were asked about their favourite hair colour, and the bar chart shows the results.

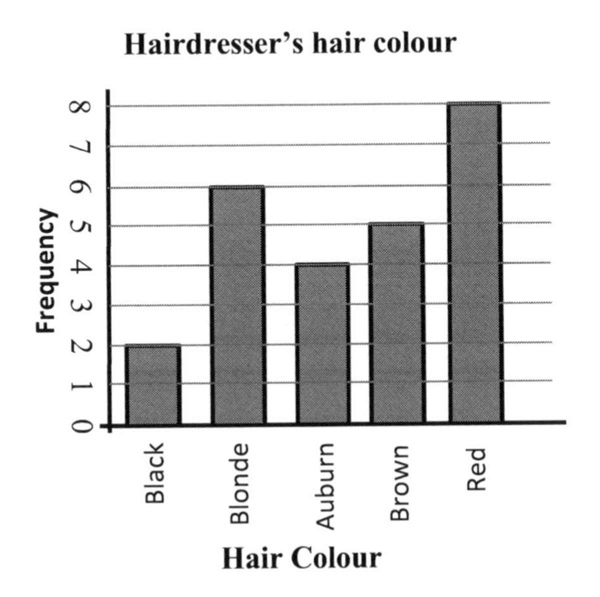

a) What is the modal hair colour?
The longest bar is red. Therefore, ***red*** is the modal colour.

b) How many adults took part in the survey?

$$2+6+4+5+8 = 25$$

25 adults took part in the survey.

c) How many adults have blonde hair?
6 adults

2) Study the bar chart shown below.

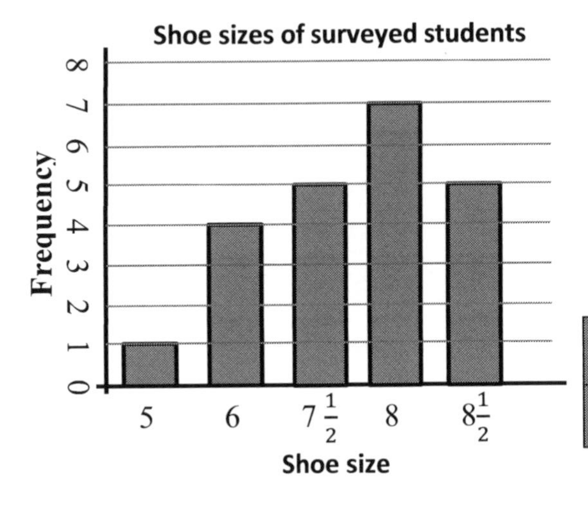

a) How many students were surveyed?

$1 + 4 + 5 + 7 + 5 =$ **22 students**

b) What is the modal shoe size? The longest bar has shoe size of 8. Therefore, **8** is the modal shoe size.

c) Complete the frequency table below.

Shoe size	5	6	$7\frac{1}{2}$	8	$8\frac{1}{2}$
Frequency	1	4	5	7	5

d) Work out the median shoe size.
$1 + 4 + 5 + 7 + 5 = 22$…….$(22 + 1) \div 2 = 11.5$. The median lies on the 11.5^{th} value which is 8 (Refer to section 20.2 example 1). Therefore, the median shoe size is **8**.

e) How many more students wear size eight shoes than size 6?
$7 - 4 =$ **3 more students**

EXERCISE 20 F

1) Emeka stood in front of his house and counted the number of children in each of the cars that passed for 20 minutes. He drew a bar chart to represent his findings as shown below.

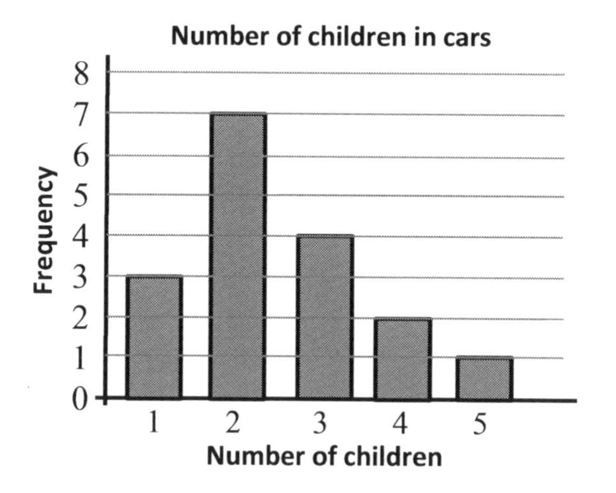

a) How many cars passed Emeka's house in total?
b) Find the modal number of children in the car.
c) Work out the range for the data.
d) Calculate the median number of children in the car.

2) The pie chart shows types of pets owned by some students.

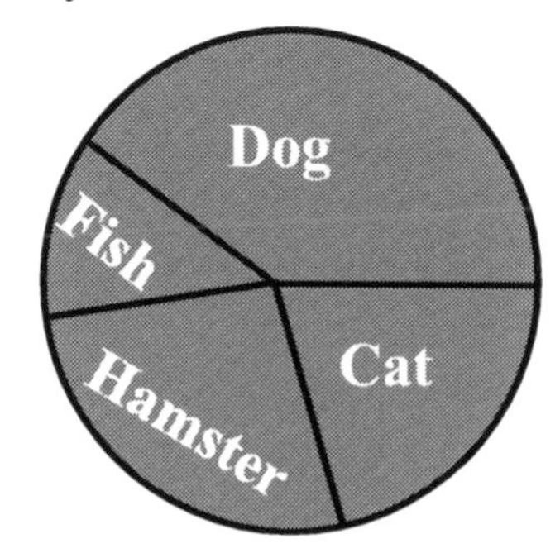

Write the mode of this data.

3) The pie chart below shows information about mock grades of some SSS3 students in a school.

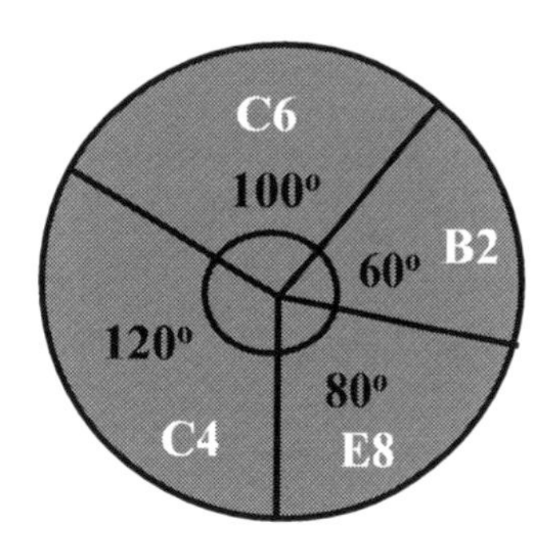

a) What fraction of the students achieved grade E8? Simplify your answer to its lowest form.
b) What grade was the mode?
c) If 12 students achieved grade B2,
i) how many students took part in the survey?
ii) how many students achieved grades C6 and C4?

4) The pictogram represents the number of apples sold each day by Tunji.

Key: ⊕ Represent 8 apples

Monday	⊕ ⊕
Tuesday	⊕ ⊕ ⊕ ◖
Wednesday	⊕ ⊕ ◔
Thursday	⊕ ⊕ ⊕ ⊕
Friday	⊕ ⊕ ⊕ ⊕ ⊕ ◖

a) Work out the range.
b) How many apples were sold on Tuesday?
c) On which day were the most apples sold?
d) How many apples were sold altogether?

5) The bar chart represents the number of horror books owned by four boys.

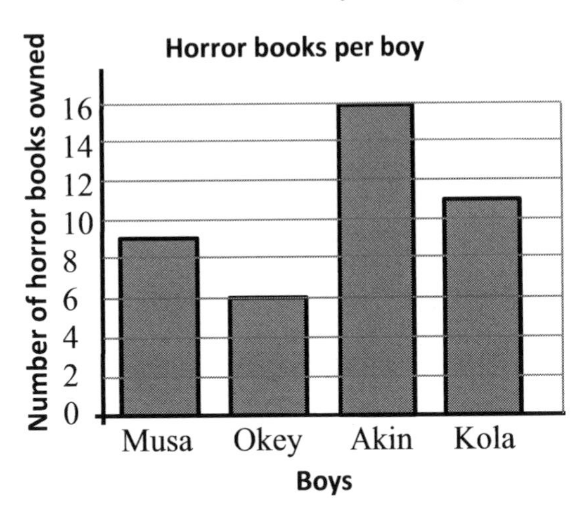

a) Who owned the most horror books?
b) Work out the range of the number of horror books owned by the boys.
c) Work out the mean number of horror books.
d) Musa and Kola owned how many horror books?

6) An advert in the Sun Newspaper reads:

Profession	**Wages per month(₦)**
Carpenter	40 000
Electrician	65 000
Handy person	35 000
Cleaner	45 000
Plumber	55 000

a) What is the median wage?
b) Work out the mode of these wages.
c) Work out the mean wage.

The advert further says "Carpenter needed, average salary more than ₦53 000 per month."

d) Show why the advert is deceptive.

7) Name three things that are wrong with this bar chart.

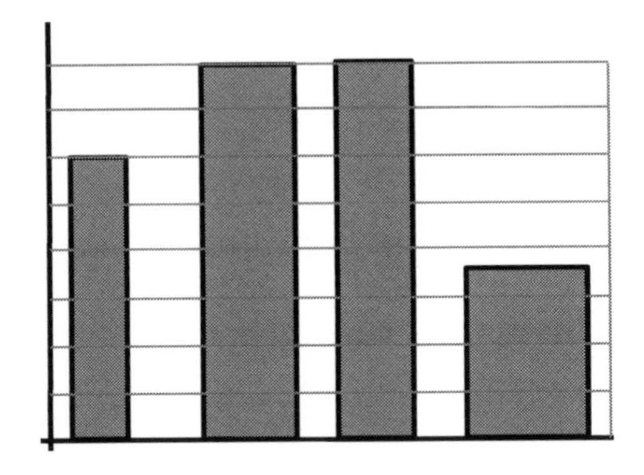

8) The numbers of students living in 50 households in **two** different cities in Nigeria are shown in the tables below.

Lagos

No. of Students	Frequency
0	7
1	15
2	20
3	5
4	2
5	1

Enugu

No. of students	Frequency
0	12
1	16
2	14
3	6
4	2
5	0

a) Draw bar charts for the two cities.
b) Work out the mean, median, mode and range for Lagos and Enugu.
c) Comment on the two cities.

Chapter 20 Review Section

Assessment

1) Work out the *mode*, *median*, *range* and *mean* of the amounts below.

₦200, ₦600, ₦120, ₦350, ₦600
...........**5 marks**

2) Write down a set of data that has no mode. ....**1 mark**

3) Write down a set of data that has two modes. ....**1 mark**

4) The following table shows some 14-year-old females show sizes.

Shoe size	**14-year-old females**
6	3
7	5
7½	10
8	3
8½	5

a) Draw a bar chart for the data.
..........**2 marks**

b) Work out the modal shoe size.
............**1 mark**

c) How many females were surveyed?
............**1 mark**

d) Find the range of the shoe sizes.
............**1 mark**

e) What type of data is presented in the table? ....**1 mark**

5) Nneka counted the number of food items in some containers and recorded her results in the table below.

Food items	**Frequency**
3	2
4	3
5	7
6	6

a) Work out the mode............ **1 mark**

b) Work out the median number of food items. ....**2 marks**

c) Calculate the mean number of food items to the *nearest whole number*.
...........**3 marks**

6) The pie chart below shows information about mock grades of some GCE students.

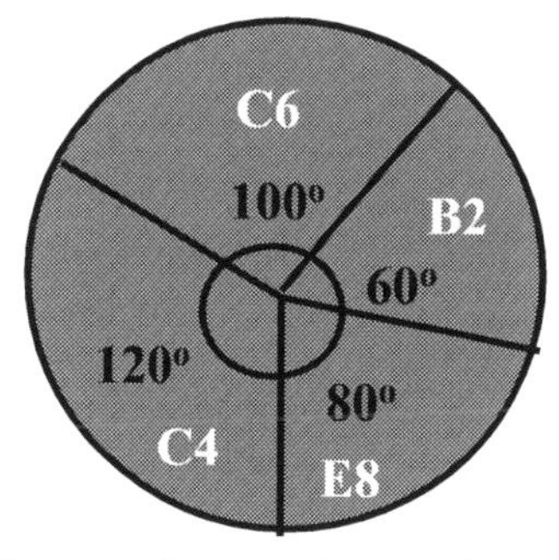

a) What grade was the mode?....**1 mark**

b) If 50 students achieved grade C6,

i) how many students took part in the survey?...................................**3 marks**

ii) how many students achieved grades B2 and C4?..........................**2 marks**

21 Coordinates

This section covers the following topics:

- Coordinates of a point
- Coordinates and quadrants

LEARNING OBJECTIVES

By the end of this unit, you should be able to:

a) Locate a point using coordinates
b) Read coordinates
c) Plot coordinates in all the four quadrants

KEYWORDS

- Coordinates
- x-axes
- y-axes
- Quadrants

21.1 UNDERSTANDING COORDINATES

Coordinates describe the position of a point and are written inside brackets. The first number in the bracket represents the horizontal axes (x) while the second number represents the vertical axes (y).

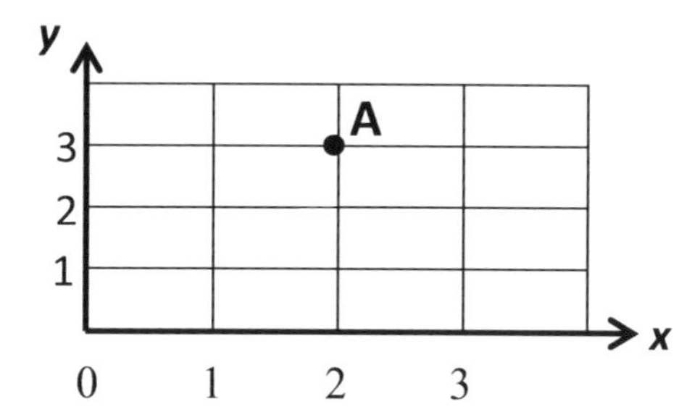

Coordinates are written in pairs inside a bracket. The *x*- coordinate (left or right) written first, while the *y*-coordinate (up or down), written second with a comma separating them. Point **A** can be located and described by the coordinate (2, 3).

Example 2: Write down the coordinates of points A to F.

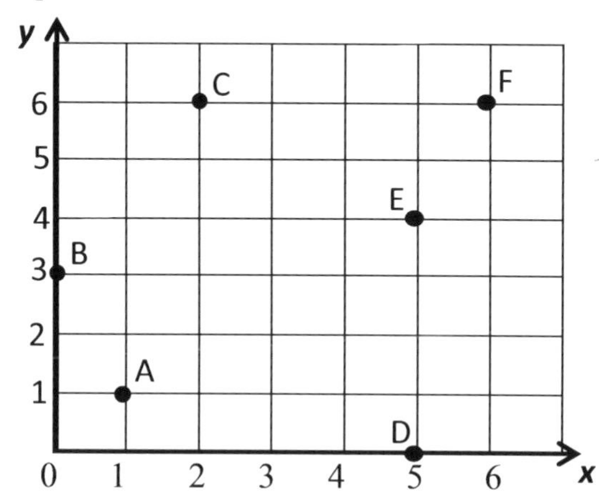

Answers: A = (1,1), B = (0,3), C = (2,6)
D = (5, 0), E = (5, 4), F = (6, 6)

EXERCISE 21A

1a) Write down the coordinates of points A, B, C, D, E and F.

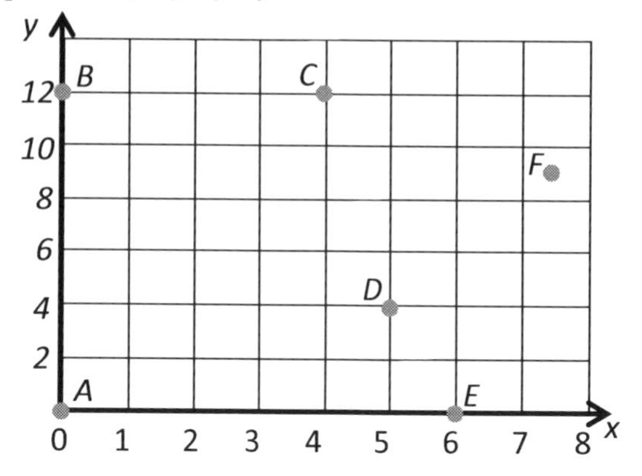

1b)
i) Join points A, F and C with a ruler.
ii) What is the mathematical name for the shape formed?

1c)
i) Join points A, B, C and E and back to A, in that order.
ii) What is the mathematical name for the shape formed? Explain.

2)

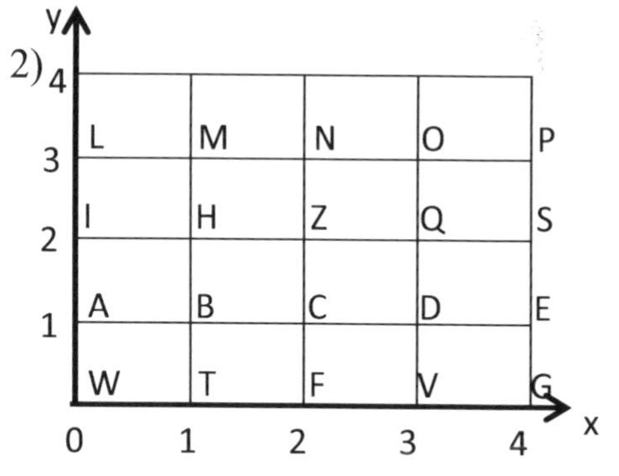

a) Write down the coordinates of points
i) C ii) G iii) D

b) Form a sentence given by the letters with the coordinates give below.
(0,2) (0,3) (3,3) (3,0) (4,1) (1,3)
(0,1) (1,0) (1,2) (4,2)

QUADRANTS

A graph can be split into four quadrants as shown below.

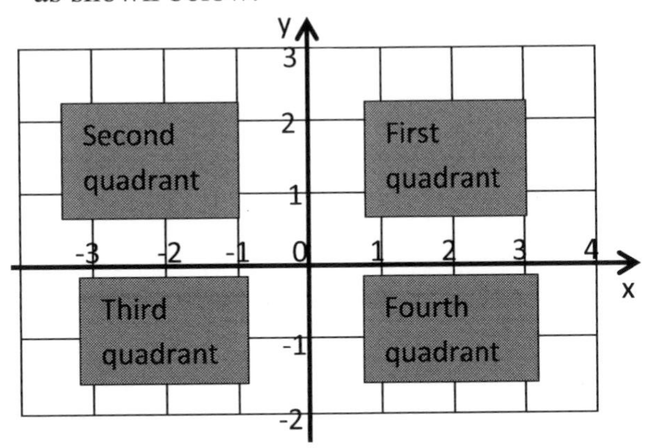

First quadrant: all coordinates are positive. For example (3, 2)
Second quadrant: coordinates are negative and positive. E.g. (-1, 2)
Third quadrant: all coordinates are negative. For example (-2, -1)
Fourth quadrant: coordinates are positive and negative. E.g. (4, - 2)

EXERCISE 21 B

1) Write the coordinates of the points, A, B, C, D, E, F and G.

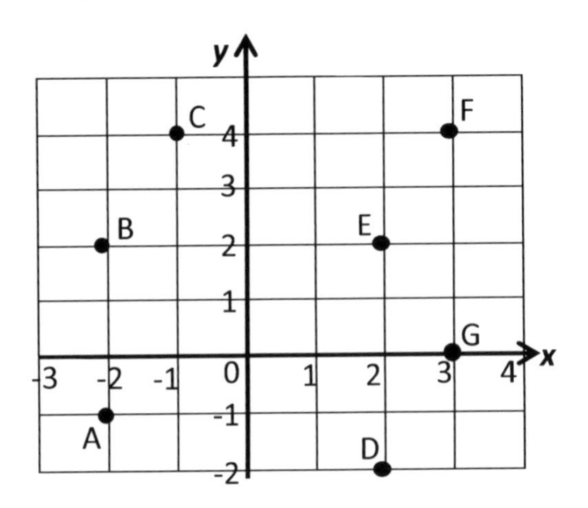

2) From question one above, join points BCFEB in that order. What is the mathematical name for the shape formed?

3) John says "When you join points ABEDA in that order, you have a rectangle" Is John correct? Explain.

4a) Draw a coordinate grid with x axes going from -4 to 4 and y axes going from -4 to 5.

4b) Plot the points below on the grid: A(0,3), B(2,5), C(4,3), D(2,0).

4c) By using a ruler, join points A, B, C, D in that order and back to A.

4d) What shape ABCD?

4e) Also, plot points E (-4, 2), F (-4,-2), G (-1,-1), H (-1, 1) and join the points in that order and back to E using a ruler.

4f) What is the mathematical name for shape EFGH?

4g) Charles says "Shape ABD is an isosceles triangle" Is Charles correct? Explain fully.

5) In question 1, join points EFG with a ruler. What type of triangle is EFG? Explain fully.

6) What quadrant is the coordinate (4, -5)?

22 Sequences

This section covers the following topics:

- Sequence of numbers
- Finding terms from the nth term of a sequence
- Graph of linear sequence
- Finding the nth term of a number sequence

LEARNING OBJECTIVES

By the end of this unit, you should be able to:

a) Produce sequence of numbers
b) Understand terms and nth terms
c) Find a formula for the nth term of a linear sequence
d) Draw graphs to show linear sequences

KEYWORDS

- Sequence
- Terms
- Nth terms
- Patterns
- Rule

22.1 LINEAR SEQUENCE

A number **sequence** is a list of numbers that follow a particular rule (pattern). It is linear if it can be represented on a straight line graph. It increases or decreases in equal-sized steps.

Each number in a sequence is called a **term**.

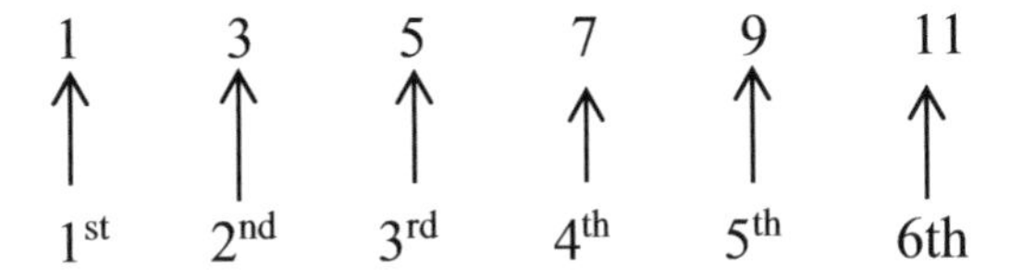

Consecutive terms are terms that are next to each other. Commas usually separate them. In the example above, 1 and 3, 3 and 5, 5 and 7, 7 and 9, 9 and 11 are consecutive odd numbers.

There are lots of different number patterns. A sequence can continue forever (infinite). For example, 2, 4, 6, 8, 10…

Sequences follow a **rule**.

The sequence 4, 8, 12, 16, 20 will continue if we keep on adding 4.

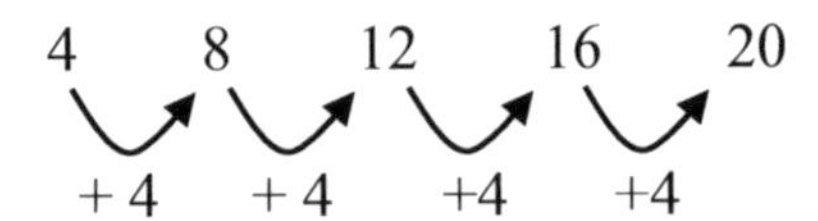

The rule for this sequence is to **add 4** each time. It is often called **term-to-term** rule.
Therefore, 4, 8, 12, 16, 20 is a linear sequence because it has a common difference of 4.

Some common sequences to remember are:

1, 3, 5, 7, 9…odd numbers
1, 4, 9, 16, 25… square numbers
2, 4, 6, 8, 10 … even numbers
10, 100, 1000, 10000… powers of ten
1, 3, 6, 10, 15… triangular numbers
1, 1, 2, 3, 5, 8… Fibonacci sequence

TRIANGULAR NUMBERS

Triangular numbers are formed by using consecutive numbers. The first triangular number is 1.

1
1 + 2 = **3**
1 + 2 + 3 = **6**
1 + 2 + 3 + 4 = **10**
1 + 2 + 3 + 4 + 5 = **15**
1 + 2 + 3 + 4 + 5 + 6 = **21**

FIBONACCI NUMBERS (SEQUENCE)

Fibonacci was an Italian mathematician regarded as the "Greatest European mathematician of the middle ages." His full name was Leonardo Pisano.

Source: Wikipedia

Fibonacci numbers are:1,1,2,3,5,8….

Each term in the Fibonacci sequence is the sum of the previous two terms.

1
1 + 0 = **1**
1 + 1 = **2**
2 + 1 = **3**
3 + 2 = **5**
5 + 3 = **8**
8 + 5 = **13**

Example 1: For each of the sequence, work out the missing terms.

a) 4, 7, 10, 13, …, …, …
b) 1, …, 13, …, …, 31, 37

Answers:
a) The rule is to add 3 each time. The missing numbers are 16, 19 and 22.

b) Look at the 6th and 7th terms. The difference between 31 and 37 is 6. The rule is + 6 each time. The missing numbers are: 7, 19, 25

EXERCISE 22A

1) Find the next two terms in each sequence.
a) 4, 5, 6, 7, ___ , ___
b) 7, 9, 11, 13, ___, ___
c) 1, 6, 11, 16, ___, ___
d) 4, 7, 13, 25, ___, ___
e) 3, 7, 11, 15, ___, ___
f) 22, 16, 10, 4, ___, ___
g) 5, 6, 8, 11, ___, ___
h) 7, 6.9, 6.8, 6.7, ___, ___
i) 2, 4, 8, 16, ___, ___
j) 1, 4, 8, 13, ___, ___
k) 3, 9, 27, ___, ___
l) $\frac{1}{2}, \frac{2}{6}, \frac{3}{18}, \frac{4}{54},$ ___, ___

2) Write the term-to-term rule for questions 1a, b and c.

3) The term-to-term rule is +7. Write two different sequences that fit the rule.

4) For the sequence 4, 9, 14, 19, write down
a) the term-to-term rule
b) the first term
c) the 10thterm

5) You are given the first term and the rule of different sequences. Write down the first four terms of each sequence.

First term	Rule
4	Add 6
47	Add 15
2	Subtract 7
3	Triple
200	Divide by 2

6) Copy and complete the sequences.

3 × 88 = 264
4 × ☐ = 352
☐ × 88 = 440
6 × ☐ = 528
7 × 88 = ☐
. . .
. . .
19 × 88 = ☐

7) Find i) the rule ii) the missing numbers for each sequence

a) 7, 15, 31, 63, ___
b) 225, 180, 135, ___ , ___
c) ___, 80, 77, 74, 71
d) 0.3, ___, 30, 300, ___
e) 8, 9, 11, 14, ___

8) Look at the sequence below.

$5^2 =$ 25
$55^2 =$ 3025
$555^2 =$ 308025
$5555^2 =$ 30858025

What is the value of 55555^2?

TERM NUMBER FROM A FORMULA

From the sequence, 4, 7, 10, 13, 16, a formula can be written as $3 \times n + 1$ or $3n + 1$ where n is the number of terms. Using the formula $3n + 1$, you could find any number of terms. For example, the 2000th term will be $3 \times 2000 + 1 =$ **6001**

Example 2: Write down the first three terms of the formula (nth term) $8n + 5$.

First term is when $n = 1$,
$(8 \times 1) + 5 =$ **13**.
Second term is when $n = 2$,
$(8 \times 2) + 5 =$ **21**.
Third term is when $n = 3$,
$(8 \times 3) + 5 =$ **29**
Therefore, the first three terms are:
13, 21 and 29

Example 3: Find the first three terms of the formula 5n.
First term: $5 \times 1 =$ **5**, second term: $5 \times 2 =$ **10**, third term: $5 \times 3 =$ **15**

Example 4: Find the 20th term of the nth term $2n - 3$.
20th term $= (2 \times 20) - 3 =$ **37**

Example 5: Write down the first five terms of the nth term n - 5

First term: $1 - 5 =$ **- 4**
Second term: $2 - 5 =$ **-3**
Third term: $3 - 5 =$ **-2**
Fourth term: $4 - 5 =$ **-1**
Fifth term: $5 - 5 =$ **0**
The first 5 terms are: -4, -3, -2, -1, 0

Example 6: Find the 10th term of the nth term formula, $3n^2$.
Note: $3n^2 = 3 \times n^2$. Therefore, the 10th term will be $3 \times 10^2 = 3 \times 100 =$ **300**

Example 7: Write down the first three terms of the nth term $4n^2 + 1$.

1st term: $4 \times (1)^2 + 1 = 4 \times 1 + 1 =$ **5**
2nd term: $4 \times (2)^2 + 1 = 4 \times 4 + 1 =$ **17**
3rd term: $4 \times (3)^2 + 1 = 4 \times 9 + 1 =$ **37**

A common mistake would be to multiply 4 by n and then square it. It is wrong.
Example: To find the 1st term of $4n^2 + 1$, a common mistake will be $(4 \times 1)^2 + 1 = 4^2 + 1 = 17$. ✖
This is wrong.

Example 8: For the sequence 11, 14, 17, 20, write down a) $T_{(1)}$, b) $T_{(2)}$, c) $T_{(10)}$

Solution: $T_{(1)}$ means the 1st term, $T_{(2)}$ means the 2nd term and $T_{(10)}$ means the 10th term.

$T_{(1)} = 11$
$T_{(2)} = 14$
$T_{(10)} = 38$.......from continuing the sequence.

EXERCISE 22B

1) Write down the first **three** terms of the sequences produced by these nth term formulae.

a) $2n$	g) $3 - n$	m) $20n - 3$
b) $14n$	h) $4 + 2n$	n) $2n^2$
c) $2n + 2$	i) $10 - 5n$	o) $3n^2 - 1$
d) $3n - 1$	j) $8 + 3n$	p) $4n^2 + 10$
e) $4n + 1$	k) $n^2 + 7$	q) $13 - 2n$
f) n^2	l) $9 + n^2$	r) $\frac{1}{n^2}$

2) Write down the first **five** terms of the sequences below.

a) First number is 40; the rule is add 4
b) First number is 7; the rule is add 5
c) First number is 13; rule is minus 2
d) First number is 2; rule is to subtract 3
e) First number is -7, the rule is to add 9

3) The third term of a sequence is 16, fourth term, 22 and fifth term, 28. Find the first and second terms.

4) For the sequence -14, -9, -4, 1…
a) Find the term-to-term rule
b) Find the next term in the sequence
c) Find the 100th term in the sequence.

5) For the sequence below
7, 11, 15, 19, 23…,
Write down
a) $T_{(1)}$ b) $T_{(2)}$ c) $T_{(97)}$

6) Each number in this sequence is -3 times the previous number.
-4, 12, -36, …
Write down
a) the 4th term b) the 6th term

7) Fill in the missing numbers

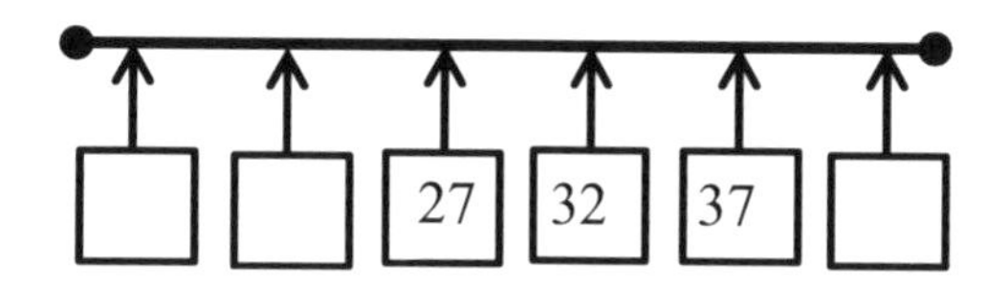

8) Write down the first three terms of the sequences where $T_{(n)}$ is:

a) $2n + 11$ c) n^2
b) $n - 3$ d) $3n^2 + 3n$

SEQUENCES AND GRAPHS

Number patterns can be drawn on a graph. For example, the number pattern 3, 6, 9, 12, 15 can be shown on a graph as follows.

Term	1	2	3	4	5
Magnitude	3	6	9	12	15

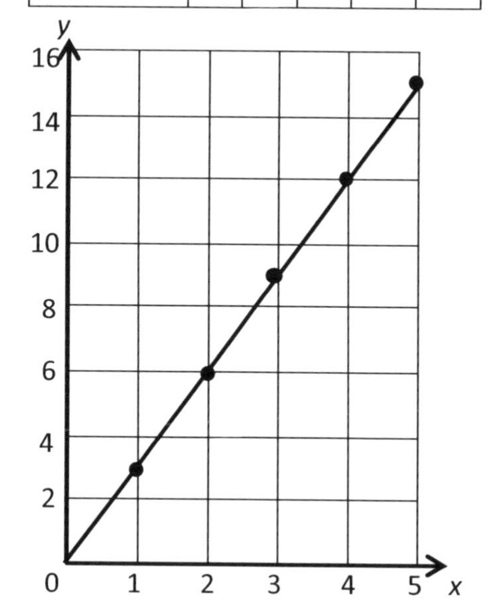

The graph of the sequence is a straight line indicating linear sequence.

22.2 NTH TERM FORMULA

For most mathematical sequences, there are formulae (nth terms). The nth term is very important in predicting the number of terms in a sequence or the term number.

A general term in the sequence is the **nth term**, where *n* stands for any number of terms.

For a linear sequence, the difference is constant. The sequence 4, 6, 8, 10 …have a constant difference of 2, all through. The sequence is linear. To find the nth term of a linear sequence, try to memorise **DNO**.

D = difference between the terms
N = Number of terms
O = **Zero** term
(Term before the 1st term)

Example 1: Find the nth term of the sequence 9, 14, 19, 24, 29…

Solution:

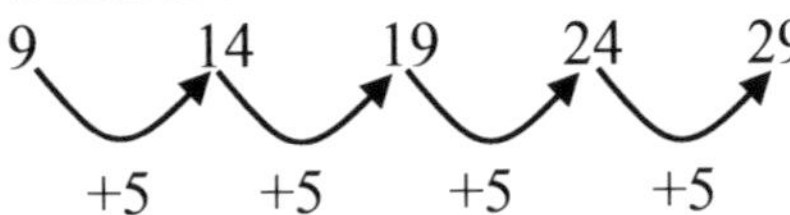

Difference between the terms is 5.
N, which is the number of terms, is written beside the difference to give 5n.
0, which is the zero term is the number before 9.
So we work backwards to get the zero term. 9 – 5 = **4**
Our sequence would look like this:

4, 9, 14 19, 24, 29
↗
Zero term
Using DNO
↓ ↓ ↓
5 n +4
The nth term is now **5n + 4** ✓

CHECK YOUR ANSWER
Using the nth term 5n + 4 obtained, find the first 2 or three terms. If they equal the original sequence, then the nth term is correct.
1st term: 5 × **1** + 4 = 9
2nd term: 5 × **2** + 4 = 14
3rd term: 5 × **3** + 4 = 19

Since the numbers obtained are the original numbers in the sequence, the nth term 5n + 4 is correct.

Example 2: Find the nth term formula for the sequence 2, 5, 8, 11, 14,..
Solution
Difference = 3, which implies a 3n term
Zero term = 2 – 3 = -1.
Nth term using DNO = **3n - 1** ✓

Example 3: Find the nth term formula for the sequence 20, 18, 16, 14, 12…
Solution
Difference = 18 – 20 = -2
which implies -2n term
Zero term = 20 – (-2) = 22
Nth term = **-2n + 22** ✓

Example 4: Find the nth term of the sequence $\frac{5}{7}, \frac{6}{9}, \frac{7}{11}, \frac{8}{13}$
For numerator, nth term = n + 4
For denominator, nth term = 2n + 5
Nth term = $\frac{\mathbf{n+4}}{\mathbf{2n+5}}$ ✓

EXERCISE 22 C

1) Find the nth term of each of the sequences below.

a) 5, 7, 9, 11, 13…
b) 11, 14, 17, 20, 23…
c) 4, 5, 6, 7, 8, 9 …
d) 30, 25, 20, 15, 10 …
e) 1, 5, 9, 13, 17 …
f) 9, 12, 15, 18, 21 …
g) 7, 14, 21, 28, 35 …
h) 0, 4, 8, 12, 16 …
i) -1, 1, 3, 5, 7 …
j) 17, 14, 11, 8, 5 …
k) 500, 550, 600, 650, 700 …

2) Copy and complete the mapping diagram below.

Term number (n)	4n	Term
1	4	7
2	8	11
3		
4		19
70		

3) Write down each sequence and select the correct expression for the nth term.

a) 4, 6, 8, 10, 12…
b) 33, 30, 27, 24…
c) 2, 4, 6, 8…
d) 2, 5, 8, 11, 14…
e) 3, 4, 5, 6…
f) $1^2, 2^2, 3^2, 4^2$…

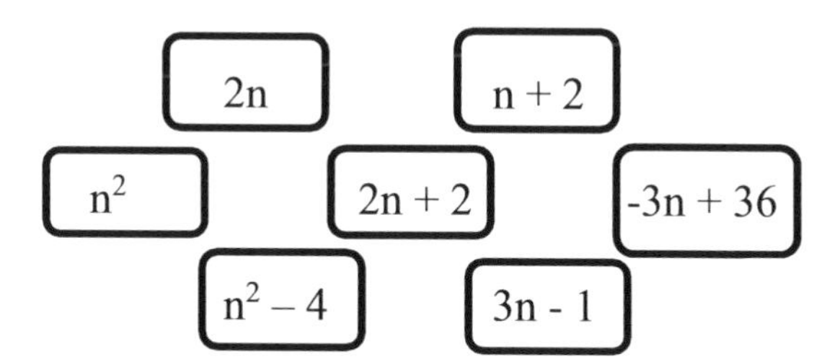

4) This sequence of patterns is made from regular pentagons with sides of 1 centimetre.

Pattern **1** Pattern **2** Pattern **3**

a) What is the perimeter of pattern number 3?
b) Draw pattern numbers 4 and 5
c) Copy and complete the table below.

Pattern number	1	2	3	4	5	6
Perimeter (cm)						

d) Find a formula for the perimeter of the nth pattern.
e) Find the perimeter of the 100th pattern.
f) Which pattern number would have a perimeter of 3150?

5) Rods are placed to form a pattern.

Pattern 1 Pattern 2 Pattern 3

a) How many rods will there be in pattern number 5?
b) How many rods are needed for the nth pattern?
c) What pattern number will have 59 rods?
d) Copy and complete the table below.

Pattern number	1	2	3	50
Number of rods				

ANSWERS

EXERCISE 1A

1)
a) 20
b) 40
c) 20
d) 100
e) 4000
f) 8000
g) 800
h) 30
i) 500
j) 300
k) 7
l) 300
m) 9000
n) 2
o) 8
p) 50
q) 5
r) 300
s) 6000
t) 30

EXERCISE 1B

1)
2 ⟶ 2 000
5 ⟶ 500
3 ⟶ 30
7 ⟶ 7

2)
4 ⟶ 400
3 ⟶ 30
2 ⟶ 2

3)
1 ⟶ 1000
2 ⟶ 200
6 ⟶ 60
3 ⟶ 3

4)
3 ⟶ 3000
4 ⟶ 400
5 ⟶ 50
1 ⟶ 1

5)
1 ⟶ 10000
5 ⟶ 5000
8 ⟶ 800
7 ⟶ 70
2 ⟶ 2

6)
1 ⟶ 1000
2 ⟶ 200
3 ⟶ 30
1 ⟶ 1

7)
7 ⟶ 700
5 ⟶ 50
3 ⟶ 3

8)
a) 24
b) Largest: 9765
Smallest: 5679
c) 700

9)
a)

Tens	Ones
4	5

b)

H	T	O
1	3	5

c)

Thousands	H	T	O
5	6	7	8

d)

H	T	O	.	1/10	1/100
1	3	4	.	9	8

e)

0	.	Tenths	1/100
0	.	8	9

f)

Ten ths	Thousand	H	T	O
1	2	3	4	5

10)
For 134.98
1 = 100
3 = 30
4 = 4
0.9 = 9/10
0.08 = 8/100

For 0.89
0.8 = 8/10
0.09 = 9/100

For 12345
1 = 10 000
2 = 2000
3 = 300
4 = 40
5 = 5

EXERCISE 1C

1) 600
2) 40
3) 30
4) 2
5) 6000 and 50
6) 1

EXERCISE 1D

1) Forty-three
2) Six
3) One hundred and forty-seven
4) Two hundred and twenty-two
5) Forty-two thousand one hundred and thirty-five
6) Eight thousand and twenty
7) Twenty thousand
8) Four thousand six hundred and nineteen
9) Five hundred and seventy-one thousand two hundred and thirty-five
10) Two million three hundred and fifteen thousand two hundred
11) Eight thousand three hundred and fifty-nine
12) Seven hundred and fifty-one million three hundred and fifty thousand two hundred and ten
13) One thousand one hundred and eleven
14) Eighty-eight thousand eight hundred and eighty-eight
15) Nine hundred and ninety-nine thousand nine hundred and ninety-nine

16) Eight thousand nine hundred and ninety-seven
17) Three hundred and forty-four thousand fifty-seven
18) One million eight hundred and eighty-eight thousand three hundred
19) Eight hundred and nineteen billion two hundred and seventy million
20) Eight trillion four hundred billion

EXERCISE 1E

1) 2
2) 11
3) 25
4) 205
5) 803
6) 1 001
7) 10 950
8) 23 600
9) 501 265
10) 1 703 950
11) 607 400 000
12) 4 290 000 000

EXERCISE 1F

1) 4 567
2) 45 678
3) 1 000
4) 987 654
5) 250 000
6) 2 000 000
7) 12 000 000
8) 1 000 000 000
9) 345 000 000 000
10) 1 000 000 000 000
11) 3 490 034
12) 7 623
13) 4 000 000 000
14) 3 500 000
15) 200 000 000
16) 6 789 743
17) 300 000 000
18) 5 000 000 000 000
19) 1 564 566
20) 7 500 000 000

21) Students/ teacher's responses

22)
Nigeria: 180 000 000
UK: 70 000 000
USA: 320 million
Japan: 140 million

EXERCISE 1G

1) 400
2) 170
3) 1 800
4) 4 000
5) 13 000
6) 876 000
7) 1 300
8) 33 000
9) 2 900
10) 33 000
11) 2 000
12) 55 000
13) 45 000
14) 270
15) 19 000
16) 412 000
17) 160
18) 1 300
19) 82 500
20) 169 800

EXERCISE 1H

1) 20
2) 0.04
3) 0.005
4) 4
5) 0.9
6) 0.0015
7) 0.00865
8) 50

EXERCISE 1I

1) 500
2) 2 000
3) 15 000
4) 14 000
5) 4 000
6) 6 000
7) 20 000
8) 100 000
9) 7 000
10) 9 500
11) 450
12) 16 000
13) 6 900
14) 4
15) 1.7
16) 0.318
17) 0.417
18) 1.2
19) 8.7
20) 4
21) 2
22) 8.888
23) 8 880
24) 0.865
25) 9 000
26) 630

EXERCISE IJ

1a) 400 g
b) 500 g
c) 80 g

2) 1 200 g or 1.2 kg
3) 4.2 kg
4) 200 000 g

5)

	Litre (l)	(ml)
Water	2	2 000
Milk	0.3	300
Coca Cola	1	1 000
Petrol	30	30 000

6a)
A ≈ 4cm
B ≈ 2 cm
C ≈ 3 cm
D ≈ 7 cm

6b)
A = 4.2 cm
B = 2.2 cm
C = 3.5 cm
D = 6.8 cm

6c) **A** 0.2 cm
B 0.2 cm
C 0.5 cm
D 0.2 cm

6d) 116.5 cm

EXERCISE 1K

1) 24
2) 48
3) 52
4) 365/366
5) 180
6) 2

7) 5
8) 3
9) 300
10) 4
11) a) Wednesday
b) Monday
c) Sunday
d) Saturday
12) 31 days
13) 12
14) 4
15a) 56 b) 280
c) 63 d) 0.25 or 1/4
16a) 300 b) 1200
c) 210 d) 5
17a) 3 b) 5 c) 1 d) 1
18) 18 weeks 13 days
19) 1 800 seconds
20) 156 weeks
21) 7 hours 42 mins
22) Friday
23) 13 weeks 4 days

EXERCISE 1L

1a) 2 hours 1 minute
b) 5 hours
c) 17 hours 25 minutes
d) 12 hours
e) 7 hours
f) 1 hour 25 mins
g) 4 hours 21 mins
h) 19 hours 30 mins

2a) 05:25 b) 18:45
c) 07:30 d) 10: 55
e) 09:00 f) 17:00
g) 23:45 h) 16:45
i) 06:15

EXERCISE 1M

1a) 08:30 bus b) 4 hours
2a) 11:25
b) 3 hours 32 mins
c) No, he is incorrect. Ikenna's journey was longer by 28 mins.
3a) 5 mins b) 14:20
c) 3 hours 15 mins
d) 6 hours 15 mins

4a)

	A	B	C	D	E
BS	07:00	07:10	07:20	07:30	07:40
Sch	07:15	07:25	07:35	07:45	07:55

4b)
i) A ii) B iii) C iv) D

5a) Asia World
b) Are You There?
c) Sports
d) Africa World

6a) 45 mins b) 30 mins
c) 13 mins d) 35 mins
7) Weather

CHAPTER 1 REVIEW SECTION ASSESSMENT

1)
a) 1 567
b) 7 651
c) 60
d) 0

2a) 80 b) 1 c) 7
3a) 150 b) 5 c) 44
4) 9
5a) 4 b) 35
6a) 09:45 b) 10:33
c) 68 mins or 1hr 8 mins
7a) 7:10 am or 7:10 pm
b) 07:10 or 19:10
8a) 06:45 or 18:45
b) 10:15 or 22:15
c) 09:00 or 21:00
9) September, April, June, November
10a) 7 b) 8.7 c) 400
d) 760 e) 70 f) 67.5
11) 9
12) 79 days
13a) 1330 millimetres
b) 133 Centimetres
14) 10
15a) 31 days
b) Friday
c) 17th day
d) Friday 29th
16a) 365 b) 940
c) 4687 d) 440 900
17a) One million nine hundred and sixty-seven thousand five hundred
b) 7 945 942
c) Seven million nine hundred and forty-five thousand nine hundred and forty-two
d) No, Bimbo is incorrect. 510456 × 2 = 1 020 912 which is not the population of Latvia

EXERCISE 2A

1)
a) 10
b) 10
c) 30
d) 50
e) 90
f) 430
g) 530
h) 330
i) 590
j) 180
k) 220
l) 410
m) 590
n) 890
o) 410
p) 140
q) 200
r) 220
s) 920
t) 660
u) 90
v) 140
w) 5460
x) 1230

2)
a) 100
b) 800
c) 200
d) 700
e) 500
f) 600
g) 300
h) 200
i) 900
j) 1400
k) 1400
l) 2200
m) 1000
n) 11900
o) 6000
p) 1100

3)

a) 4000
b) 3000
c) 3000
d) 9000
e) 1000
f) 2000
g) 1000
h) 7000
i) 42000
j) 77000
k) 14000
l) 619000
m) 506000

4) 23 200
5) ₦450 300
6a) 2 b) 24
c) 104 kg d) 350

7a) 10 pens
b) 65, 66, 67, 68, 69, 70, 71, 72, 73, 74

EXERCISE 2B

1) 1
2) 2
3) 2
4) 3
5) 2
6) 3
7) 3
8) 2
9) 3
10) 3
11) 3
12) 3
13) 1
14) 4
15) 4
16) 6
17) 1
18) 3
19) 2
20) 4
21) 3
22) 7
23) 3
24) 3

EXERCISE 2C

1)
a) 2.3 b) 0.4
c) 4.5 d) 0.1
e) 0.2 f) 23.8
g) 154.0 h) 32.1
i) 100.5 j) 1234.8
k) 1.0 l) 72.1
2)
a) 2.32 b) 0.38
c) 4.53 d) 0.08
e) 0.16 f) 23.76
g) 154.00 h) 32.06
i) 100.54 j) 1234.77
k) 1.04 l) 72.12

3)
a) 3.752 b) 2.635
c) 10.238 d) 11.986
e) 0.126 f) 17.990
g) 879.123 h) 0.763
i) 10.088 j) 76.130
k) 100.675 l) 9.651

4)
a) 456.8 cm^3
b) 456.78 cm^3
c) 456.783 cm^3
d) 456.7833 cm^3

5)
a) 153 786.5 m^2
b) 153 786.45 m^2
c) 153 786.454 m^2
d) 153 786.4536 m^2

EXERCISE 2D

1)
a) 2 b) 3 c) 5 d) 9
e) 4 f) 3 g) 5 h) 5

2)
a) 4 b) 8 c) 6 d) 0
e) 0 f) 8 g) 0 h) 9

3)
a) 70 b) 800
c) 3 d) 2
e) 0.4 f) 70
g) 40000 h) 0.08
4)
a) 4000 b) 9
c) 5.7 d) 3.2
e) 8.91 f) 12.6

5a)
D Africana 1000 m
Hi-Impact 900 m
T.W.Bunch 1000 m
Klubdelag 800 m
Fun Factory 1000 m
Omu Resort 500 m

b) 1230 m
c) 200 m
6) 65
7

a) 40
b) 55
c) 0.9
d) 90
e) 679
f) 60
g) 800
h) 0.88
i) 543
j) 1.1
k) 400
l) 4.9

EXERCISE 2E

1

a) 70
b) 200
c) 400
d) ₦45
e) 300

2)
₦50 000 to ₦52 000

3) ₦45 000
4a) 30 b) 4
c) 80 d) 20
5a) 2 b) 40
c) 250 d) 3
6a) 12 b) 0.2
c) 105
7a) 8 b) 81
c) 14 d) 70
e) 50 f) 10
g) Chukwudi should round 95 pence to £1. Then multiply £1 by 8 to give £8. Since he rounded to £1, he must expect more than £2... (From £10 – £8).

EXERCISE 2F

1)
a) 14 b) 14 c) 100
d) 25 e) 53 f) 91
g) 84 h) 145 i) 402

j) 575 k) 1112 l) 648
m) 1582 n) 1061
2)
a) 55 b) 102 c) 174
d) 1020 e) 964 f) 1096
g) 1398

3)
a) 468 b) 93 c) 1308
d) 1107 e) 2342
f) 8442 g) 8802

4)
a) ₦300 b) ₦1 030
c) ₦1 000 d) ₦1 100
e) ₦750 f) ₦2 100

5)
a) 4 536 b) 2 796
c) 68 181

6) 7 521
7) ₦754 325
8a) 68 b) 3767 c) 394

9a)

6	**2**
1	9
8	1

b)

2	6	**5**
1	4	9
4	1	4

c)

1	2	0	**0**
2	4	3	7
3	6	3	7

EXERCISE 2G

1a) 4 b) 7 c) 57
d) 44 e) 12 f) 52
g) 13 h) 30 i) 36
j) 31 k) 25 l) 19
m) 25 n) 68 o) 24
p) 49 q) 26 r) 75

2a) 50 b) 141 c) 41
d) 72 e) 420 f) 756
g) 520 h) 212 i) 448
j) 208 k) 168 l) 447
m) 727 n) 748 o) 642
p) 16860 q)10036
r) 635

3) 46
4a) APGA b) 21554

5) 524 6) 254
7a) 423 b) 1112
c) 5286
8) 125 pens

9a)

3	7
1	**2**
2	5

b)

6	**1**
1	9
4	2

9c)

8	6	**5**
1	**8**	4
6	8	1

10a) 122 b) 615
c) 1098 d) 6806
e) 46 f) 551

EXERCISE 2H

1a) 24 b) 126 c) 176
d) 427 e) 380 f) 168
g) 160 h) 228 i) 84
j) 490 k) 77 l) 603
m) 408 n) 741 o) 7068

2a) 300 b) 240 c) 60
d) 120 e) 180 f) 150
g) 125 h) 210 i) 210
j) 80 k) 180 l) 560
m) 60 n) 72 o) 220

3) 390 4) 3318 5) 75
6) 2280
7a) 240 b) 3819
c) 49725 d) 9432
e) 10680 f) 6732

8) 1820
9) ₦306 000
10)

×	3	6	7	9	10
2	6	12	14	18	20
8	24	48	56	72	80
7	21	42	49	63	70
11	33	66	77	99	110

11) 1998 m^2 12) 200
13) ₦109 200
14a) 7 b) 10 c) 3015
15a) ₦2 700 b) ₦6 750
c) ₦63 450 d) 67 500
16) 300 eggs
17a) 49152 b) 128
c) 32, 128, 4096
d) 16 or 4096
18) ₦15 000
19) 1645 seats

20)

×	4	7	8	9
2	8	14	16	18
5	20	35	40	45
7	28	49	56	63
8	32	56	64	72

EXERCISE 2I

1a) 20 b) 3 c) 4
d) 20 e) 12 f) 13
g) 10 h) 34 i) 129
j) 16 k) 75 l) 400
m) 700 n) 74 o) 2222

2) ₦2 400 3) 24 kg
4) 120 5) ₦650
6) 749 7) ₦987
8a) 6 coaches
b) 9 coaches
9) Students own answers that add up to ₦9 250 e.g. ₦4 000, ₦3 000 and ₦2 250.

EXERCISE 2J

1a) 30 b) 430 c) 770
d) 3190 e) 4000
f) 5190 g) 760
h) 4310 i) 7800
j) 2340 k) 9800
l) 1340 m) 5550
n) 11110 o) 789650

2a) 3 000 b) 4 200
c) 600 d) 7 500
e) 31 900 f) 5 000
g) 41 900 h) 91 700
i) 700 j) 4 700
k) 928 800 l) 111 200
m) 97 600 n) 5 900
o) 8 669 700

3a) 3 000 b) 5 000
c) 77 000 d) 318 000
e) 96 000 f) 7 000
g) 431 000 h) 96 000
i) 419 000 j) 60 000
k) 8 776 000 l) 92 000
m) 1112000 n) 59 000
o) 486 000

4a) 3 b) 7 c) 4.6
d) 9.6 e) 1 f) 46.5
g) 543.21 h) 67.8
i) 128 j) 40 k) 831
l) 5700 m) 11 n) 5
o) 50

5a) 9 b) 86 c) 32.450
d) 100 e) 654 f) 4590
g) 7000 h) 234 i) 870
j) 31.8 k) 0.83 l) 8.9
m) 0.452 n) 98.74
o) 40.8

FUNCTIONAL MATHS
1) £8.97 2) £9.49
3a) £11.98 b) £8.02
4) 240 5a) 800 g
5b) £2.04 6) £45

CHAPTER 2 REVIEW SECTION

1) 35, 57, 76, 89, 789
2) 123, 245, 621, 705, 947
3a) 40 b) 50 c) 80
d) 140 e) 280 f) 310
g) 460 h) 790 i) 840
j) 1460 k) 4570
l) 5560
4a) 9000 m b) 7000 m
c) 6000 m d) 8200 m

5) No, Mbakwe is wrong. The digit, 4 is not up to 5.
3490 would round to 3000 to 1 significant figure and not 4000.

6a) 50 b) 87
c) 432 d) 733

7a) 50 paintings
b) ₦81 750
c) ₦81 750

8a) 991 tins
b) £51.45

9a) 2300 b) 768000
c) 900890 d) 540
e) 6867 f) 23576

10) ₦551 200

11)

×	2	5	6	7	9
3	6	15	18	21	27
2	4	10	12	14	18
7	14	35	42	49	63

12)

11	6	13
12	10	8
7	14	9

13) ₦12 154 214
14) ₦1 219 104
15a) ₦31 050
b) ₦11 610
c) ₦62 660

16a) £41.75
b) £275.50 c) £30
d) £42 e) £25.10
17) 289 students
18) 49
19a) Port Harcourt ➡ Warri ➡ Benin route is the quickest. 198 + 97 = 295 km.

It takes longer to take Port Harcourt ➡ Aba ➡ Owerri ➡ Benin route.
66 + 69 + 227 = 362 km

b) 361 km c) 657 km
d) Yes, Achike is correct. 69 + 227 = 296 km through Owerri.

Through PH would take longer: 66 + 198 + 97 = 361 km
20a) 100 b) 600 c) 500
d) 400 e) 300 f) 2000
21a) 120 b) 580 c) 530
d) 350 e) 330 f) 1600
22a) 1.3 b) 23.5 c) 1.0
d) 897.3 e) 3.1 e) 98.6
23a) 70 b) 1400 c) 5
24a) 45 b) 24
c) 358 d) 658

EXERCISE 3A

1) 5.68
2) 10.5
3) 7.54
4) ₦35.46
5) 5.4
6) 15.74
7) 157.04
8) 85.25
9) 57.55
10) 18.94
11) 9.7
12) 11.93
13) 2
14) 102.89
15) 20.91
16) 21.534
17) 845.72
18) 667.326
19) £36.62
20) 145.59 cm
21) a) 102.476 km
b) 126.64 km
22) 21.16 m
23) 2.088 kg

EXERCISE 3B

1) 2.11
2) 0.71
3) 27.41
4) 4.07
5) 4.11
6) 0.043
7) 0.14
8) 4.46
9) 71.1
10) 9.4
11) 5.86
12) 699.088
13) 74.8
14) ₦623.44
15) 134.11 kg
16) 23.8
17) ₦168.85
18) ₦969.44
19)
a) 19.71
b) 131.07
c) 153.44
d) 1.076
e) 42.154
f) 232.105

Any calculation(s) that is correct e.g.
20) 9.75 – 2.5
21) 105.87 – 3.3
22) 71.36 – 4.7
23) 999.6 – 5.8
24) ₦11 331.09

EXERCISE 3C

1) 12.6
2) 0.38
3) 0.272
4) 9.5
5) 18.396
6) 39.88
7) 27.75
8) 912.084
9) 0.04
10) 0.16
11) 17.376
12) 50.4
13) 91.59
14) 2.1
15) 0.81
16) 31.996
17) 42.752
18) 141.183
19) 702.1
20) 20.355

EXERCISE 3D

1) 32 8) 5
2) 45.6 9) 18.97
3) 17634 10) 1839
4) 5123
5) 8.2
6) 13900
7) 80034

EXERCISE 3E

1) ₦41 888
2) £253.75
3) £18.45
4) £25.80
5) £31
6) £72.01
7) £7.99
8) £967.25
9) 230.268 m^2
10) 28.14
11) 0.00653 tonnes
12
   a) 20.25
   b) 0.49
   c) 0.1609
   d) 41.87

EXERCISE 3F

1) 7800
2) 3478
3) 907000
4) 43128900
5) 100000000
6) 54
7) 1214500
8) 7
9) 888
10) 456540

EXERCISE 3G

1a) 87 b) 50 c) 80
d) 5 e) 0.32 f) 13
g) 490 h) 5706
i) 0.425 j) 50

2a) 0.02 b) 0.5
c) 5.14 d) 2.6
e) 9.22 f) 0.099
g) 0.748 h) 21.75
i) 2.4 j) 3.45
3) 3 4) 2.1 5) 0.3
6) 0.002 7) 3.2
8) ₦658 9) 5.6
10) 186 11) 18.6
12) 1860 13) 1860
14) 186 15) 1860
16) 1.86 17) 18.6

EXERCISE 3H

1) 0.41, 2.14, 2.5, 4.01
2) 4.05, 4.054, 4.504, 40.54
3) 0.24, 2.4, 4.2, 42.4
4) 6.90, 9.6, 9.69, 96
5) 0.53, 0.953, 1.53, 5, 5.35
6) 0.43, 0.5, 0.57, 0.6, 0.7
7) 2.6, 6.2, 20.6, 21.62, 26.2
8) 4.31, 3.1, 1.43, 1.3
9) 86.5, 80.65, 8.65, 6.85

10) 2.7, 1.5, 1.3, 0.9

11) 9.88, 8.9, 8.8, 8

12) 30.1, 13.3, 10.3, 1.3

13) 0.32, 0.23, 0.032, 0.023

14) 8.17, 1.87, 0.78, 0.187

15) <
16) >
17) =
18) >
19) <
20) <

CHAPTER 3 REVIEW SECTION

1a) 6.5 b) 17.4
c) 32.3

2a) 5.37 b) 7.6 c) 0.9
d) 632.45
3a) 0.07, 0.7, 0.712, 7.2
b) 4.6, 4.67, 4.7, 4.76
c) -6, -0.55, -0.4, -0.3
d) 2.38, 2.83, 3.08, 3.83
4)

3	.	**4**	5
5	.	2	**5**
8	.	7	0

5a) £14.74 b) £2.76
c) £26.25 d) £23.75
6) 13.89 seconds
7a) 4390 b) 7.67
c) 78 d) 2.3934
8a) 30.1 b) 10.5
c) 40.8 d) 39.4
9a) 500 b) 4
c) 2.268 d) 0.0759
10a) £22.50 b) £1.25
11) 71.17m
12a) 5.6 kg b) 11.1 kg
13a) 409.96 b) 70.13
c) 8.94 d) 31.5 e) 8.9
14) £162.87
15a) ₦600 b) ₦1 400
c) ₦400 d) ₦70
e) ₦2 600

EXERCISE 4A

1a) 1, 2, 3, 4
b) 2, 4, 6, 8
c) 5, 10, 15, 20
d) 7, 14, 21, 28
e) 9, 18, 27, 36
f) 10, 20, 30, 40
g) 13, 26, 39, 52
h) 17, 34, 51, 68
i) 20, 40, 60, 80

2a) 2, 12, 16, 22, 30, 36, 54, 70
b) 12, 30, 36, 54
c) 12, 16, 36
d) 5, 30, 70
e) 7, 49, 70

3) Any number that fits the requirements e.g. 9, 15, ...

4) None
5) Examples: 10, 20…
6) 49 and so on
7) 8 and 20
8a) 33 and 55
b) 27 and 63
c) 16
9) YES, because 6 x 3 = 18
10) NO, Sanusi is incorrect. The first multiple of a number is the number itself. Therefore, the first five multiples of 4 are: 4, 8, 12, 16, and 20

11a) 7, 14, 21, 28, 35
b) 3, 6, 9, 12, 15, 18, 21, 24, 27, 30
c) e.g. 21
12) 12 and 30
13) 18, 36, 63

EXERCISE 4B

1a) 4 b) 6 c) 35
d) 70 e) 24 f) 30
g) 60 h) 20
2a) 12 b) 210 c) 240
d) 56 e) 315 f) 84
g) 90 h) 300

3) 360 minutes

4a) $2 \times 2 \times 3 \times 3 \times 3$
b) $2 \times 2 \times 3 \times 3 \times 5 \times 5$
c) $2 \times 2 \times 2 \times 3 \times 3 \times 3 \times 5 \times 5$
d) $5 \times 5 \times 7 \times 7 \times 7 \times 11 \times 11$
f) $2 \times 2 \times 3 \times 3 \times 3 \times 3 \times 4 \times 4 \times 5 \times 5 \times 5$

5) Jude is wrong. LCM of 3 and 6 is 6 and not ($3 \times 6 = 18$)
6) 560
7) 40
8) 4 cups

EXERCISE 4C

1)
- a) 1, 3
- b) 1,2,4
- c) 1,2,7,14
- d) 1,2,3,4,6,8,12,24
- e) 1,2,3,5,6,10,15, 30
- f) 1,5,7,35
- g) 1,2,3,4,6,9,12, 18,36
- h) 1,2,3,4,6,8,12,16,24, 48
- i) 1,2,4,7,8,14,28, 56
- j) 1,2,3,4,6,7,12,14,21,28,42,84
- k) 1,2,4,5,10,20,25,50, 100

l) 1,2,3,4,5,6,8,10,12,15, 20,24,30,40, 60,120

2)
a) 3, 6
b) 3, 7
c) 3, 6, 9
d) 3, 9
e) 3, 5, 6
f) 3, 6, 9

3) 1 or 5
4) Could be 2, 6 or 18
5) 5, 10 and 25
6) 1 and 5
7) 4 and 18
8) Yes, Anthony is correct. $7 \times 8 = 56$
9) 72

EXERCISE 4D

1)
- a) 3, 7, 19
- b) 2, 5, 13
- c) 31, 41
- d) 37, 59, 97
- e) 11, 17

2)
- a) 2
- b) 2 and 3
- c) 17 and 2
- d) 2 and 5

3) 2 or 3
4) 2 or 3
5a)

1	2	3	4	5	6	7
8	9	10	11	12	13	14
15	16	17	18	19	20	21
22	23	24	25	26	27	28
29	30	31	32	33	34	35
36	37	38	39	40	41	42
43	44	45	46	47	48	49

5b

1	2	3	4	5	6	7
8	9	10	11	12	13	14
15	16	17	18	19	20	21
22	23	24	25	26	27	28
29	30	31	32	33	34	35
36	37	38	39	40	41	42
43	44	45	46	47	48	49

5c) 71
5d) Yes, 71 is a prime number because it has only two factor, 1 and 71.

EXERCISE 4E

1a) $2 \times 3 \times 5$
b) 5×11
c) $2 \times 5 \times 7$
d) $2 \times 2 \times 2 \times 2 \times 3 \times 3$
$= 2^4 \times 3^2$
e) $2 \times 2 \times 2 \times 2 \times 3 \times 3 \times 5 = 2^4 \times 3^2 \times 5$
f) $2 \times 2 \times 5 \times 47$
$= 2^2 \times 5 \times 47$
2a) 2, 5 and 7
b) $2 \times 2 \times 5 \times 7$
$= 2^2 \times 5 \times 7$
3a) $2 \times 3 \times 5^2$
b) $2^2 \times 5^3$
c) $2 \times 3 \times 5 \times 17$
4a) 2×3
b) 3×7
c) 11×13
5a) 3, 3, 5
b) 3, 7, 2, 5
6a) $2^3 \times 3^3$
b) $2^2 \times 5 \times 7^4$
c) $3^4 \times 5^2$
7) NO, he is incorrect. For example;

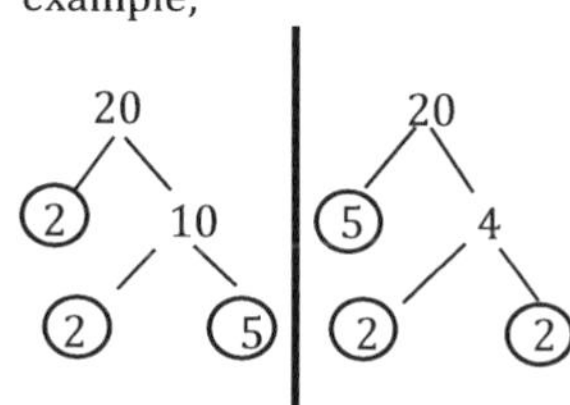

$2^2 \times 5$ $2^2 \times 5$

8a) $2^2 \times 3 \times 5^2 \times 11$
b) $2^2 \times 5 \times 7 \times 11$

EXERCISE 4F

1a)
6: 1 2 3 6
10: 1 2 5 10

HCF = 2

b) 12: 1 2 3 4 6
18: 1 2 3 6 9 18

HCF = 6

c) 5: 1 5
25: 1 5 25
HCF = 5

d) 27: 1 3 9 27
49: 1 7 49
HCF = 1

e) 30: 1,2,3,5,6,10,15,30
76: 1,2,4,19,38,76
HCF = 2

f) 36: 1,2,3,4,6,9,12,18,36
48: 1,2,3,4,6,8,12,16,24,48
Common factors are 1,2,3,4,6 and12
HCF = 12

2a) 20 b) 30 c) 140

3a $3 \times 5 \times 5 \times 5 \times 7 \times 7$
$= 3 \times 5^3 \times 7^2$
3b $2 \times 2 \times 3 \times 3 \times 3$
$= 2^2 \times 3^3$
3c $5 \times 7 \times 13$
4a) 6 b) 5
5a) 8 b) 9 c) 30 d) 5
6a) 1,2,4 b) 1,2,4
c) 1,3,9 d) 1,2,4
7) 35 cm by 35 cm
8) No, Kenechukwu is incorrect. The HCF of 25 and 50 is 25.

EXERCISE 4G

1a) YES, last digit is even
b) YES, last digit is even
c) NO, last digit (7) is not 0 or even
d) NO, last digit (9) is not 0 or even
e) YES, last digit is 0
f) NO, last digit (5) is not 0 or even

2a) YES, sum of digits is divisible by 3. 1 + 2 = 3 and 3 ÷ 3 = 1
2b) YES, same reason as above
2c) NO, 3 + 4 = 7 and 7 is not divisible by 3
2d) NO, 4 + 6 = 10 and 10 is not divisible by 3
2e) YES, 1 + 8 + 0 = 9 and 9 is divisible by 3
2f) YES, 6 + 5 + 4 = 15 and 15 is divisible by 3

3) YES, last digit is 5.
$5 \times 2 = 10$. $28794 - 10 = 28784$ and $28784 \div 7 = 4112$

4a) Yes, b) Yes, c) No
d) No e) Yes, f) Yes
5) NO, the last digit is not 0 or 5
6) No, the last three digits (451) is not divisible by 8.
7) Yes, last digit is 0 and even
8a) Yes b) No

CHAPTER 4 REVIEW SECTION

1a) 3,6,9,12,15
b) 4,8,12,16,20
c) 14,28,42,56,70
2a) 1, 13
b) 1,2,3,4,6,9,12,18,36
c) 1,2,4,8,13,26,52,104
3a) 7 e) None
b) 7,9
c) 6, 72
d) 2, 7, 23
4) $2 \times 5 \times 5 \times 5$
$= 2 \times 5^3$
5a) 9 b) 100
6a) 14 b) 350 c) 5460
7a) LCM = 132
HCF = 66
b) LCM = 1680
HCF = 140
8a) $2 \times 3^3 \times 5$ b) 7
9) Azubuike is wrong. LCM of 2 and 4 is **4**
10) 9
11) 165 seconds
12a) $2 \times 3 \times 3 \times 3$
b) 2×3^3
c) 18
13) 60
14) Henry is not correct. 1 is not a prime number since it has only one factor. Also, 2 is the only even prime number with factors 1 and 2
15) 37, 41, 43, 47, 53
16a)
16 → 2, 8; 8 → 2, 4; 4 → 2, 2

16b)
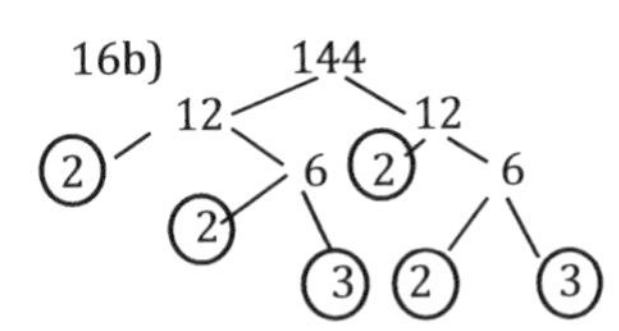

17) Yes.
1+8+ 9 + 3 + 5 + 7 = 33
33 ÷ 3 = 11 (Whole number)

18) $2 \times 3 \times 5 \times 5 \times 5$ which implies 2, 3 and 5
19a) 47 b) 54
c) 16 or 32
d) 12 or 16
e) 16

EXERCISE 5A

1) ¼ 2) 2/4 3) 7/12
4) 4/6 5) 1/5 6) 5/6
7)

8)
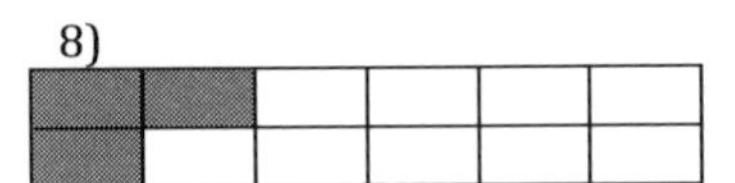

9)
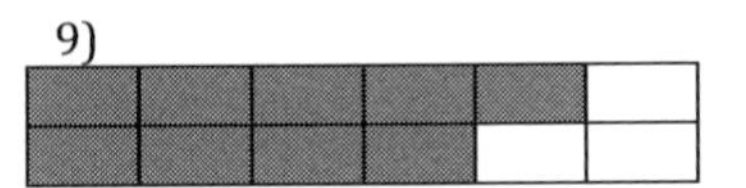

10) $1\frac{1}{2}$ 11) $2\frac{2}{6}$ or $2\frac{1}{3}$
12) $3\frac{2}{11}$ 13) $3\frac{2}{40}$ or $3\frac{1}{20}$
14) $\frac{8}{5}$ 15) $\frac{7}{3}$ 16) $\frac{32}{5}$
17) $\frac{68}{7}$ 18) 12 19) 14
20) 6 21) 14 22) 21
23) 5 24) 35 25) 5

26) 1 27) 2 and 4
28) 3, 4, 5

EXERCISE 5B

1a) $\frac{1}{2}$, b) $\frac{2}{3}$ c) $\frac{1}{3}$
d) $\frac{2}{11}$ e) $\frac{1}{4}$ f) $\frac{2}{9}$
g) $\frac{5}{24}$ h) $\frac{5}{8}$ i) $\frac{1}{3}$
j) $\frac{3}{10}$ k) $\frac{4}{15}$ l) $\frac{11}{20}$
m) $\frac{1}{5}$ n) $\frac{2}{5}$ o) 1

2a) $\frac{12}{15}$ and $\frac{10}{15}$
b) $\frac{5}{30}$ and $\frac{24}{30}$
c) $\frac{6}{8}$ and $\frac{4}{8}$
3a) $1\frac{3}{5}$ b) $4\frac{1}{7}$
c) $18\frac{1}{3}$ d) $1\frac{5}{6}$
e) $1\frac{4}{5}$ f) $4\frac{1}{4}$
g) $10\frac{12}{13}$ h) $13\frac{5}{15}$

4a) $\frac{4}{9}$ b) $\frac{3}{7}$ c) $\frac{5}{6}$
d) 0 e) $\frac{7}{12}$ f) $\frac{14}{45}$
g) $\frac{13}{15}$ h) 0 i) $\frac{26}{45}$
j) $\frac{34}{40}=\frac{17}{20}$ k) $\frac{9}{14}$
l) $\frac{47}{70}$ m) $1\frac{1}{12}$
n) $1\frac{1}{10}$ o) $1\frac{3}{20}$
5a) $\frac{2}{3}$ b) $\frac{8}{9}$ c) $\frac{1}{3}$
d) $\frac{2}{17}$ e) $\frac{7}{13}$ f) $\frac{4}{19}$

6a) $\frac{5}{6}, \frac{2}{3}, \frac{1}{2}$
b) $\frac{7}{8}, \frac{1}{2}, \frac{1}{4}$
c) $3\frac{2}{3}, 2\frac{2}{3}, 2\frac{1}{3}$

7a) ¼ b) ¼ c) ¼
d) ¼
8) $\frac{7}{15}$
9a) $1\frac{1}{19}$ m b) $\frac{12}{19}$ m
10) $\frac{5}{45}$. In its lowest form, $\frac{5}{45}$ = $\frac{1}{9}$ but the rest are $\frac{4}{15}$ to their lowest terms.
11a) 3 b) $2\frac{1}{2}$ c) 1
d) 1 e) $2\frac{1}{2}$

EXERCISE 5C

1) $3\frac{1}{3}$ 2) $4\frac{1}{15}$ 3) 8
4) 3 5) $6\frac{1}{2}$ 6) 5
7) $1\frac{37}{105}$ 8) $3\frac{13}{21}$ 9) 0
10) $4\frac{13}{15}$ 11) $5\frac{5}{12}$

EXERCISE 5D

1a) $\frac{2}{15}$ b) $\frac{4}{35}$ c) $\frac{18}{28}=\frac{9}{14}$
d) $\frac{10}{54}=\frac{5}{27}$ e) $\frac{9}{25}$ f) $\frac{7}{32}$
g) $\frac{40}{63}$ h) $\frac{3}{96}=\frac{1}{32}$ i) $\frac{12}{63}$
j) $\frac{12}{105}$ k) $\frac{24}{40}=\frac{3}{5}$
l) $\frac{4}{400}=\frac{1}{100}$

2) 12 m 3) $\frac{12}{55}$ m² 4) $\frac{1}{6}$
5) P: $\frac{3}{21}$ m² Q: $\frac{15}{63}$ m²
R: $\frac{5}{63}$ m² S: $\frac{25}{189}$ m²

6a) $\frac{1}{27}$ b) $\frac{6}{60}=\frac{1}{10}$
c) $\frac{36}{143}$
7a) Nnaemeka is wrong.
Correct answer should be $\frac{5}{18}$
b) Nnaemeka added the numerators and denominators instead of multiplying them.

EXERCISE 5E

1)
a) 1 b) 4 c) 9 d) 21
e) 8 f) 10 g) 50 h) 15
i) 40 j) 55 k) 350 l) 1

2a) ₦40 b) ₦50 c) $50
d) 6 kg e) 40 f) ₦1200
g) £15.50 h) 1020 kg

3a) £40
b) Arinze is correct. $\frac{4}{5}$ of £40 = £32 and £32 is more than £30

EXERCISE 5F

1a) $\frac{7}{4}$ b) $\frac{28}{5}$ c) $\frac{103}{9}$
d) $\frac{52}{3}$
2a) $1\frac{14}{25}$ b) $3\frac{31}{35}$ c) $4\frac{14}{25}$
d) $4\frac{24}{50}$ e) $5\frac{1}{4}$ f) $17\frac{1}{7}$
g) $110\frac{1}{4}$ h) $6\frac{19}{25}$
3a) $1\frac{10}{21}$ b) $6\frac{3}{25}$ c) $17\frac{2}{3}$
d) $60\frac{3}{4}$
4) $3\frac{1}{10}$
5a) $27\frac{2}{5}$ kg b) 29 kg

EXERCISE 5G

1a) 2 b) $1\frac{1}{5}$ c) $1\frac{1}{6}$
d) $4\frac{2}{7}$ e) 14 f) $1\frac{2}{3}$
g) 21 h) $7\frac{21}{45}$ i) $11\frac{1}{9}$
j) $1\frac{97}{153}$ k) $2\frac{8}{81}$ l) $3\frac{9}{77}$
2a) 15 b) 21 c) 27
d) 45
3a) 60 b) 80 c) 110
d) 200
4) $1\frac{1}{2}$ m 5) $12\frac{3}{5}$ m
6) 9 7) $2\frac{26}{235}$

CHAPTER 5 REVIEW SECTION

1a) $\frac{2}{3}$ b) $\frac{1}{3}$ c) $\frac{1}{3}$
2) $\frac{3}{12}=\frac{1}{4}$
3) $\frac{1}{3}$
4a) $\frac{10}{40}$ b) $\frac{15}{40}$ c) $\frac{32}{40}$
d) $\frac{36}{40}$
5a) $\frac{15}{7}$ b) $\frac{13}{8}$ c) $\frac{37}{3}$
d) $\frac{229}{11}$
6a) $\frac{5}{7}$ b) $1\frac{5}{21}$ c) $1\frac{17}{42}$
d) $\frac{65}{77}$
7a) $\frac{1}{2}$ b) $\frac{1}{3}$ c) $\frac{3}{5}$
d) $\frac{6}{7}$
8a) 12 b) 60 c) 72 d) 12
9a) $2\frac{1}{3}$ b) $5\frac{3}{4}$ c) $8\frac{3}{7}$
d) $10\frac{4}{20}=10\frac{1}{5}$
10a) $\frac{8}{17}$ b) $\frac{5}{9}$ c) $6\frac{3}{5}$
d) $3\frac{1}{3}$ e) $1\frac{5}{9}$ f) $\frac{5}{6}$
g) $5\frac{29}{54}$ h) $1\frac{44}{81}$
11a) 12 b) 27 kg c) ₦300
d) 140
12a) Q = $\frac{6}{77}$ m² R = $\frac{8}{77}$ m²

$S = \frac{36}{77}$ m²

12b) 1 m²   12c) Square
Same lengths for the four sides. I metre each

13a) 2 b) 3 c) $9\frac{9}{35}$

14a) $19\frac{5}{6}$ kg b) $38\frac{1}{2}$ kg
c) $5\frac{1}{2}$ kg d) ₦115 500

EXERCISE 6A

1a) 0.6 b) 0.8 c) 0.3 d) 0.1
e) 0.28 f) 0.125 g) 0.25
h) 0.625 i) 0.06 j) 0.76
k) 0.75 l) 0.8 m) 0.35
n) 0.48 o) 0.6 p) 0.5 q) 0.1
r) 0.875 s) 0.02 t) 0.8

2a) $\frac{1}{50}$ b) $\frac{3}{100}$ c) $\frac{1}{10}$ d) $\frac{3}{20}$
e) $\frac{9}{50}$ f) $\frac{1}{5}$ g) $\frac{1}{4}$ h) $\frac{3}{10}$ i) $\frac{7}{20}$
j) $\frac{23}{50}$ k) $\frac{1}{2}$ l) $\frac{11}{20}$ m) $\frac{3}{5}$ n) $\frac{7}{10}$
o) $\frac{3}{4}$ p) $\frac{79}{100}$ q) $\frac{4}{5}$ r) $\frac{81}{100}$ s) $\frac{9}{10}$
t) $\frac{49}{50}$

EXERCISE 6B

1a) 25% b) 20% c) 40%
d) 30% e) 80% f) 15%
g) 60% h) $37\frac{1}{2}$% i) 35%
j) 36% k) 32% l) 60%
m) 24% n) 64% o) 65%
p) 10% q) 50% r) 50%
s) 25% t) 5%

2a)
i) $\frac{7}{10}$ ii) 70% iii) $\frac{3}{10}$ iv) 30%

2b) i) $\frac{3}{25}$ ii) 12% iii) $\frac{22}{25}$
iv) 88%

2c)
i) $\frac{1}{2}$ ii) 50% iii) $\frac{1}{2}$ iv) 50%

EXERCISE 6C

1) $\frac{5}{7}$ 2) $\frac{4}{20} = \frac{1}{5}$ 3) $\frac{5}{200} = \frac{1}{40}$
4) $\frac{30}{300} = \frac{1}{10}$
5a) $\frac{3}{13}$ b) $\frac{1}{4}$ c) $\frac{1}{30}$ d) $\frac{17}{50}$
e) $\frac{5}{48}$ f) $\frac{2}{125}$ g) $\frac{1}{150}$

EXERCISE 6D

1a) 40% b) 25% c) 20%
d) $33\frac{1}{3}$%
2a) 10% b) 50% c) 25%
d) 100%
3) $1\frac{2}{3}$% 4) 15% 5) 30%
6a) 62.5% b) 37.5%
7a) 20% b) 50% c) No, Okonkwo is wrong. Uduak scored 70%

CHAPTER 6 REVIEW SECTION

1) 27%

2a)

2b)

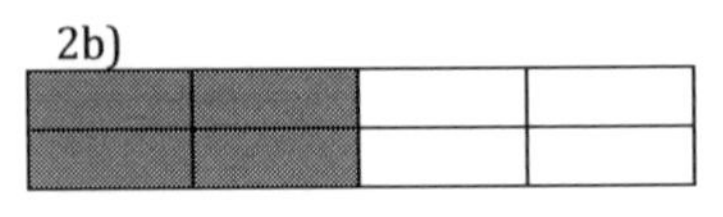

2c)

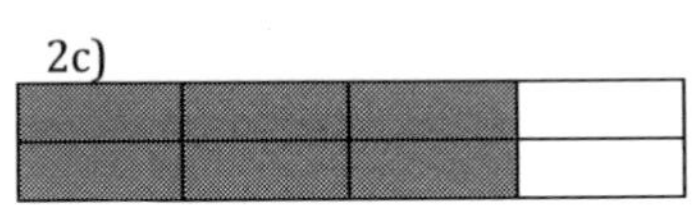

3) $\frac{1}{50}$
4a) 28.8% b) $\frac{1}{4} \times 100 = 25\%$
c) 50%
5a) 50% b) 60% c) 70%
d) 15%
6a) Maths – 68%
Economics – 5%
English - 60% Physics – 35%
6b) 5%, 35%, 60%, 68%
7)

Fractions	%	Decimals
$\frac{2}{5}$	40%	0.4
$\frac{7}{20}$	35%	0.35
$\frac{3}{5}$	60%	0.6
$\frac{6}{25}$	24%	0.24

8) 0.65, $\frac{3}{5}$, 51%, $\frac{1}{2}$
9) $\frac{8}{25} \times 100 = 32\%$
$\frac{8}{25}$ is bigger
10a) 75% b) 25%

EXERCISE 7A

1) 2 2) 9 3) 8 4) 5 5) 68
6) 7 7) 22 8) 87 9) 6 10) 3
11) 10 12) 40 13) 356
14) 0 15) 9 16) 1 17) 4
18) 20 19) 10 20) 9 21) 9
22) 4 23) 10 24) 1 25) 15
26) 12

EXERCISE 7B

1a) 11 b) 15 c) 14 d) 4
e) 10 f) 1 g) 135 h) 491
i) 81

2a) 30 b) 4 c) 15 d) 65
e) 50 f) 19 g) 3 h) 0

EXERCISE 7C

1a) a + 7 b) b + 20 c) c – 9
d) d - 2 e) e + 17 f) f + 6
g) g - 4 h) h - 13

2a) k – 5 b) k + 10
3a) 2k + 5 b) Chuka: 15 years
Okoro: 30 years
4a) y + 20 b) y – 2 c) 6y + 23
d) 3y
5a) 24w b) 24w + 72n
6a) b + 7 b) b - 4

EXERCISE 7D

1) n = 3 2) v = 4 3) m = 12
4) b = 13 5) k = 4 6) x = 11
7) y = 18 8) y = 5 9) d = 8
10) c = 35 11) d = 18
12) c = 34 13) x = 40
14) w = 10 15) d = 5
16) k = 6 17) g = 55
18) w = 96 19) t = 0
20) w = 105 21) 6 22) 10
23) 2 24) 9 25) 6 26) 3
27) 3 28) 75 29) 45
30) 110 31) y = 19 years
32) 44 cm

CHAPTER 7 REVIEW SECTION

1) 28 2) 39 3) 6 4) -3
5) 22 6) 7 7) 10 8) 40
9) 86 10) 1 11) 4 12) 8
13) 20 14) 1 15) 7
16a) a + 5 b) b + 12
c) c – 4 d) h – 4
17a) m – 3 b) m + 7
18a) 2m + 4
b) Taiwo = 6 yrs Olu = 16 yrs
19a) y + 3 b) y – 8
c) 7y + 54 d) y^2
20a) g + 7 b) g – 4
21) n = 9 22) v = 4 23) m = 64
24) b = 38 25) k = 29
26) d = 48 27) c = 37
28) x = 80 29) w = 30
30) d = 20 31) 32 32) 25
33) 7 34) 64

EXERCISE 8A

1)

Name	Edge	Vertex	Face
Cuboid	12	8	6
Square-Based Pyramid	8	5	5
Triangular-Based Pyramid	6	4	4
Triangular Prism	9	6	5

2a) some tins of milk
b) Sugar cubes
c) Most of the toothpaste packs
d) Football

3a) Ice-cream cone or mound of rice
b) Triangle ornaments
c) Toblerone chocolate (triangle-based prism)
4) A, D and H
5) B – Rectangle, C – Cylinder, E – Sphere, F – Square-based pyramid, G - Cone

6) B – a 2-d shape but prisms are 3-d
C – Curved surfaces but a prism should have flat surfaces only
E – A prism must have flat surface(s) only
F – The cross section does not go through the whole length
G – Curved surface but a prism should have flat surfaces only.

EXERCISE 8B

Accurate net drawn with the following dimensions in cm:

1)

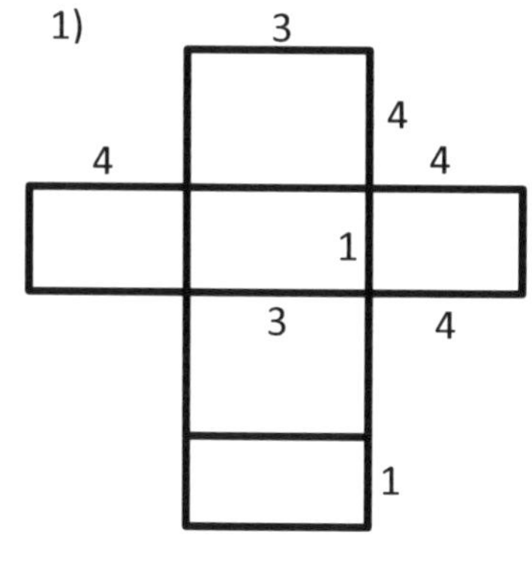

2)

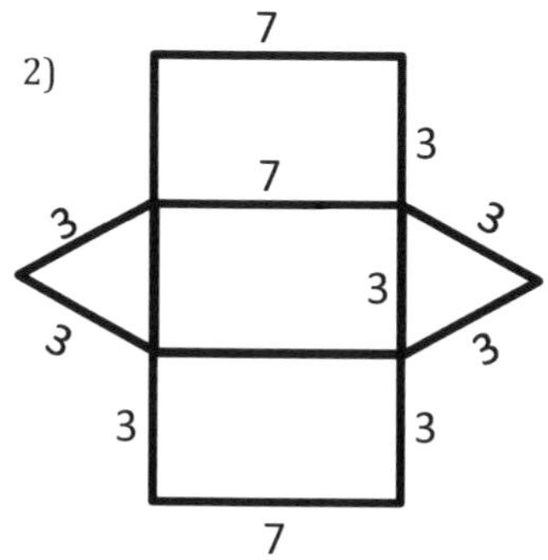

3) 3, 4
4a) 3 b) 2 c) 5 d) 9 e) 6
5a) 2, 4, 6
b) 6, c) 9 d) 5
6) A i)Cuboid
ii) Student's work
iii) Student's work
iv) 8 vertices and 12 edges

B i) Square-based pyramid
ii) Student's work
iii) Students work
iv) 5 vertices and 8 edges

CHAPTER 8 REVIEW SECTION

1a) A – Cuboid,
B – Pyramid, C – Cylinder,
D – Cone, E – Sphere
1b) 5 c) 0 d) No, it has curved surface
e)

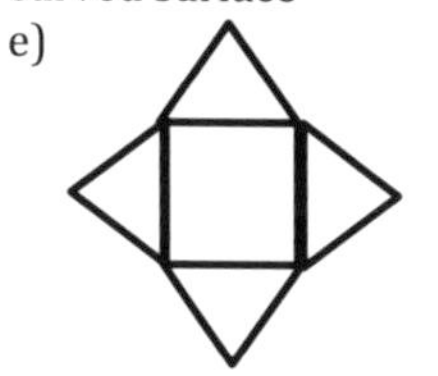

2a) Square-based pyramid
b) Vertices: 5, Faces: 5

3a) Cuboid b) Rectangle
c) WV and UV d) Rectangle e) 3

4a) Square-based pyramid
b) Triangle-based prism
c) Cuboid d) Cuboid
e) Hexagonal- based pyramid or cube or cuboid
5a) Pentagon b) 5 c) DI d) 3
6)

Name	Vertices	Faces	Edges
Triangular prism	6	5	9
Cube/cuboid	8	6	12
Square-based pyramid	5	5	8
Tetrahedron	6	4	4
Hexagonal-based pyramid	7	7	12

EXERCISE 9A

1a) p + 4 b) c – 3 c) 2n d) 3k
e) 2p + 8 f) 3q + 2 g) 6(f + 7)
h) $\frac{y-c+p}{n}$ i) $\frac{b+10}{h}$ j) r – t + w
2a) m = i + 10 b) 47 years
3a) w = 100h b) i) ₦300 ii) ₦1000
iii) ₦150 iv) 550
4a) py b) 7w c) 2500m
d) 3000b + 150k e) 13y + 25f

5a) 24d b) 24m c) 120
6a) 15y b) 15d c) 45
7a) c= $(\frac{3000}{n})$ + 5
b) i) ₦205 ii) ₦155
8a) a = 60 + 140w
b) i) ₦5 660 ii) ₦14 060
9) $\frac{1}{2}$w + 11
10a) 3n b) 4n
11) m – 5 12) f + 3

EXERCISE 9B

1a) 4 b) 1 c) 6 d) 3 e) 5
f) 7 g) 24 h) 8 i) $\frac{1}{7}$ j) $\frac{2}{3}$
k) $\frac{3}{7}$ l) $\frac{4}{9}$ m) $\frac{2}{7}$ n) $\frac{7}{11}$ o) $\frac{1}{2}$ p) $\frac{1}{8}$

EXERCISE 9C

1) 2w 2) 9x 3) 3w 4) 5p 5) 5w 6) 17x 7) 19y 8) 5y
9) 0 10) 11g 11) 9n 12) 10e 13) 11y 14) 21p 15) 7u
16) 2p 17) 6u 18) 7a 19) 3c 20) 12t 21) 6.5m 22) 6.6x
23) 6.6v 24) 9.6z 25) 29.8f 26) 11g 27) 3w 28) 7k
29) 9c 30) 76u 31) 6m 32) 3x

EXERCISE 9D

1) 2d + 3e 2) 6s + 12 3) 15t -6 4) 7y+7t 5) 3x –t
6) 7d +3x + 9 7) 5x + 2y 8) 7g + 4x + 5 9) 2p + 11e
10) 7c + 8 11) 4x + 7y 12) 2a – 3b 13) 2a – 1 14) 4p + 16
15) 4b + 3c 16) 9a + 2 17) 6s + v 18) -6y
19) 5t – 3u + 8v – 3 20) 9b + 6c - 3

EXERCISE 9E
1a) p + q
b) c)

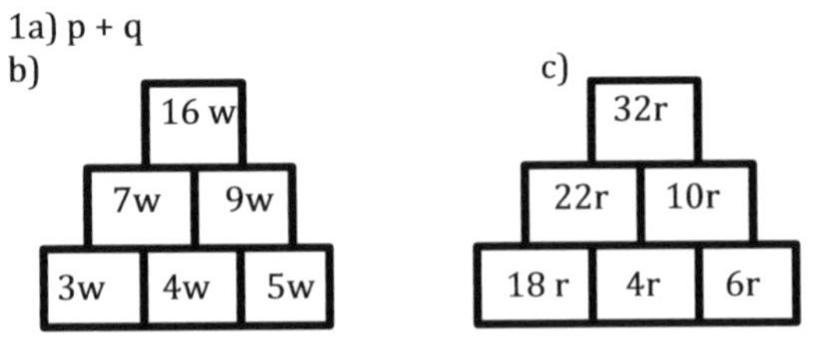

c)

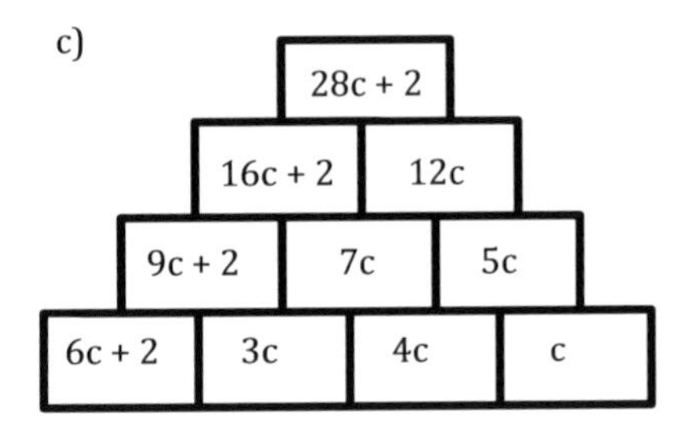

EXERCISE 9F

1a) 16x b) 26c c) 8w + 6s d) 3x + 3 e) 2t + 4v + 1
f) 3w + 8n + 4 g) 4k + 14c h) 2x + 8y -2

CHAPTER 9 REVIEW SECTION

1a) w + 8 b) x + 4 c) t – 5 d) 2c e) 4c f) $\frac{x}{5}$ g) $\frac{m}{3}$ h) $\frac{p}{x}$ i) ac
j) n + 30
2a) 7a b) 3t c) 6x d) 2a e) 10n + 6m f) 2v g) 13m + 2
h) s + 3j i) 9c + 3e + 3 j) 6 + 6y
3a) p = 3c + 7 b) p = 22 c) c = 7
4a) 8t + 12u b) 2t – 8u c) 20t + 5u d) 56t + 31u
5a) 28a b) 6b + 8 c) 10c + 10d + 3
6a) 24 b) 12 c) 14 d) 64
7) Range of answers like: 6w + 4w; 12w – 2w; 20w – 10w .....
8) No. The answer should be 6x – y – 3
9a) f = 3c + 30 b) 90ºf iii) 120 ºf

EXERCISE 10A

1a) 45° 2) 135° 3) 315°
4) 180° 5) 225° 6) 90°
7) 315° 8) 360° 9) 270°
10) 225° 11) 90° 12) 270°
13) 270°
14a) East b) 37.5°
15a) South b) 315°
16a) 45° b) 15° c) 15°
d) 67.5°
17a) 30° b) 60°
c) 300° or 60°
18) 90° clockwise or 270° anti-clockwise

EXERCISE 10B

1a) Obtuse angle b) $B\hat{C}D$
2a) Acute b) Right angle
c) Reflex d) Obtuse
e) Reflex f) Acute

3a) $A\hat{B}C$ b) $P\hat{Q}R$

c) $J\hat{K}L$ (reflex) d) $D\hat{E}F$

e) $G\hat{H}I$ (reflex) f) $X\hat{Y}Z$

4a) acute b) reflex c) obtuse
d) right angle e) acute
f) straight line

5a) obtuse b) right angle
c) obtuse d) obtuse
e) obtuse

EXERCISE 10C

1) Student's accurate construction
2) They add up to 180° (120 + 60) .... Straight line

EXERCISE 10D

1) Any ten angles between 90 and 180 degrees, exclusive.
2) Any four angles between 90 and zero degrees, exclusive.

3a) $Q\hat{P}R$ b) $S\hat{Q}P$ and $T\hat{R}P$

c) $Q\hat{S}T$ and $R\hat{T}S$
4a) 20° b) 70° c) 110°
d) 160° e) 60°
5a) obtuse b) acute c) acute
d) acute e) acute f) reflex
g) right angle h) reflex
6a) reflex b) acute c) obtuse
d) right angle e) acute
f) reflex g) obtuse h) acute
i) reflex j) reflex k) acute
l) acute m) reflex n) obtuse
7) ...+/- two degrees
a) 40° b) 100° c) 35°
d) 120° e) 90° f) 15°
8) Students own sketch using 90° as a guide
9) Students own accurate angles
10a) acute b) obtuse
c) acute d) obtuse
e) right angle f) acute
11a) obtuse b) right angle
c) obtuse

CHAPTER 10 REVIEW SECTION

1) 180° 2) 30° 3) 300° bigger angle or 60o smaller angle 4a) 30° b) 100°
c) 170° 5) Students own accurate construction
6a) 50° b) 130°
7a) reflex b) acute c) obtuse
d) obtuse e) right angle
f) obtuse g) acute

EXERCISE 11A

1a) 9,5,4,-3,-8,-10
b) 40,10,9,-2,-7,-15
c) 42,17,-10,-15,-20
d) 5,3,-1,-5,-7
e) 9,8,4,-4,-8,-9
f) 30,14,10,-12,-16,-18
2a) A : 9°C, B : 13°C,
C : -9°C, D : -2°C
3a) – 12°C b) 14°C
c) 4°C d) 9°C
4a) > b) > c) < d) =
5a) 8°C, b) 16°C c) 3°C
d) 5°C e) 3°C f) 3°C
g) 13°C h) 6°C
6a) -1°C,-3°C,-7°C,-10°C, -15°C
b) 5°C c) 5°C d) -15°C
e) 20°C
7a) Fallen by 6°C
b) Increased (risen) 15°C
c) Fallen by 10°C
d) Increase by 23°C

8a) 5 b) 1 c) -2 d) -4
e) -9 f) 4

EXERCISE 11B

1a) -10, -12 b) -2, -5
c) -5, -10 d) -4, -10
e) -1, 1 f) -9, -11
2a) -9,-4,-3, 0, 5
b) -50,-30, 20, 40, 55
c) -7,-3,-2,-1, 0, 1, 4

3)

Temp °C	Change °C	New Temp °C
-3	+4	1
-3	-3	-6
-7	-3	-10
+6	+6	12
-9	-9	-18

4a) 2°C b) 9°C c) 28°C
5a) Kano b) Essex c) Essex, Manchester, Lagos, Onitsha, Abuja, Kano d) Essex: 1°C
Onitsha: 31°C

EXERCISE 11C

1a) 11 b) 0 c) -1 d) -4 e) -4
f) 3 g) 5 h) -4 i) 3 j) -2
k) -5 l) 7 m) 3 n) 2 o) 2
p) 0 q) 0 r) 1
2a) 2 b) 2 c) -3 d) -1 e) 0
f) 0 g) -7 h) -3 i) -1 j) -7
k) -8 l) 4 m) -5 n) 7 o) -6
p) -6 q) 0 r) 3
3a) -6 b) -16 c) -10 d) -6
e) -3 f) -16 g) -13 h) -13
i) -11 j) -17 k) -11 l) -10
m) -15 n) -9 o) -5 p) -12
q) -10 r) -23
4a) -11 b) -15 c) -7 d) -12
e) -22 f) -10 g) -15 h) -15
i) -10 j) -5 k) -14 l) -16
m) -8 n) -11 o) -9 p) -6
q) -5 r) 15
5a) -2 b) -1 c) -2 d) -2
e) -2 f) -6 g) 7 h) -4 i) 7
j) 1 k) -130
6a) 13 b) 5 c) 4 d) 12
e) 19 f) 10 g) 10 h) -5

7)

×	-4	-3	-2	-1	0	1	2
+4	0	1	2	3	4	5	6
-5	-9	-8	-7	-6	-5	-4	-3
+3	-1	0	1	2	3	4	5
-2	-6	-5	-4	-3	-2	-1	0
0	-4	-3	-2	-1	0	1	2
+8	4	5	6	7	8	9	10
-10	-14	-13	-12	-11	-10	-9	-8
-7	-11	-10	-9	-8	-7	-6	-5
-6	-10	-9	-8	-7	-6	-5	-4

8a) -14 b) -4 c) -4 d) -1 9) ₦22 259 overdrawn

EXERCISE 11D

1a) 6 b) 20 c) 7 d) 30 e) 21 f) 25 g) 120 h) 45
i) 14 j) 72 k) 8 l) 56
2a) -6 b) -70 c) -15 d) -36 e) -36 f) -8 g) -30
h) -18 i) -63 j) -6 k) -16 l) -33
3a) 2 b) 3 c) 3 d) 3 e) 10 f) 4 g) -3 h) -10
i) -4 j) -4 k) -5 l) -5
4)

×	8	-2	-5	7	-3
-3	-24	6	15	-21	9
4	32	-8	-20	28	-12
5	40	-10	-25	35	-15
-10	-80	20	50	-70	30
-2	-16	4	10	-14	6
-7	-56	14	35	-49	21

5)
-3 × 7 ⟶ -21
-6 × 6 ⟶ - 36
-6 × -6 ⟶ 36
-7 × -3 ⟶ 21
-10 ÷ -2 ⟶ 5
10 ÷ -2 ⟶ -5
6a) 1 b) 6 c) 2 d) -10 e) -32
f) -27 g) 56 h) -9

CHAPTER 11 REVIEW SECTION

1) 2 2) -5 3a) -3 3b) -3 3c) -3 3d) -14
4) -9°C 5a) 5 5b) 2 5c) 9 5d) 11
6a) -5, -3, -2, -1, 7, 10 b) -7, -5, -4, 0, 3, 5, 8
c) -17, -5, -2, 1, 2, 4, 5

7)

×	-4	+3	-8	2
-2	8	-6	16	-4
-7	28	-21	56	-14
+8	-32	24	-64	16
+5	-20	15	-40	10

EXERCISE 12A

1a) 3 b) 0.4 c) 6 d) 0.8 e) 1.6 f) 4 g) ₦8 h) ₦30
i) 6 j) 0.8 k) 12 l) 1.6 m) 16 n) 18 o) 16 p) 50
2a) 1 b) 7 c) 60 d) 2 e) 1.9 f) 0.24 g) 1220
h) 28 kg

EXERCISE 12B

1a) 1.3 b) 1.05 c) 16 d) 24 e) 6.93 f) 7.7 g) 5.2
h) ₦21 i) £33 j) 675 k) 1.2 l) 28 m) 24.4 kg
n) 62.3 o) ₦460 p) 0.36
2a) £1.89 b) 1.792 c) 5.6 d) 80.5 e) 0.75 f) 9
g) 2.38 kg h) ₦325

EXERCISE 12C

1a) 22 b) 88 c) ₦220 d) ₦374 e) 924 f) 3300
2a) 23 b) 92 c) ₦230 d) ₦391 e) 966 f) 3450
3a) 32 kg b) 73.6 litres c) £98.40 d) $146.42
e) 3200 g f) 196 cm 4) ₦2500 5) £3150
6) b ⟶ 50 to 70
7) For £15, sale price = £9.75
For £32, sale price = £20.80
For £45, sale price = £29.25
8) 1.94 kg 9) 33.3% 10) ₦5610 11) ₦5220
12) 82.8 kg 13) £772.20

CHAPTER 12 REVIEW SECTION

1a) 13 b) 0.34 c) 180 d) ₦1260
2) ₦3600 3) 67.5kg 4a) 25% 4b) 4% 4c) 4.2%
5) 10% 6) ₦1068.48 7a) ₦24 500 7b) ₦350 000
7c) ₦31 500

EXERCISE 13A

1a) 12 b) 16 c) 15 d) 15 e) 6 f) 11 g) 7 h) 8
i) 27 j) 24 k) 7 l) 12
2a) p + c + e b) 3x + 4 c) m + n + o + p d) 9x + y
e) y + n – m f) 5k + 3 g) 3c – d + e -7f
h) p – q + r – s i) 2n – m j) 7g – 4c k) 10w + 6y
l) d + e – f - g

EXERCISE 13B

1a) 1 b) 3 c) 0 d) 6 e) 18 f) 3 g) 5
h) 13 i) 0 j) 9 k) 12 l) 13
2a) d – e + f b) t – h – g c) p – e + w
d) x + 3 e) 11b – 5 f) 5 g) f + 4b
h) 10 – 7y + 3 i) 9v – 2 j) c – 2d
k) -3m + 7n l) 7a – 3d + 2e

EXERCISE 13C

1a) 6n b) 12m c) 7n d) mn e) 6w
f) 15cx g) $5r^2s$ h) $9y^2$ i) 130a
j) $8a^2c$ k) 45ab l) $21v^2$ m) $40c^2$
n) $18c^2$ o) $4d^2$ p) $9m^2$ q) s^2 r) $24py^2$
s) 30d t) $42q^2$ u) 56cd v) $22e^2fg$
w) 3aw x) $4b^2c$ y) a^2b^2c z) 200gh

EXERCISE 13D

1a) c b) 4s c) x d) 3ef e) 9w f) 2ab
g) 2 h) 8y i) d j) 3m k) 5wx l) 2
m) k/8 n) 3b o) 2pq p) 5k

EXERCISE 13E

1a) 22 b) 11 c) 47 d) 8 e) 4 f) 5
g) 18 h) 22 i) 44 j) 9 k) 39 l) 3
m) 18 n) 22 o) 14 p) 10 q) 15
r) 11 s) -2 t) 300
2a) 5 + (6÷ 2) = 8 b) 21 ÷ (3+4) = 3
c) 2 × (7-4) = 6 d) 8 – (4÷4) = 7
e) 9 + (3+3) × 2 = 21
f) (40-8) × 7 = 224
3a) 12w + 10 b) 3 + 8y c) 47c
d) 5 + 3x e) 5n – 1 f) 7w – 2
g) 18 h) 12m + 10 i) 47p – 3
j) 4w + 5 k) 51x + 2 l) 10x – 2
m) 27p – 9 n) 24 – 2n o) 35k – 21
p) 10 q) 44b r) $24g^2$ – 3
s) -2x t) 28v

EXERCISE 13F

1a) 20 b) 30 c) 13 d) 7 e) 10 f) 5
g) 100 h) 60.25 i) ¼ j) 18 k) 130
l) 5
2a) 21 b) 33 c) 37 d) 49 e) 0 f) 70
g) 42 h) -6 i) 21 j) 9 k) 49 l) 20
3a) i) 22 cm ii) 10 m 3b) i) 18 cm
ii) 9m c) i) 36 cm ii) 19.5 m
d) i) 11 cm ii) 6.6 m
4a) 18 b) 116 c) 10 d) 140 e) 3
f) -24 g) 2 h) 10

EXERCISE 13G

1a) -9 b) -6 c) -2 d) 13 e) -10 f) 24 g) -103 h) -9 i) -27
j) 0 k) 1 l) -58
2a) -2 b) -11 c) 28 d) 1 e) -1 f) 8 g) 95 h) -21 i) -48 j) 4
k) 34 l) 19
3a) -18 b) -18 c) -30 d) 6 e) -6 f) -24 g) -13 h) -24 i) 9
j) 2 k) 1 l) 34

EXERCISE 13H

1a) c = 2 b) c = 13 c) c = 15 d) x = 1 e) x = 11 f) x = 91
g) x = 18 h) w = 10 i) w = 22 j) w = -9 k) u = -13 l) x = - 16
m) u = -7 n) r = 53 o) g = -5 p) w = 23 q) y = -16 r) r = 64
s) y = 30 t) n = 22 u) e = -7 v) x = 24 w) c = 10 x) x = 6.8
y) x = 7.8 z) u = 344

EXERCISE 13I

1a) x = 2 b) x = 3 c) x = 6 d) x = 7 e) x = 11 f) w = 3 g) w = 9
h) c = -2 i) c = -2 j) w = -4 k) x = 8 l) x = -10 m) x = 7 n) n = 3
o) n = 42 p) x = 30 q) w = -18 r) n = 3 s) x = 7 t) n = -9
u) n = -2 v) m = 3 w) w = 4.5 x) z = 1/3 y) n = 10 z) y = -94

EXERCISE 13J

1a) c = 4 b) c = 2 c) x = 4 d) w = 2 e) x = 1 f) x = -1 g) x = 3
h) w = 1 i) y = 6 j) x = 2 k) u = -1 l) x = -1 m) y = 2 n) y = 10
o) n = 6 p) v = 3 q) x = 2 r) x = 3 s) c = 5 t) w = 2 u) x = 8
v) n = 7 w) w = 8 x) x = 3 y) c = 4 z) x = -1

EXERCISE 13K

1a) n = 15 b) n = 40 c) n = 10 d) n = 5 e) n = -9 f) n = 300
g) n = -63 h) n = 5 i) n = 14 j) n = 120 k) n = 3 l) n = 78
m) n = 32 n) n = 4 o) n = 2 p) n = 4/5 q) n = -4 r) n = -18
s) n = 5 t) n = 4 u) n = 5 v) n = 1 w) n = 1 x) n = 4 y) n = 28
z) n = -8

EXERCISE 13 EXTENSION QUESTIONS

1) 9 cm 2) 36 cm^2 3a) d = -15 b) w = -10 c) n = 1

CHATPER 13 REVIEW SECTION

1a) 16 b) -3 c) 7
2a) c – d + e – f b) 13w + 6n
3a) f + 4b b) 7a – 3d + 2e c) $8w^2$ d) $3a^2b$
4a) 12 b) 44 c) 23
5a) 3 b) 7 c) 4 d) 9
6a) n = 2 b) n = 11 c) n = 27 d) x = 4 e) x = -9 f) x = -1
g) y = 14 h) x = 10
7a) n = 15 b) x = 70 c) x = 7 d) x = -45
e) x = -60 f) x = 18 g) y = 48 h) x = 3

EXERCISE 14A

1a) 27 cm^3 b) 64 cm^3 c) 27 cm^3
d) 15 cm^3

EXERCISE 14B

1) 24 cm^3 2) 90 cm^3 3) 36 cm^3
4) 3000 cm^3
5a) 60m^3
b) i) 60 000 litres
ii) 30 days
6) 64 cm^3 7) 264 m^3 8) 900
9)

Volume	Length	Breadth	Height
40 cm^3			
	5 cm		
210 m^3			
			6 cm

10a) 25.2 m^3 or 25,200, 000 cm^3
b) 75.6 tonnes
11) x = 6 cm y = 11 cm
12) 6m^2

EXERCISE 14C

1) 180 cm^3 2) 3000 cm^3
3) 2100 cm^3 4) 8m 5) 192 cm^3

CHAPTER 14 REVIEW SECTION

1) 24 cm^3 2) 147 cm^3 3) 343 cm^3
4) 18 m^3 5a) 168.75 cm^3
5b) 1200 cm^3 5c) 4000 cm^3
6) 5.4 m

EXERCISE 15A

1a) 30° b) 10° c) 145° d) 44° e) 60°
f) 100° g) 50° h) 141° i) a= 27°,
b = 153°, c = 153° j) x = 30° k) 125°
l) w = 90°, y = 47° m) 215°
n) n = 90°, x = 40°

EXERCISE 15B

1a) a = 100° 2) b = 40°
3) c = 42°, d = 42°
4) e = 70°, f = 70°, g = 110°
5) h = 75°, i = 30°
7) m = 32°, n = 50°, o = 82°, p = 98°
8) q = 60°, r = 60°, s = 60°, t = 120°, u = 120°
9) v = 33°
10a) scalene............three different angles
b) Right-angled..........have a 90° angle
c) Isosceles......two equal lengths
d) Equilateral......all sides the same
e) Isosceles......two base angles are equal
f) Equilateral.......three equal sides
11a) 25° b) 25° c) 59° d) 56° e) 84°
12) NO, Obiora is incorrect. 54 + 85 + 42 = 181° but angles in a triangle add up to 180°.
13a) 94° b) 30° c) 81° d) 10° e) 40° f) 60° g) 90°
h) 81° i) 39° j) 55°
14a) 9w = 180 b) w = 20° c) 100° d) 80
15a) i) 55° ii) 70° iii) 18°
15b) i) Isosceles triangle because two base angles are the same (55° and 55°).
ii) Scalene triangle because all three angles are different.

EXERCISE 15C

1a) Rhombus b) Trapezium c) Rectangle d) Square e) kite
f) Parallelogram g) Trapezium
2)

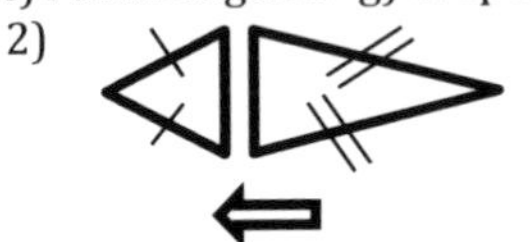

3)

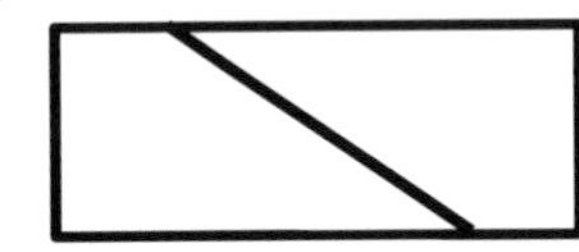

4) A square has all the corners at 90° while the corners of a rhombus are not 90°.
5) No. A quadrilateral has four sides but shape B has three sides.
6a) Trapezium b) Rhombus c) Kite d) Square e) Rectangle
f) Parallelogram
7) No, Kunle is incorrect. A trapezium has only a pair of parallel sides. The shape drawn is a parallelogram.

EXERCISE 15D

1a) p = 141° b) w = 65°, x = 95° c) a = 93°, b = 75°, c = 92°
d) f = 90° e) d = 43° 2) 140° 3a) 3n = 156 , n = 52°
3b) 8x = 280, x = 35° 3c) 6x = 210, x = 35° 3d) 10c = 360, c = 36°

EXERCISE 15E

1) NO. A polygon has straight sides but a circle is curved.
2) Any four quadrilaterals, for example; Pentagon – 540°, Hexagon – 720°, Heptagon – 900°, Octagon – 1080°
3) Dodecagon – 1800°
For questions 4 - 7): Students to construct circles accurately, using the radii given. They must use a pencil, ruler and a pair of compasses for the construction. Teachers/parents to check the accuracy of the circles drawn.
8) No, Okoro is incorrect. A circle with a diameter of 20 cm must have a radius of 10 cm. *2 × radius = diameter*

6) j = 60°, k = 60°, l = 60°

EXERCISE 15F

1a) a = 47°........ Alternate angles are equal
b) b = 107°...... alternate angles are equal
c) c = 38° .....Angle on a straight line where 142° lies is 38° (180 – 142). Therefore, Angle C = 38° ....alternate angles. Other Mathematical reasons are also encouraged.
d) d = 102°... corresponding angles are equal
e) e = 113° ....alternate angles
f = 70° ....corresponding angles
f) g = 96° ....corresponding angles
g) h = 30° ....alternate angles
i = 47° ....angles on a straight line
j = 133° ....corresponding angles
h) k = 110°
Angle next to K is 70° ....Corresponding angles.
Angle on a straight line = 180 – 70 = 110°
i) l = 40° ....angles on a straight line
n = 14° ....corresponding angles
m = 166° ... from (180 – 14) = 166°, then corresponding angles are equal.
m = 40° ......alternate angle to l
j) p = 58° ...co-interior angles add up to 180°

CHAPTER 15 REVIEW SECTION

1a) x = 85° b) a = 109°, b = 71°, c = 71°
d = 109° c) e = 63°, f = 117°, g = 63°, h = 54°
i = 63°
2a) j = 92°, k = 110°, l = 70° b) m = 95°
c) n = 48°, o = 89°, p = 48°
3a) 3 b) CD c) AB d) Regular hexagon
e) 6 f) obtuse
4a) 76.5°....vertically opposite angles are equal
b) g = 115° .....Angles on a straight line add up to 180°
a = 115° ....corresponding to g
b = 65° ....corresponding to 65°
f = 115° ....vertically opposite to g
d = a = 115° ....vertically opposite angles
c = b = 65° ....vertically opposite angles
e = 65° ....vertically opposite angles

EXERCISE 16A

1a) 9 b) 10 c) 10 d) 15 e) 17 f) 18 g) 7
h) 8 i) 24 j) 20 k) 36 l) 41 m) 312 n) 48
o) 209
2a) × 2 b) -1 c) × 3 d) ÷ 11 e) ÷ 10 f) × 10
g) -16 h) +50 i) ÷4

EXERCISE 16B

1a) 15 b) 3 c) 120 d) 10 e) -8 f) 1

2)

Input	2	5	0	10	7	3	5	15
Output	10	16	6	26	20	12	16	36

3)

Input	10	20	5	15	80	100
Output	6	12	3	9	48	60

4)

Input	2	10	7	13	20	65
Output	4	12	9	15	22	67

5a) 9 b) 100

6)

Input	16	25	100	64	144
Output	400	500	1000	800	1200

EXERCISE 16C

1a)

8	3	4
1	5	9
6	7	2

b)

6	5	10
11	7	3
4	9	8

c)

13	4	7
2	8	14
9	12	3

d)

9	4	11
10	8	6
5	12	7

e)

7	13	4
5	8	11
12	3	9

f)

11	6	7
4	8	12
9	10	5

EXERCISE 17A

1a) 6 cm b) 4.8 cm c) 8.8 cm d) 3.9 cm e) 5cm
f) 5.5 cm g) 5 cm h) 5.5 cm
2a) 8 cm b) 10 cm c) 15.4 cm d) 14.8 cm

EXERCISE 17B

1a) 22 cm b) 16.5 cm c) 35 cm d) 25.2 cm e) 10.1 cm
f) 40 cm

2) 440 km

3)

Perimeter	Length	Width
22 m		
30 cm		
	7 mm	
		7 m
300 cm		
23.4 km		

EXERCISE 17C

1) Teacher's supervision
2a) 9.4 cm b) 7.1 cm c) 8.1 cm d) 5.1 cm
3a) 22 cm b) 44 cm c) 110 km d) 132 cm
e) 176 m f) 6.3 km
4a) 25 cm b) 90 cm c) 3.6 m d) 94 cm
e) 35.7 cm
5) 131.9 cm 6) 502.4 cm 7) 2.2 cm
8) 159 rev.

EXERCISE 17D

1a) 21 cm^2 b) 24 cm^2 c) 22 cm^2 d) 4 cm^2
2) A = 3cm^2, B = 6 cm^2, C = 2 cm^2, D = 3.5 cm^2
3) P = 2 cm^2, Q = 11.5 cm^2
4) Could be 4 cm and 2 cm or 8 cm and 1 cm.
5) 6 m 6) 127 cm^2 7a) 22 m^2 b) ₦37 300
8)

Area	Length of side
	1 cm
	3 m
64 cm^2	
18.49 cm^2	
	9 cm

EXERCISE 17E

1a) 15 cm^2 b) 6 cm^2 c) 24 cm^2 d) 14 cm^2
e) 17.5 cm^2 f) 18 cm^2
2) 63.5m^2 3) 6 cm
4) For example; 14 cm and 4 cm or any two lengths that will multiply to give 56.
5a) 18 cm^2 b) 12 cm^2 c) 14 cm^2
d) 240 cm^2 e) 40 cm^2
6a) 18 cm^2 b) 20 cm^2 c) 28 cm^2
d) 75 cm^2 e) 36 cm^2 f) 175 cm^2
7a) 5 m b) 5 cm c) 5 m d) 2 m
8a) 3.52 cm^2 b) 171 cm^2 c) 30 cm^2
d) 153 cm^2 e) 67.5 cm^2 f) 6.3 cm^2

EXERCISE 17F

1a) 80 cm^2 b) 45 cm^2 c) 19.5 cm^2 d) 45.5 cm^2 e) 21.2 cm^2
f) 17.5 cm^2 g) 48 cm^2 h) 117 cm^2
2a) 6 m b) 5 m c) 7 m d) 20 m

EXERCISE 17G

1a) 28 cm^2 b) 36 cm^2 c) 126.5 cm^2 d) 148 cm^2 e) 24 cm^2
f) 58 cm^2 g) 90 cm^2 2) 66 cm^2 3) $4 \times 12 = 48\ m^2$, $5 \times 6 = 30\ m^2$...........48 + 30 = 78m^2 is the area.
4a) 85 m^2 b) 88 m^2 c) 125 m^2

EXERCISE 17H

1a) 129 m^2 b) 41.5 m^2 c) 51.5 m^2 d) 81 m^2 e) 88 m^2
f) 40.5 m^2

EXERCISE 17I

1a) 12.6 cm^2 b) 78.6 cm^2 c) 95.1 m^2 d) 314.3 cm^2
e) 201.1 km^2 f) 113.1 mm^2
2a) 140 m b) 15386 m^2
3)

Diameter (m)	Radius (m)	Area m^2
	7	154
	21	1386
42		1386
28		616

4a) 14.1 m^2 b) 50.2 m^2 c) 100.5 m^2 d) 379.9 m^2
5a) 218.5 m^2 b) 74.1 m^2

EXERCISE 17J

1a) 13.8 cm^2 b) 109.9 cm^2 2a) 19.63 m^2 2b) 58.88 m^2
2c) 16.13 m^2 3) 21.5% 4) 314 cm^2 5a) 471 cm^2
5b) 66.67% 6) 36.25 m^2 7) 176.7 m^2

CHAPTERS 16 AND 17 REVIEW SECTIONS

1a) 7 b) 25 2) times 10
3)

Input	9	30	6	15	45	39
Output	15	50	10	25	75	65

4a)

8	3	10
9	7	5
4	11	6

4b)

9	4	11
10	8	6
5	12	7

g) 19.22 cm²
5) 11.4 cm 6) 37.84 am
7a) 66 cm b) 264 cm 8) 237 cm²
9a) 15 cm² b) 24 cm² c) 27 cm²
d) 100 cm² 10) 6 cm
11a) 3 cm² b) 52.5 cm²
c) 92 cm² d) 420 cm²
12a) 7 m b) 5 m c) 4 m d) 20 m
13a) 142 cm² b) 72 cm²
14a) 7.1 cm² b) 314.3 cm²
c) 173.3 cm² d) 44.2 cm² e) 285.2 cm²
15a) 294.6 cm² b) 60%
16) ₦3 142 8 57.14 17) 7626 m²

EXERCISE 18A

1a) 65 or more b) 30 to 39 c) 18 to 24
d) Student's rational explanation like older people are more aware of their sugar intake by consuming low-sugar drinks.

EXERCISE 18B

1a) Qualitative b) quantitative c) quantitative
d) qualitative e) quantitative f) qualitative
g) quantitative h) quantitative i) quantitative
j) quantitative
2a) Primary b) Secondary c) Secondary
d) Primary e) Secondary/primary depending on the context. F) Secondary

EXERCISE 18C

1a)

Colour	Tally	Frequency
B	~~////~~	5
G	////	4
R	///	3
W	~~////~~ ///	8
Y	~~////~~ //	7

b) red c) white d) 27 times e) 1/9

2a)

No. of shoes	Tally	Frequency
1	//	2
2	~~////~~ //	7
3	~~////~~ ~~////~~ /	11
4+	///	3

b) 23
c) No, Joe is incorrect. Number of people who own two pairs of shoes are 7, which is more than 3 for the over 4.

3a)

Marks	Tally	Frequency
12	//	2
13	///	3
14	//	2
15	~~////~~ //	7
18	/	1
20	//	2
25	/	1
40	/	1
42	////	4
45	/	1

b) 19 students c) 1 d) 24 e) 15 f) 26%

4a)

Travel method	Tally	Frequency
Cycle	//	2
Bus	~~////~~ ///	8
Taxi	////	4
Walk	~~////~~ ~~////~~ ~~////~~ //	17

b) 31 pupils c) Walk d) 9

EXERCISE 18D

1a) i) Overlapping responses –
(₦100 - ₦200 and ₦200 - ₦300)
ii) How much pocket money do you receive each week? Please tick a box.

☐ Under ₦100 ☐ ₦100 - ₦500 ☐ ₦501 - ₦1000 ☐ More than ₦1000

1b) i) Too personal
ii) Have you ever lost your freedom? Please tick a box

☐ Never ☐ Once ☐ Twice ☐ More than twice

1c) i) No boxes or time frame
ii) How much time do you spend on homework each week? Please tick a box.

☐ Less than 1 hour ☐ 1 – 3 hours ☐ More than 3 hours

1d) i) Does not cover all the possible answers.
Tick one

☐ Fried rice ☐ Beans ☐ Fish and chips ☐ others/specify

e) i) Vague response

ii) Do you eat breakfast every day? Please tick a box.

Yes No Not sure

f) i) Overlapping responses

ii) How many uncles do you have?

0 1 – 3 4 – 6 More than 6

g) i) Leading question

ii) Do you think smoking should be banned? Please tick a box.

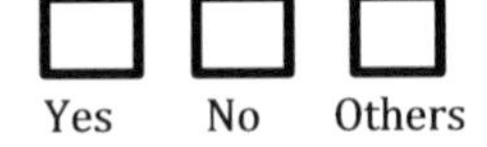

Yes No Others

EXERCISE 19A

1a)

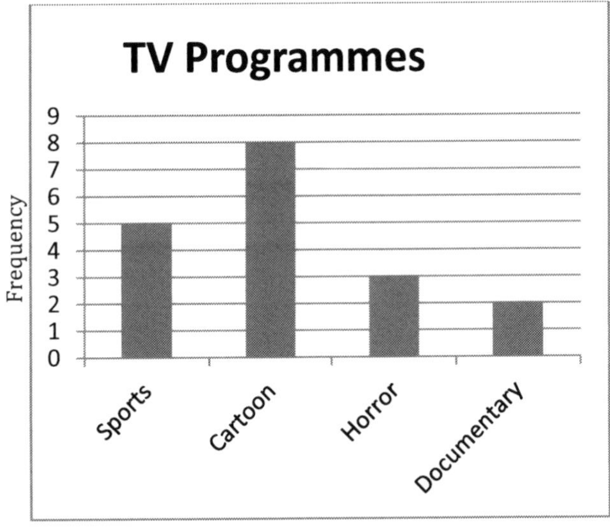

b) Cartoon c) 8

2a) 8

2b) 25

2c) Black

3a)

Number of eggs	Tally	Frequency
0	//	2
1	/	1
2	////	4
3	/	1
4	~~////~~	5
5	//	2
6	///	3

3b)

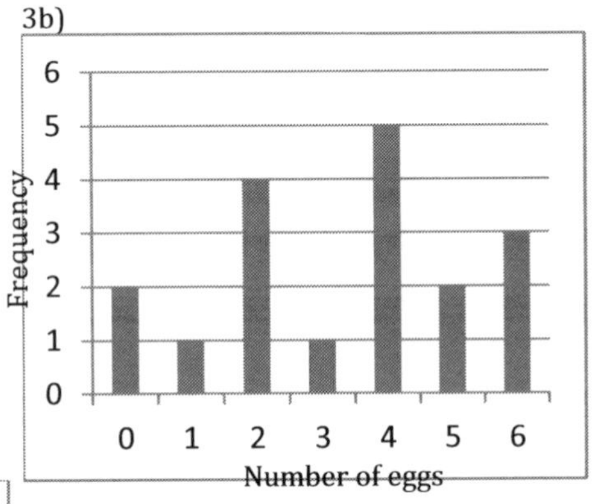

4a)

Colours	Tally	Frequency
B	~~////~~ //	7
P	~~////~~	5
R	///	3
Y	~~////~~ //	7

4b)

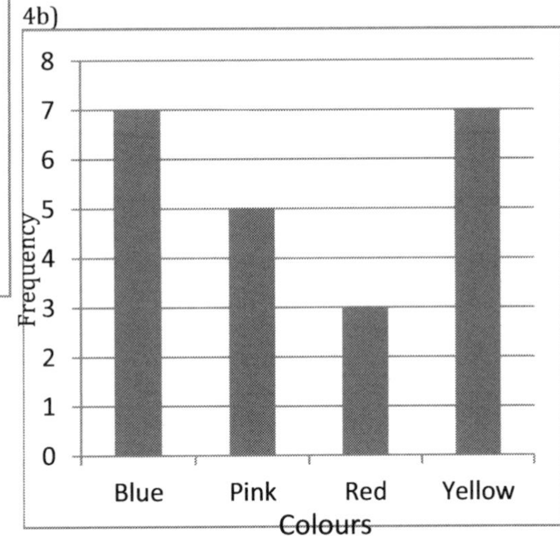

c) Blue and yellow

d) Red

5a) 20
b) Ngozi might not have gone to work on Thursday due to sickness or AWOL. Accept any other reasonable explanation.
c) 6
d) ₦1 000

EXERCISE 19B

1) Key: O represents 3 people/persons

Colour of shirts	
Yellow	O
Red	O O O O O
Blue	O O O O
Pink	O O

2a) 6 b) 9 c) 1 d) 36

3) Key: Δ represents 4 people

Favourite State	
Anambra	Δ Δ Δ
Ondo	Δ Δ
Rivers	Δ Δ Δ Δ Δ Δ

EXERCISE 19C

1a) To find what the angle would be in degrees.

Mode of transport	Number of pupils	Angle
Private car	4	4 × 18 = 72°
Walk	10	10 × 18 = 180°
Cycle	1	1 × 18 = 18°
Taxi	2	2 × 18 = 36°
Others	3	3 × 18 = 54°
Total	20	360°

c)

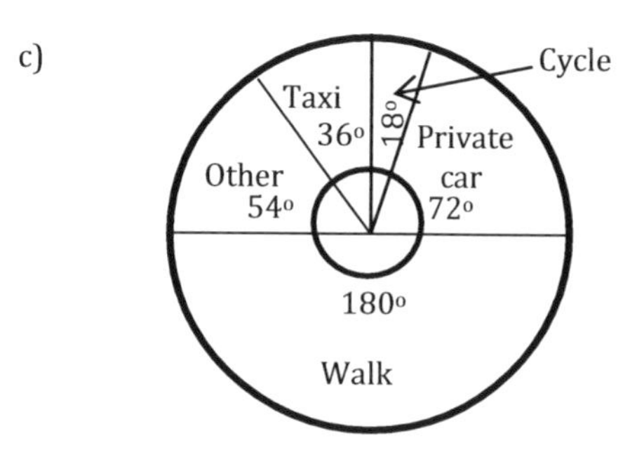

2)

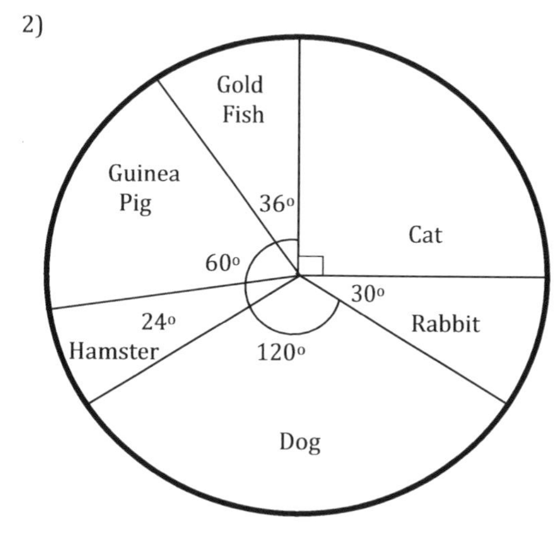

Grade 9 class Pet Owners

3) People who chose yellow $\frac{1}{4} \times 78 = 19.5$
This is not possible as we **cannot** have any fraction of a person.

4a) $\frac{1}{12}$ b) Nigeria c) 44 students d) 6 students

CHAPTERS 18 AND 19 REVIEW SECTIONS

1a) What gender are you? Please tick one box

☐ Male ☐ Female

b) How many hours of TV do you watch in a day?

☐ 0 Hours ☐ 1 – 3 hours ☐ 4 – 6 hours ☐ More than 6 hours

2a)

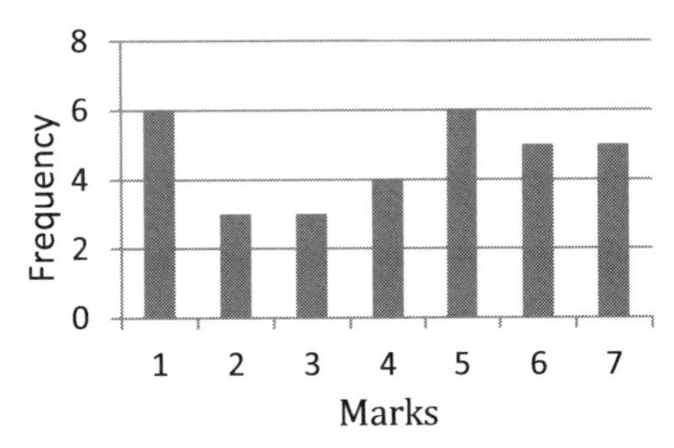

2b) 5 and 1

3)

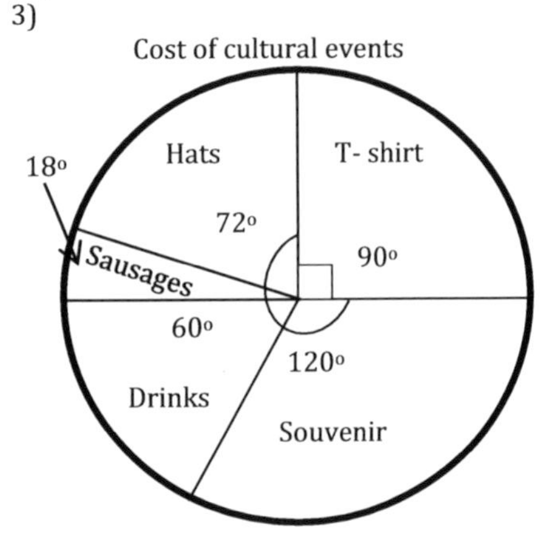

4a) No late comers on Wednesday
b) 22 students
5a) 22.5 °C b) May

EXERCISE 20A

1a) 4 b) 9 c) 6 d) 4 e) 13.25
2a) ₦3 600 b) ₦240
3a) 1270 g b) 423.33 g
4) 287.5 g 5a) 6 b) $13\frac{1}{3}$ c) $5\frac{3}{4}$
d) 5 e) 6 m f) $1\frac{1}{2}$ 6) 37%
7a) 27 b) 8 8) 7 9a) 5.2 b) 5.8
c) 5.3 d) 2.7

EXERCISE 20B

1a) 6 b) 2.5 c) 4 d) 5 e) 40 f) 4.5
2a) 19 b) 20 c) Chuba's marks were more consistent.
3) 20°C 4) 10 5) 0.9 6a) 5.75 b) 60

EXERCISE 20C

1a) 1 b) 3 c) 7 and 12 d) 9 e) No mode 2) 8
3) Blue 4) No mode 5a) 13 b) -8 c) 3.5 d) $\frac{2}{5}$
6a) Blue b) π c) Dog and Cat

EXERCISE 20D

1a) 9 b) 12 c) 30 d) 80 2) 2 3) 36 cm
4a) 40 b) 16 c) 5.5 d) 0.4

EXERCISE 20E

1a) 115 b) 110 c) 15 d) 110
2a) 4 b) 5 c) 4 d) 3.6 3a) 1.2 b) 40 matches
4a) 61 b) 56 c) 99 d) 2 e) 6
5a) 0 b) 1.9 c) 5 d) 1

EXERCISE 20F

1a) 17 b) 2 c) 4 d) 2 2) Dog 3a) $\frac{80}{360} = \frac{2}{9}$ 3b) C4
3c) i) 72 students ii) C6 : 20 Students C4: 24 students
Altogether 44 students
4a) 28 b) 28 c) Friday d) 138 apples
5a) Akin b) 10 c) 10.5 d) 20
6a) ₦45 000 b) No mode c) ₦48 000 d) The average salary is ₦48 000 which is less than ₦49 000 stated in the advert. Also, the Carpenter's salary was advertised as ₦40 000.
7) i) No title ii) No label on both axes iii) Bars are of different widths.
8a) **LAGOS**

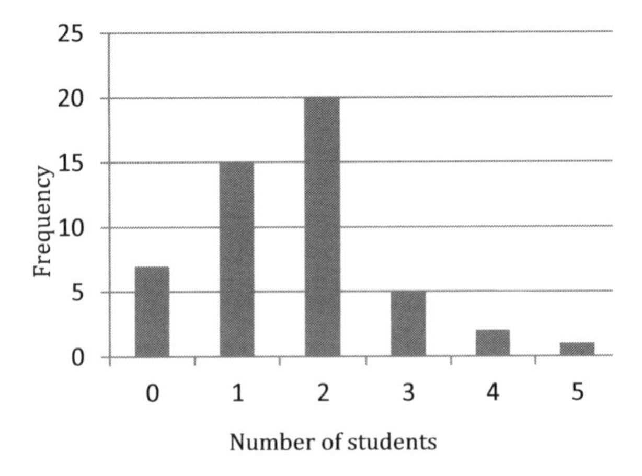

ENUGU

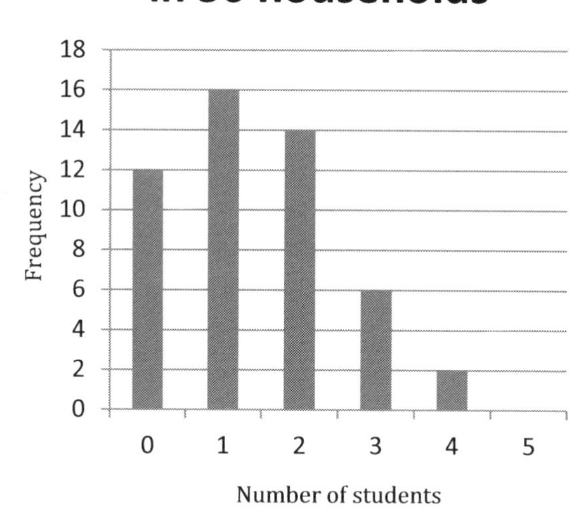

b) For Lagos:
Mean = 1.7 ≈ 2 students
Median = 2 students
Mode = 2 students
Range = 5 students

For Enugu:
Mean = 1.4 ≈ 1 student
Median = 1 student
Mode = 1 student
Range = 4 students

c) From the averages calculated, it shows that there were fewer students living in households in Enugu than in Lagos. Also, the spread of students per household is bigger in Lagos than in Enugu (From their ranges).

CHAPTER 20 REVIEW SECTION

1) Mode = ₦600, median = ₦350
Range = ₦480, mean= ₦374
2) For example, 4, 5,6,7,8
3) For example 8, 9, 9, 13, 13

4a)

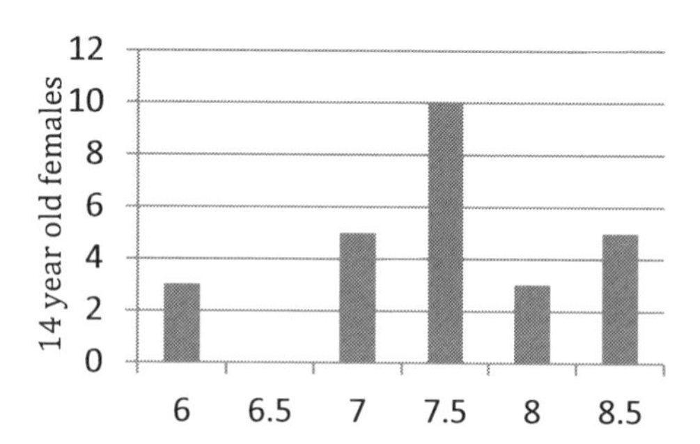

4b) $7\frac{1}{2}$ c) 26 females d) $2\frac{1}{2}$ e) Discrete data (quantitative) 5a) 5 b) 5 c) 5
6a) C4 b) i) 180 students ii) B2: 30 students, C4: 60 students. Altogether, 90 students

EXERCISE 21A

1a) A = (0, 0) B = (0, 12) C = (4, 12) D = (5, 4)
E = (6, 0) F = (7.5, 9)
1b) i) Student's diagram ii) Triangle
1c) i) Student's diagram ii) Trapezium
2a) i) C = (2, 1) G = (4, 0) D = (3, 1)
2b) I LOVE MATHS

EXERCISE 21B

1) A = (-2, -1) B = (-2, 2) C = (-1, 4) D = (2, -2)
E = (2, 2) F = (3, 4) G = (3, 0)
2) Parallelogram
3) No, John is wrong. The shape formed is a trapezium because it has only one pair of parallel sides.

4a)

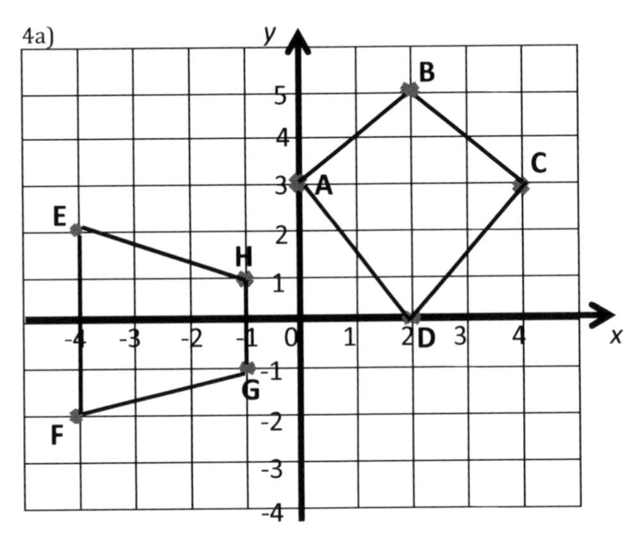

4b) see 4a 4c) See 4a 4d) Kite
4e) See 4a 4f) Trapezium
4g) No, Charles is wrong. ABD is scalene since all the lengths are of different sizes, but an isosceles triangle has two sides that are equal.
5) Isosceles triangle EF = EG
6) 4th quadrant

EXERCISE 22A

1a) 8, 9 b) 15, 17 c) 21, 26 d) 49, 97
e) 19, 23 f) -2, -8 g) 15, 20 h) 6.6, 6.5
i) 32, 64 j) 19, 26 k) 81, 243 l) $\frac{5}{162}, \frac{6}{486}$
2a)add 1 each time
b) add 2 each time
c) add 5 each time
3) Could be
i) 1, 8, 15, 22, 29 ii) 5, 12, 19, 26, 33
4a) add 5 each time b) 4 c) 49
5a) 4, 10, 16, 22 b) 47, 62, 77, 92
c) 2, -5, -12, -19 d) 3, 9, 27, 81
e) 200, 100, 50, 25
6) 88, 5, 88, 616, 1672
7a) i) multiply difference by 2 and add to get the next term. ii) 127
7b) i) Subtract 45 ii) 90, 45
7c) i) Subtract 3 ii) 83
7d) i) Multiply by 10 ii) 3, 3000
7e) i) Difference increases by 1 each then, then add to the next number ii) 18
8) 3 086 358 025

EXERCISE 22B

1a) 2, 4, 6 b) 14, 28, 42 c) 4, 6, 8
d) 2, 5, 8 e) 5, 9, 13 f) 1, 4, 9
g) 2, 1, 0 h) 6, 8, 10 i) 5, 0, -5
j) 11, 14, 17 k) 8, 11, 16 l) 10, 13, 18
m) 17, 37, 57 n) 2, 8, 18 o) 2, 11, 26
p) 14, 26, 46 q) 11, 9, 7 r) $1, \frac{1}{4}, \frac{1}{9}$

2a) 40, 44, 48, 52, 56 b) 7, 12, 17, 22, 27
c) 13, 11, 9, 7, 5 d) 2, -1, -4, -7, -10
e) -7, 2, 11, 20, 29
3) 4 and 10
4a) add 5 each time b) 6 c) 481
5a) 7, b) 11, c) 391
6a) 108 b) 972
7) 17, 22, 42
8a) 13, 15, 17 b) -2, -1, 0
c) 1, 4, 9 d) 6, 18, 36

EXERCISE 22C

1a) 2n + 3 b) 3n + 8 c) n + 3 d) -5n + 35 e) 4n – 3 f) 3n + 6
g) 7n h) 4n – 4 i) 2n – 3 j) -3n + 20 k) 50n + 450
2)

Term number	4n	Term
1	4	7
2	8	11
3	12	15
4	16	19
70	280	283

3a) 2n + 2 b) -3n + 36 c) 2n d) 3n – 1 e) n + 2 f) n^2
4a) 15 cm
4b)

Pattern 4

Pattern 5

4c)

Pattern number	1	2	3	4	5	6
Perimeter (cm)	5	10	15	20	25	30

4d) 5n
4e) 500 cm
4f) 630
5a) 14 rods
5b) 3n - 1
5c) 20
5d)

Pattern number	1	2	3	50
Number of rods	2	5	8	149

INDEX

T

V

Printed in Poland
by Amazon Fulfillment
Poland Sp. z o.o., Wrocław

58998229R00222